Dryland Farming
■ A SYSTEMS APPROACH ■

Dryland Farming

■ A SYSTEMS APPROACH ■

AN ANALYSIS OF DRYLAND AGRICULTURE IN AUSTRALIA

Edited by VICTOR SQUIRES and PHILIP TOW

SYDNEY
UNIVERSITY PRESS

This book has been brought to publication with the
generous assistance of the Key Centre

SYDNEY UNIVERSITY PRESS
in association with
OXFORD UNIVERSITY PRESS AUSTRALIA

National Library of Australia
Cataloguing-in-Publication data:

Dryland farming.

  Bibliography.
  Includes index.
  ISBN 0 424 00169 1.

  I. Dry farming – Australia. I. Squires, Victor.
  II. Tow, P. G.

Edited by Lee White
Designed by Lynn Twelftree
Typeset by Syarikat Seng Teik Sdn. Bhd. Malaysia
Printed by Macarthur Press
Published by Sydney University Press in association
with Oxford University Press,
253 Normanby Road, South Melbourne, Australia

Contents

Contributors

Dr Jock R. Anderson
Professor of Agricultural Economics
University of New England
Armidale, NSW
*Presently on extended leave with the World Bank,
Washington, USA.*

Mr Malcolm Bartholomaeus
Rural Management Consultant
Callum Downs, Clare, SA

Mr John Barton
Western Australian Department of Agriculture
Perth, WA

Mr Andrew Bathgate
Western Australian Department of Agriculture
Perth, WA

Dr Richard J. Bawden
Professor and Dean
Faculty of Agriculture and Rural Development
University of Western Sydney—Hawkesbury
Richmond, NSW

Mr Chris M. Boast
Lecturer in Farm Management
University of Adelaide—Roseworthy Campus
Roseworthy, SA
Farm Management Consultant
Boast Rural Management
Gawler, SA

Dr Kevin G. Boyce
Division of Plant Industries and Natural Resources
Department of Agriculture
Adelaide, SA

Mr Barry Bull
Chief Agronomist
Hi-Fert Pty Ltd
Glenside, SA

Mr Andrew Campbell
National Landcare Facilitator
Faculty of Agriculture and Forestry
University of Melbourne, Vic.

Dr Peter S. Cornish
Principal Research Scientist
New South Wales Department of Agriculture
Gosford, NSW

Mr Peter Cregan
Faculty of Science and Agriculture
Charles Sturt University (Riverina)
Wagga Wagga, NSW

Mr Trevor Dillon
Senior District Agronomist
South Australian Department of Agriculture
Kadina, SA

Dr Ian S. Ferguson
Professor of Forest Science and
Head of Faculty Section
University of Melbourne, Vic.

Dr Ian R. Fillery
Principal Research Scientist
CSIRO Dryland Crop and Soil Research Unit
Floreat Park, Wembley, WA

Mr Geoff G. Fosbery
Regional Agronomist
Elders Pastoral
Northam, WA

Mr Reg J. French
Former Chief, Land Use and Protection Division
Department of Agriculture
Adelaide, SA

Mr James Gaffney
Agricultural Economist
Queensland Department of Primary Industries
Toowoomba, QLD

Dr Robin Graham
Department of Plant Science
University of Adelaide—Waite Campus
Glen Osmond, SA

Dr Peter J. Gregory
Principal Research Scientist
CSIRO Dryland Crops and Soils Research Unit
Floreat Park, Wembley, WA

Mr Iain Grierson
Department of Soil Science
University of Adelaide—Roseworthy Campus
Roseworthy, SA

Mr John B. Griffiths
Officer in Charge, Mallee Research Station
Victorian Department of Agriculture and Rural
 Affairs
Walpeup, Vic.

Dr Avaz Koocheki
College of Agriculture
Ferdowsi—University of Mashad
Iran

Mr Tom L. J. Mann
Department of Agricultural Technology
University of Adelaide—Roseworthy Campus
Roseworthy, SA

Dr Robert L. McCown
CSIRO Division of Tropical Pastures
QDPI/CSIRO Agricultural Systems Research Unit
Toowoomba, QLD

Dr Greg M. McKeon
Principal Scientist
Queensland Department of Primary Industries
Brisbane, QLD

Dr David Morrison
Principal Research Economist
Western Australian Department of Agriculture
Perth, WA

Dr Roger G. Packham
Graduate Programme Director and Associate
 Professor
Faculty of Agriculture and Rural Development
University of Western Sydney—Hawkesbury
Richmond, NSW

Dr Pam Pittaway
Department of Plant Production
University of Queensland—Gatton College
Lawes, QLD

Dr Jim E. Pratley
Associate Professor of Agronomy and Director
Centre for Conservation Farming
Faculty of Agriculture
Charles Sturt University
Wagga Wagga, NSW

Dr Ken G. Rickert
Department of Plant Production
University of Queensland—Gatton College
Lawes, QLD

Dr Brian R. Roberts
Head, Land Use Study Centre
University of Southern Queensland
Toowoomba, QLD

Mr Jeff E. Schultz
Senior Research Officer
Land Management Section
Northfield Research Laboratories
South Australian Department of Agriculture
Adelaide, SA

Dr Victor R. Squires
Director
Key Centre for Dryland Agriculture and Land Use
 Systems
University of Adelaide—Roseworthy Campus
Roseworthy, SA

Dr Philip G. Tow
Department of Agricultural Technology
University of Adelaide—Roseworthy Campus
Roseworthy, SA

Mr Steve J. Trevenen
Operations Manager
Division of Extension and Regional Operations
Western Australian Department of Agriculture
Perth, WA

Dr David H. White
Bureau of Rural Resources
Commonwealth Department of Primary Industries
 and Energy
Canberra, ACT

Dr Allan D. Wilson
Senior Research Fellow
Centre for Farm Planning and Land Management
Faculty of Agriculture and Forestry
University of Melbourne
Parkville, Vic.

Preface

Farming is the practice of agriculture, an activity that produces food and fibre by the deliberate and controlled use of plants and animals. This book deals with farming systems of dryland farming areas, that is, where moisture limits production for at least part of the growing season. Geographically, this fits the cereal zone of temperate and subtropical Australia. It describes the characteristics of dryland farming systems, examines their operation and management and considers procedures used to evaluate performance on dryland farms. In order to emphasize certain principles, some examples will also be taken from North America, Europe, Africa or the Middle East.

Most Australian dryland farms are based on rotations of crops and often pasture species. Dryland agriculture based on integrated sheep and cropping farms is important in most mainland States of Australia. On these farms crops and pastures are grown in rotation and sheep graze pastures and eat stubbles and some grain. These interdependencies mean that it is important to look at the whole farm rather than just at separate enterprises. A systems approach allows this. A farming system can be regarded as a group of interacting components that operate as a whole for a common purpose and react to external stimuli within a boundary that includes all the significant feedbacks.

The traditional textbook on agriculture tends to be compartmentalized along discipline lines, for example, soils, climate, crop production, pastures, livestock, farm machinery, economics. This book aims to approach agriculture as the farmer must in terms of the relationships of the various system components, how the operation and management of each section affects that of the others and how it affects the resource base of soil, water and biological assets. Farming should be both profitable and sustainable. This book explains how. It deals with the principles and practice.

After an introduction to systems thinking in agriculture, to goals and to classification of systems, the book shows how environmental, technological, economic and social factors have shaped the composition of farming systems. Case studies are presented from each of the five States in which dryland agriculture is a major land use. Various types of systems are analysed to show their strengths and weaknesses in terms of productivity, profitability, stability and sustainability.

The book is designed for students at tertiary level, educated farmers and their advisers, and for specialist researchers who wish to think about whole systems.

V. R. Squires
P. G. Tow

Faculty of Agricultural and Natural Resource Sciences
Roseworthy Campus, University of Adelaide

How to Use This Book

The layout of the book reflects a view of dryland farming as a system and its hierarchical nature. There are different routes through the book. Readers must choose a route suited to their needs. The matrix (shown below) is our attempt to identify the chapter sequences that might best suit various users. Lecturers and students intending to use the book as a course text can devise the chapter sequence best suited to their requirements in terms of subject matter and level. Hopefully, the majority of users will wish to attain a breadth of view that will enable them to gain an insight into a system of farming (the Australian Dryland Farming System) that is truly comprehensive.

Two broad treatments are possible. First, the holistic approach can be preserved but the level varied. Second, particular interests can be satisfied by following particular components—such as sheep or cropping—of the system. Case studies from each state where dryland farming systems are important provide a starting point for practical analysis.

Each chapter has a brief synopsis which outlines the scope of the chapter. A list of major concepts provides the reader with a set of salient points. Most contributors sum up in a conclusion section and list further readings to allow users to pursue topics in greater depth. The *Glossary of Terms in Dryland Agriculture* is a further source of information, and is available from the Key Centre for Dryland Agriculture and Land Use Systems, University of Adelaide, Roseworthy College, SA, 5371.

POSSIBLE PATHS THROUGH THE BOOK

APPROACHES	1	2	3	4	5	6	7	8	9	10	11	12	13	14	15	16	17	18	19	20	21	A	B	C	D	E
Section / Chapter	I					II							III					IV				V				
Holistic/ ecosystem approach	●	●	■	■	●						■	■	■	■	■	■		●		■	■	▲	▲	▲	▲	▲
Agri-business approach			●											●			●			■		●	●	●	●	●
Farmer's perspective	■		●	■																■	■	●	●	●	●	●
Livestock sub-system		■	▲	■	■	■	■	●	■	■		■	■	■	●		■					■	■	■	■	■
Pasture sub-system				■	●	■	■	■	■	■		■	■	■	■	■	■	■	■			■	■	■	■	■
Cropping sub-system			■		■	■	■	■	■	■		■	■	■		■	■	■	■			■	■	■	■	■

● core reading   ■ only relevant sections of chapter   ▲ optional related reading

Characteristics of Farming Systems

CHAPTER 1

A Systems Approach to Agriculture

V. R. Squires

SYNOPSIS This chapter defines dryland farming systems, examines their geographic distribution in Australia (and elsewhere), and explains why a systems approach is of value to those who wish to study dryland farming systems.

MAJOR CONCEPTS

1 Dryland farming is dependent on natural rainfall during the growing season and on moisture stored during fallow periods. Dryland farming is practised in regions which are seasonally arid.

2 A system of farming which involves crops, pasture and livestock has been developed to best utilize the soil resources and climate sequences in seasonally arid regions. Climate and soils dictate which farming options are used in a particular region.

3 Agricultural systems are complex. The farm manager operates at the interface between the biophysical and the socio-economic environment.

4 Agroecosystems are modified ecosystems. They exhibit a number of important properties — stability, resilience, sustainability.

5 Productivity of an ecosystem is not determined simply by yield potential of the particular crop or livestock that is used. It is a consequence of the functioning of the total interactive agricultural–environment–social system.

6 The performance of a system must be studied in its complexity but the size and complexity of agricultural systems generally precludes experimentation on system performance.

7 Agriculture is changing rapidly almost everywhere. The student of agriculture needs to be aware of this and master the methodologies which allow an integrated approach.

Dryland farming, also called rain-fed agriculture, is practised in regions where lack of moisture limits crop and/or pasture production to part of the year. This period is called the growing season and varies from a few months to about nine months in a 'normal' year (Chapter 3). Rainfall is often low and variable but the more successful dryland farming systems have been established where rainfall (however low) is reliable and the start and finish of the growing season are fairly predictable. Successful systems are not restricted to the wetter end of the spectrum. For example, in Western Australia wheat can be grown successfully where rainfall is as low as 200 mm. Figure 1.1 is a map of the cereal/livestock region of Australia (the so-called wheat/sheep zone). Rainfall distribution varies from winter-dominant in the south, through a region of non-seasonal rainfall to summer-dominant in the north.

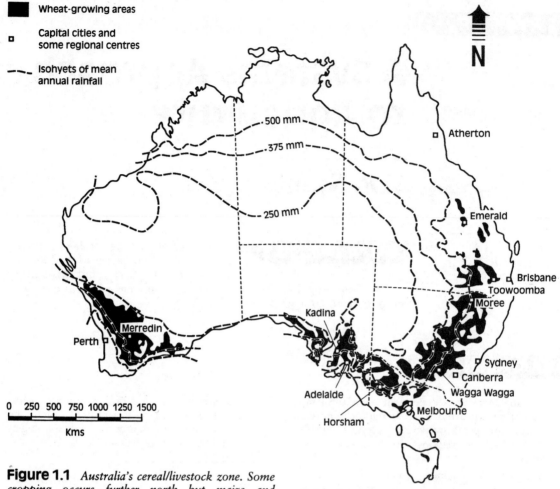

Figure 1.1 *Australia's cereal/livestock zone. Some cropping occurs further north but maize and sorghum replace wheat as the principal cereal.*

Dryland farming systems

Dryland farming systems combine crops, pastures and fallow periods for the fundamental purpose of making efficient use of the limited water. Lack of water is a characteristic of all rain-fed systems (Chapter 3). Moisture is usually the deciding factor in the success of cereal cropping and the use of fallows is a key factor. Fallowing can be an important strategy to store and conserve water for the establishment and maturation of the crop. Winter crop production can depend on fallowing over summer. Fallows may be short or long (Chapter 7).

Much of Australia's agricultural income comes from production of food and fibre on dryland farms. Success in agriculture can be, in large measure, attributed to technological innovations for which Australian farmers and scientists are deservedly renowned. A succession of new technologies has helped transform Australian society over the last century from predominantly rural to urban.

Agriculture is changing rapidly almost everywhere. Since the turn of the century a dramatic downward trend has occurred in the use of non-purchased farm inputs — those produced on the farm — while purchased inputs, such as fertilizers, pesticides, equipment, machinery, and contract labour, have increased. Total production expenses have grown since 1890 from 45 per cent to more than 80 per cent of gross farm income. Years ago the 'inputs' to farm production were themselves produced, by and large, on the farm. Those days are over. From an economic or value-added point of view, the contribution of farming to agriculture is small and is getting smaller. Farming has changed from a productive process which originated most of its own inputs and converted them into outputs, to a process that passes materials and energy from an external supplier to an external buyer (see below).

Agriculture in the next few decades will undergo further major change. The likelihood is that global changes in climate, combined with increasing population, natural resource depletion plus changing attitudes of consumers towards food quality and changing society attitudes towards rural land use, will make the future very different to the present (Chapter 21). This will call for changes in the way we do things. Managing change is stressful; it often involves a journey into the unknown.

Agriculture comprises many small businesses and hence many independent decision makers. Most farm decisions are tactical in nature and are made in response to on-farm situations (Chapter 14). The comparative advantage of Australian agriculture depends on quick and flexible responses to changing markets. Farmers choose their management strategies on the basis of many factors but principal among these are market forces. However, farm businesses must be particularly aware of environmental issues, social change and the agricultural policies of other countries (Chapter 21).

Complexity of dryland farming

Farm managers are not merely concerned with practising husbandry skills, nor with keeping records, allocating resources, or keeping up with new knowledge as a basis for innovation. When stripped to the basics, farming becomes a way of thinking based on the management process of observation, problem identification, goal setting, planning, decision making, implementation, monitoring and evaluation. Farming is all this plus more. It is the sum of all the processes of integration of unrelated resources into a total (farm) system (Figure 1.2) comprising operative, allocative and innovative sub-systems (Chapter 4).

Agriculture was once thought of only as a production system which involved the assembly of numerous inputs such as land, labour, fertilizer and machinery, the combining of these inputs and the collection of the various agricultural outputs. Today, farming has become so much more. Like farming in Australia as a whole, agricultural man-

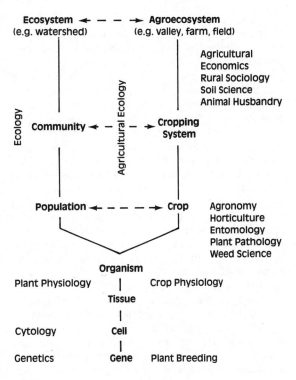

Figure 1.2 *Dryland farming systems are hierarchical.*

agement in Australia's dry farming zone has become more complex. The complexity increases as more farmers recognize the need to manage to avoid problems of soil erosion, acidification and salinity. Improved management of resources should be the first priority (Chapter 11).

Awareness of this has only recently re-emerged. Because of the accelerating rates of soil erosion, salinity and loss of tree cover, other ways of managing our natural resources must be explored if agriculture is to survive. Australians need to better respond to the several alarming land degradation problems of existing agricultural production systems and the increased sensitivity of the public about land conservation (Chapter 11).

The practice of substituting non-renewable resources for steadily diminishing renewable resources can be explained to a large extent by the perceived cost effectiveness of that strategy. Quite simply, it has been seen by many as the most profitable option to pursue. The practice of maintaining or improving the renewable resource base (soils and other biophysical support systems) in some areas of Australia was seen as relatively too expensive.

Yet the pressures brought on farmers to change their cultural practices to exploit natural resources less have to be reconciled with the need to increase productivity to cope with increasing economic pressures. The length of time that a farmer will struggle to maintain a system with declining yields is determined chiefly by the socio-economic position and the related availability of options. The period of time is the measure of economic sustainability. Complementary to the notion of economic sustainability is that of biological sustainability. In a system with declining output there is often a period during which it is biologically possible (but not necessarily economically acceptable) to maintain the system, a period which ends because the system's behaviour and the nature of its components change so radically that from the farmer's viewpoint the original system has collapsed. It is the length of this period which serves as the measure of biological sustainability.

Managers of dryland agricultural enterprises have long known that they deal in their everyday work with complex systems, systems that exhibit rich and varied behaviour governed by complex networks of causation. In studying the problems of agriculture we should invariably start with the whole situation (the 'big picture') before reducing it into smaller bits for closer study. It is not surprising, therefore, that we have come to be interested in systems science, a discipline devoted to the study of such complexities.

Systems science

The notion of applying a systems approach to agricultural problems was an outgrowth of the development of General Systems Theory and its applications in industry in the 1960s. The systems approach is characterized less by its goals than by its methods. The systems approach is a methodology for dealing objectively, and, as often as practicable, scientifically, with the complexity of systems. A more explicit systems approach can substantially improve efficiency.

Systems thinking is a way to broaden the analysis in the direction of holism. Systems thinking is characterized by being multi-faceted with respect to the kinds of consequences considered. Instead of limiting the analysis to one sector or viewpoint all aspects influenced by the issue at hand are considered. Systems thinking is a way of broadening the analysis from one-sector thinking to multisector analysis. Changing from one

perspective to another is like changing spectacles. The focus of attention and generally the kind of thing that we see or perceive will change; the conventional spectacles do not help us to see many of the things that have become increasingly important. An approach that looks at a particular agricultural problem from many different angles and disciplines is more likely to produce better long-term results because it takes the entire agricultural vista into consideration.

The cutting edge in farming is at the agricultural/environmental boundary (Chapter 21). To meet this challenge, we must ask a broader set of questions and change the way we organize ourselves. The major problems and breakthroughs will occur at the disciplinary boundaries. We have learned that everything is connected to everything else, therefore we must find a balance between the narrow and specific, and the broad and general truths. A systems approach will help accomplish this. It has the advantage of putting all the components together so that systems knowledge is primary and components become a subset of the system. Many of the continuing problems in agriculture cannot be solved by a single discipline. The best hope for agriculture today might come from embracing a systems approach. There is a big step to be taken, however, between aggregations of multi-disciplinary approaches to problem solving and holistic or systems approaches.

The systems approach involves approaching problems or situations with a sense of their interactiveness and their wholeness. Farming systems, with their many species and individuals, all interact in a network of relationships too complex for contemplation by the unaided human intellect. The systems approach provides a mechanism whereby complex biological, physico-chemical, social and economic factors can be thought about simultaneously.

The systems approach to management of dryland farming involves using the techniques and philosophy of systems analysis: that is, the methods and tools developed, largely in engineering, for studying, characterizing and making predictions about complex entities, that is, systems. Systems scientists are not just developing an esoteric theory of systems for its own sake. Many of them are becoming aware of the problems of the environment and of the potential of systems science for offering superior insights and, possibly, solutions (Chapter 20). Managers of dryland farms and their advisers are also beginning to realize the relevance of systems thinking, so that real possibilities for synergistic mixing of the two fields (agriculture and systems science) are materializing.

Systems agriculture is based on the idea of enhancing people's capacity to manage change by developing their ability to learn, to improve problem situations and to communicate effectively. It draws on the concept of experiential learning and on systems thinking and practice as well as scientific method, and encourages intuitive, creative activity as well as logical, systematic thinking. It envisages agriculture as a complex interaction between natural and social phenomena (Chapter 20).

Humans as systems agents

Many of the problems of management of agricultural systems, especially at the level of strategic planning and innovational adjustments to changing circumstances, are less than precise, and their solution is more akin to 'improving situations' than to 'solving problems'. The essence of agriculture is its interactiveness. A systems approach helps the manager and any advisory technologists cope with these complex and often messy situations.

Humans have a great potential to modify and dramatically change the environment because they are both an integral part and the most active component of agroecosystems; managers need to consider sociological and socio-economic factors.

The systems view of agriculture represents the dynamic interactions of two basically dissimilar systems. It is 'the interface between people and their environments' (Figure 1.3). Changes in any part of one system will have repercussions throughout the rest of that system — and often beyond into adjoining ones. That is one of the important messages of the systems approach to agriculture.

Only systems studies provide the framework for the prevention of these adverse conditions by concerning themselves with all of the important interactions. There is, of course, a variety of interactions in agriculture, and not just between people and their natural environment as represented by the practices of farming systems (Chapter 20) but people/people interactions are also of fundamental importance. The goals the farm managers impose on farming systems are often fundamentally inconsistent with the evolutionary forces acting on the natural environment. The performance of farming systems (called agroecosystems) can be assessed in the context of how the agroecosystems are organized, how they function, and how agroecosystems interact with the social systems of people who practice agriculture.

Figure 1.3 *Interactions between agroecosystems and human social system. 'Natural resources' in the diagram are soil, water and biological resources from which agroecosystems are constructed.*
(SOURCE: Rambo, T., 1982. 'Human ecology research on tropical ecosystems in Southeast Asia', *Singapore J. Trop. Geog.* 3, 87–99)

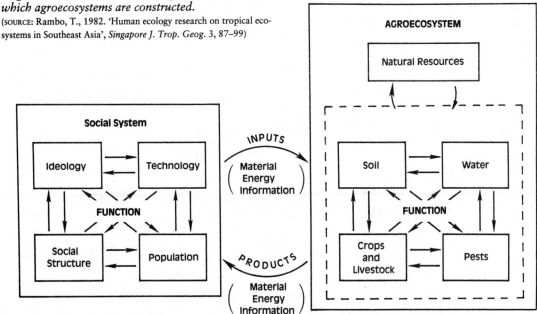

Agroecosystems

Agroecosystems are ecological systems modified by human beings to produce food, fibre or other agricultural products. In agricultural development, natural ecosystems are transformed for the purpose of food or fibre production.

An agroecosystem is a complex of air, water, soil, plants, animals, micro-organisms and everything else in a bounded area that people have modified for the purposes of agricultural production. Although human activities can make agroecosystems very complex in some ways, the ecological intricacy of agroecosystems is generally less than natural ecosystems. The major components of agroecosystems (i.e. crops and livestock) have been selected by humans over a relatively short time. Agroecosystems usually contain fewer plant and animal species than the natural ecosystems they have replaced. The co-adaptation of components (i.e. living organisms) is less complete and there is less redundancy (i.e. overlap of func-

tion). Most agroecosystems are therefore less adaptive and less equipped to persist on their own (Figure 1.4).

Agroecosystem organization can be clarified by comparing them to another kind of system, a television set. A television set has some properties in common with agroecosystems — properties that are held in common by all systems — but there are additional properties that are unique to agroecosystems because of their biological organization.

One of the most conspicuous characteristics of a television set is its design. It consists of many components, but each component is suited exactly to the components to which it is connected. This is essential for the television to function properly. Agroecosystems also have a design — co-adaptation of their biological components — which contributes to their continued persistence. An agroecosystem is not a random collection of plants, animals and micro-organisms. All the species are adapted to the physical environment and co-adapted to one another.

Both television sets and agroecosystems derive

Figure 1.4 *The systems properties of agroecosystems.* (SOURCE: Conway, G. 1985. 'Agro–economic analysis', *Agri. Admin.*, 20, 31–55)

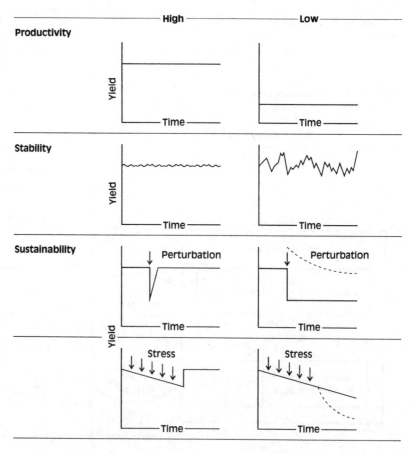

their whole-system behaviour, including their persistence, from the fact that the behaviour of each component in the system is constrained by the actions of other components. The outputs of television components depend upon inputs they receive from other components, a constraint that prevents them from damaging the other components.

Similarly, the population of every species of plant, animal or micro-organism in an agroecosystem is held in check by food or nutrient supply, natural enemies and pesticides. There are some important differences between agroecosystems and television sets. Television sets have only one component for every function. If a component is removed the television ceases to function. In agroecosystems there is a considerable overlap of function between different organisms. If a single species is removed the system may continue to function almost as though nothing has happened. Furthermore, the biological components of an agroecosystem are adaptive — they can change the way they function with respect to other parts of the agroecosystem as the need arises, e.g. disease and pest resistance.

An agricultural ecosystem such as a wheat field may be more productive than the natural forest that once occupied the same land, and it will channel a much greater percentage of its production to human consumption than did the forest, but it is not self-sustaining and may be relatively vulnerable to external disturbances. Regular inputs of energy and nutrients are required to maintain production levels (Figure 1.5). The soil may be too infertile for continuous crop growth unless it is fortified with fertilizers, or pests may destroy the crops unless they are held in check by pesticides. Agroecosystems therefore tend to be less sustainable than natural ecosystems unless sustainability is very carefully designed into the agroecosystems.

The numerous ecological processes that tie people, crops, weeds, animals, micro-organisms, soil and water together into a functioning ongoing ecosystem are so intricate that they can never be fully described, nor can they be fully comprehended. Despite this there are four simplifying assumptions :
- It is not necessary to know everything about an agroecosystem in order to produce a realistic and useful analysis.
- Understanding the behaviour and important properties of an agroecosystem requires knowledge of only a few key functional relationships.
- Producing significant improvements in the performance of an agroecosystem requires changes in only a few key management decisions.

- Identification and understanding of these key relationships and decisions require that a limited number of appropriate key questions are defined and answered.

Simplification is essential for effectively communicating agroecosystem concepts to agricultural practitioners. The dilemma is how to simplify them without losing the essence of key relationships in the agroecosystems as a whole. Agroecosystem analysis could address questions such as:
- How can new crop varieties be bred to produce not only higher yields but also more stable yields?
- How can agricultural inputs (e.g. fertilizers and pesticides) be employed to attain not only higher and more stable yields but also secure, long-term sustainability?
- Under what circumstances is a monoculture more advantageous than mixed cropping, and vice versa?
- How can crops be sequenced through time to make wisest use of the resources available for agricultural production?
- How can agriculture be intensified, and is it possible (desirable) to extend it to marginal lands, without land degradation or other environmental consequences that threaten its long-term success?
- How can the agriculture in an entire region be co-ordinated so it meets a variety of human needs on an equitable basis?

There is no established recipe for agroecosystem analysis. Agroecosystem structure and function are connected. Changes in the human social system can affect agroecosystem function through a chain of interactions between the social system, agroecosystem structure, and agroecosystem function.

Agricultural technology — its role and function

It is necessary to start with the distinction between agroecosystems and agricultural technology systems. An agroecosystem (as defined above) can be of any specified size. It can be a single field, it can be a farm, or it can be the agricultural landscape of a region or nation. An agricultural technology system is the blueprint for an agroecosystem. It is a 'design', 'plan' or 'mental image' — the total package of technology which a farmer or community uses to mould a given area into an agroecosystem.

An agricultural technology system specifies all the crops (and/or livestock) to be employed, the

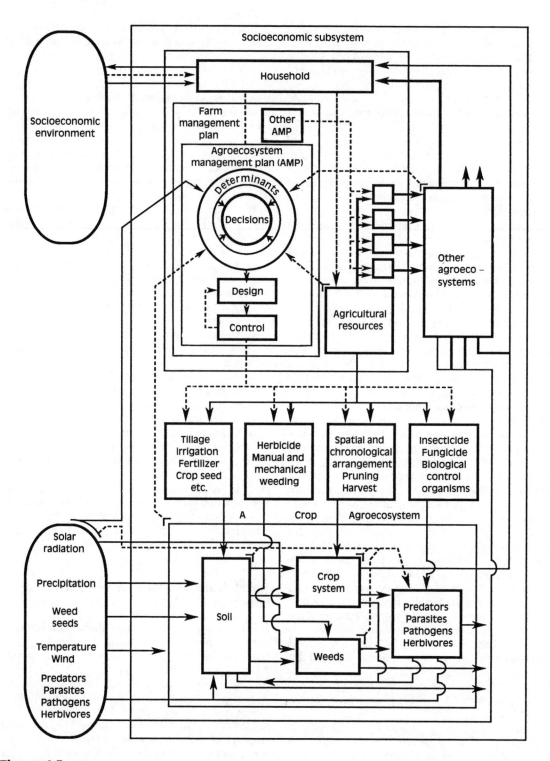

Figure 1.5 *The flow of energy and materials (———) and information (– – – –) in a farm system. The management of an agroecosystem is conceptualized as a series of decisions based on different types of determinants.*

spatial arrangement and temporal sequence of the crops, and all inputs to modify the environment so that crops, pastures and livestock produce as they should. Agricultural technology systems embrace all that is customarily included in the concept of farming systems, but agricultural technology systems are broader in the sense that they include everything that is done to shape an agroecosystem, including parts of the system that are not directly related to the production of crops and livestock, e.g. marketing, and post farm-gate activities.

Agricultural technology systems are important to farmers as their point of departure for moulding the agroecosystems in which they work, but the technology systems are particularly important to agricultural scientists. When scientists try to improve agriculture, they are seeking better designs for technology systems, and it is through technology systems that scientists communicate the fruits of their efforts to farmers.

The 'technology' can be any form of agricultural knowledge, including traditional and informal knowledge as well as technology associated with modern science. Analysis of agroecosystems can help agricultural scientists to develop agricultural technology that is both productive and sustainable (Chapters 18 and 21).

Agricultural technology systems can be at any level of generality. Just as the structure of a house is a consequence of not only an architect's blueprint, but also the particular site on which it is built, the specific materials available for construction, and the carpenter's skills and personal style with regard to details of construction, the same applies to agroecosystems.

The structure of an agroecosystem is a consequence of its agricultural blueprint (i.e. the agricultural technology system) and also of its environmental setting and the farmers and their social setting. The distinction between agricultural technology systems and agroecosystems is important because evaluations of agroecosystem performance may be directed toward one or the other. Another distinction to keep in mind is that between agroecosystem structure and agroecosystem function.

Agroecosystem structure

Agroecosystem structure is how the agroecosystem is organized. It is a consequence of both an agricultural technology system and the environmental and social setting in which the technology is applied. Farmers structure their agricultural ecosystems through the crops and livestock they select and the technology and inputs they employ to care for them. This structuring is represented by the flow of materials, energy and information from the human social system to the agroecosystem. Decisions that farmers make on structuring their agroecosystems are highly specific to the actual environmental and social conditions under which each farm must operate.

Agroecosystem structure includes all elements of the ecosystem and how they are connected functionally to one another, i.e. all species of crops, livestock, weeds, pests, soil, animals and decomposer organisms — as well as all other plants, animals or micro-organisms that are present. It includes details of soil status and everything about inputs that shape the agroecosystem: the annual calendar of human activities in the fields, sources of labour (e.g. family labour or hired labourers), how much capital and energy (e.g. petroleum) are employed, and where they come from (e.g. bank loans).

Many parts of agroecosystems are structured primarily by nature. This includes living organisms that are regarded as pests because they compete with crops for light, water or mineral nutrients or feed upon crops or livestock as herbivores, predators, parasites or pathogens. Some animals and micro-organisms are natural enemies of pests, preventing the pests from becoming abundant enough to cause serious crop damage. Other organisms are essential to sustaining crop production on a long-term basis; for example, soil animals, fungi and bacteria are essential to maintaining soil fertility because of their role in biological decomposition that releases mineral nutrients into the soil (Chapter 9).

Agroecosystem structure is a consequence of the particular crops and other components (weeds, animal pests, soil animals, micro-organisms etc.) in an agroecosystem, the way those components are structured by farm management practices, and the way those components are related functionally to one another.

Agroecosystem function

Agroecosystem function is a consequence of agroecosystem structure, it consists of (a) movements of materials, energy and information from one part of the agroecosystem to another and (b) movements of materials, energy and information in and out of the agroecosystem. The numerous plants, animals and micro-organisms in all ecosystems, including agroecosystems, interact in a complex way. They display a diversity of biological activities (i.e. flows of matter, energy and information) that are the foundation of agroecosystem function. These flows may be summarized in terms of two major agroecosystem processes: production and consumption.

Production is the growth of green plants — the formation of biomass (i.e. carbon chains) from conversion of solar energy. Consumption includes the metabolic activities of animals and microorganisms as they break down and rearrange those carbon chains while using other organisms as food for their own growth. Human beings are major consumers in agroecosystems, their crops and livestock are major producers. The part of agroecosystem function that is most important to people is its production of goods and services, expressed by the flow of materials, energy and information from agroecosystem to social system.

Agroecosytem productivity

Materials that leave the agroecosystem for human use are regarded as products. The total quantity of these products is called production. A characteristic-like production should be simple and unequivocal, but in fact it is highly multidimensional because agroecosystems have a variety of products for a variety of uses. Incidental outputs of agroecosystems, such as sediment in the water runoff from a field, can also be regarded as 'products', and each of these outputs has its own appropriate measure.

The productivity of an agroecosystem is not determined simply by the yield potential of the particular crop or livestock that is employed. The yield that actually occurs depends upon the climatic and nutritional (soil) environment which is, in part, a consequence of how farmers manage. Productivity is therefore a consequence of the functioning of the total interactive agricultural-environmental-social system. In the course of production and consumption, mineral nutrients circulate within an agroecosystem in a cyclic manner (Chapter 10). In addition, farmers move nutrients in and out of agroecosystems when they apply chemical or organic fertilizers (e.g. manure or compost) or remove the harvest from a field (Chapter 17). Water can carry mineral nutrients and other materials downhill from one agricultural field to another, causing a flow of nutrients on a watershed scale.

Energy enters an agroecosystem as sunlight and undergoes numerous physical transformations. Basically agriculture is a means of capturing and converting solar energy to food or fibre. Biological energy flow is the transfer of energy into plants by photosynthesis (production) and from one organism to another (consumption) as they use their food for metabolism and growth. The movement of energy is not cyclic like the movement of mineral nutrients.

The significance of these different measures is that production of a single agroecosystem may be relatively high for one measure and relatively low for another. Comparison of the production of different agroecosystems is therefore meaningful only when the unit of production is explicitly defined. Monetary value is the most universal measure of agroecosystem production, but no single measure — not even monetary value — is of universal significance. Productivity is defined as the output of valued product per unit of resource input. Common measures of productivity are yield or income per hectare, or total production of goods and services per household or nation, but a large number of different measures are possible. Productivity is usually expressed per unit area of land, but it can also be expressed per person, per unit of labour input, per unit of energy input, per unit of capital investment or per unit leaf area of the crop. Productivity can be interpreted more broadly to include the production of human nutrients (i.e. carbohydrates, vitamins, minerals and amino acids), building materials or fuel, as well as services such as soil conservation, water management, aesthetic functions and provision of a favourable environment for social interaction.

An equally significant source of the multidimensional character of productivity is that productivity is only meaningful when expressed as production per unit of input. Inputs take a variety of forms — land, labour, energy, cash — and a single agroecosystem's productivity can be quite different with respect to each of the inputs. In general, productivity is high with respect to inputs that limit production and low with respect to inputs that are in excess.

The functioning of agroecosystems is complex. Several attributes are particularly relevant (Figure 1.4). Two of the most important attributes are stability and sustainability.

Stability — the reliability or consistency of farm production — is important because people depend on a certain amount of production year after year. Nonetheless, fluctuation is a normal part of all ecosystems and agricultural ecosystems are no exception. Agricultural production often fluctuates from year to year, particularly on marginal land where periodic fluctuations in rainfall, pests and a variety of other short-term stresses may increase or decrease yields. Although the risks of partial or complete crop failure are an unavoidable part of farming, farmers place a high priority on minimizing those risks.

Highly productive agriculture often entails risks that reduce its stability. For example, high-yield crop varieties may fail if nutrient inputs, water supplies or protection from pests are not adequate. Planting an entire large area to the single most productive or profitable crop can make the area particularly vulnerable to large-scale pest outbreaks (Chapter 9). Highly productive agriculture also may place a drain on ecosystem resources, reducing the sustainability of the agroecosystem (Chapter 11). For example, highly productive crops may remove large quantities of mineral nutrients from the soil. These nutrients leave the field when crops are harvested and therefore no longer remain in the agroecosystem to be recycled for crop production. Finally, highly stable agriculture may not be resilient to unexpected disturbances because the capacity for dealing with such disturbances is not used. Resilience is the ability to withstand severe and unexpected disturbances such as a prolonged drought, the introduction of a new agricultural pest or disease, a significant change in markets, or an increase in the cost of inputs.

Sustainability concerns whether a given level of productivity can be maintained over time. Sustainability is defined as the ability of an agroecosystem to maintain productivity when subject to a major disturbing force. Salinity, toxicity, erosion, indebtedness or declining market demand are examples of such forces. Sustainability thus determines the persistence or durability of an agroecosystem's productivity under known or possible conditions. Sustainability has a variety of measures associated with various measures of productivity. Some measures of sustainability can be high while others are low for the same agroecosystem.

Sustainability of the system and its capacity to continue producing on a long-term basis is a problem when human activities cause ecological changes that undermine agroecosystem function. A lack of sustainability may be associated with undesirable changes in the soil, or pests can build up to a point where crops no longer function as the farmer needs. Sustainability may also be in jeopardy if the agroecosystem lacks resilience.

The multidimensionality of sustainability derives in large part from the fact that it may be necessary to increase certain inputs with successive crops to maintain yields at the same level. A lack of sustainability may be due to internal processes that cumulatively undermine agroecosystem productivity, e.g. soil degradation or an increasing dependence on expensive pesticides as pests develop increasing resistance. An agroecosystem can lack sustainability because it fails to produce satisfactorily under the impact of traumatic external disturbances such as unusually severe drought, the appearance of a pesticide-resistant pest biotype, an increase in the cost of inputs (e.g. fertilizers), or the collapse of an export market.

Agroecosystems span a scale from single fields to the entire globe. Productivity, sustainability and stability can span the same scale. Higher productivity can be associated with lower sustainability if production is at the expense of soil resources (e.g. by generating erosion, reducing soil organic matter or exporting soil nutrients), if the production is due to heavy inputs leading to major alterations in the ecosystem that eventually undercut production (e.g. pesticides leading to the loss of natural enemies and the emergence of secondary pests), or if higher production is a consequence of labour inputs that place a strain on social institutions underlying the organization of agricultural production.

Although it is desirable for an agroecosystem to be high in all three qualities — productivity, stability and sustainability — the three may conflict with one another. An agroecosystem's productivity, stability and sustainability are a consequence of its structure. Farmers' activities that structure agroecosystems affect soil fertility, pests, water supply and any of the numerous other aspects of the biophysical ecosystem that determine how high yields are and whether they can be maintained. Sometimes these effects are immediate, but in other instances because of the complex ecological processes that intervene, the effects are displaced in space or in time from the farming activity that generated them. A natural use of the productivity, stability and sustainability concepts is to evaluate alternative agricultural systems with regard to these three properties. However, extreme care is necessary in communicating the results of such evaluations because each of these properties is multidimensional and highly situation-specific.

An agroecosystem may be characterized in many other ways. Examples include energy and materials conservation, diversity, autonomy, market penetration and some measure of cultural acceptability. Each of these can be shown to contribute to social value. An early step in the analysis of a given agroecosystem is to identify the important factors and processes that affect the primary system properties. It is generally accepted today that productivity, stability and sustainability are distinct and interrelated agroecosystem properties.

Within a farm, high stability and sustainability may depend on a complementary diversity of crop fields and livestock systems, each of which produces less than its maximum potential and is more variable in yield and individually less sustainable than is the total farm.

The increased agricultural productivity has also been at the expense of the amenity, recreation and conservation values of the countryside. Pollution from agriculture is also increasing. Just how sustainable this high level of production is remains open to debate. Although yields have generally increased, particularly under monoculture, productivity is sometimes less stable.

Commercial farms are biotically simpler than traditional polycultural systems and thus potentially more unstable (Figure 1.6). Modern agriculture thus requires unprecedented management precision to avert serious production shortfalls. Stable yields and sustainable agriculture are as important as raising the yield ceiling of crops and livestock. Upholding yield gains is a core concept and it applies to all improved crops and livestock (Chapter 21).

Agroecosystem research can help to design locally adapted agroecosystems that are reasonably high in yields, low in risks, and sustainable under changing conditions. One approach is to mimic natural ecosystems by establishing the correspondence between elements of natural ecosystem in the area and analogous agroecosystem elements. We then substitute agroecosystem elements for analogous elements in the natural ecosystem. The design has three major parts :

1 The crop (and livestock) species as components of the agroecosystem.
2 The arrangement of the components in space and time.
3 The quantity and nature of inputs and outputs.

The design need not be static. It can simulate natural succession, where each successional stage provides the environmental conditions for a subsequent stage. Details of design should give particular attention to co-adaptation and constraint among component crops (i.e. mutual compatibility of crops and restraint from depleting ecosystem resources), redundancy and adaptiveness. When combined, these properties provide the basis for reliable and sustainable agroecosystem function with a minimum of dependence on external inputs.

There are numerous other ways that farmers can take advantage of natural ecological processes for controlling pests. Again, the theme is co-adaptation and constraint, redundancy and adaptive-

Natural Ecosystem

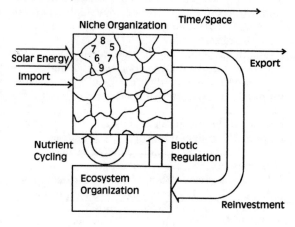

Mechanized Agroecosystem

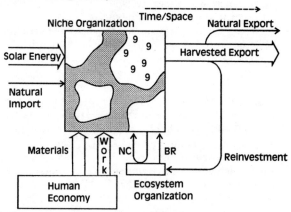

Figure 1.6 *Natural ecosystems differ from mechanized agroecosystems in several ways. The biotic community of the natural ecosystem is more diverse (indicated by the number of cells in the niche space box) than that of the agroecosystem and exploits more fully the available niche space. The characteristics of individuals (genetics, age, health) within a species (indicated by numbers within one species' cells) tend to be varied in natural ecosystems, but nearly uniform in agroecosystems. Natural ecosystems are more continuous in space and time, and they reinvest the bulk of their production in their own ecosystem organization. The export of food from agroecosystems limits such reinvestment, and makes these systems dependent on materials inputs and work from the human economy.*

ness. One way is to rotate crops so the carrying capacity of each field for particular pests is fluctuating rather than constant (Chapter 6). This may take the form of a succession of crops, or a fallow where natural processes of succession eliminate the pests (Chapter 9). It may be possible to plant the crop when pest numbers are very low and harvest it before pests attain sufficient numbers to do serious damage. Pests can be tolerated in low abundance, in fact, mild pest activity may even increase crop production. Farmers also can make full use of natural enemies for controlling pest populations by using pesticide in a way that does not destroy the natural enemies or use up high amounts of energy in the form of fossil fuel.

Conclusions

Agricultural systems are complicated. To deal with this complexity, the concepts and techniques of the systems approach have been found useful. A systems approach is a holistic approach and points to a knowledge of agroecosystems as the best basis for dryland farming.

Agroecosystems are ecological systems modified by human beings to produce food, fibre and other agricultural products. Materials that leave the agroecosystem for human use are regarded as products. We refer to the quantity of these products as production.

There is a distinction between agroecosystems and agricultural technology systems. The total package of technology which a farmer or community uses to mould a given area into an agroecosystem is a blueprint, plan or mental image — in short, the agricultural technology system.

Further reading

Lowrance, R., Stinner, B. J. and House, G. J. (1984). *Agricultural Ecosystems: Unify Concepts*, Wiley-Interscience, New York.

Spedding, C. W. (1988). *An Introduction to Agricultural Systems*, 2nd edn., Elsevier Applied Science, Amsterdam.

Wilson, J. (1988). *Changing Agriculture: An Introduction to Systems Thinking*. Kangaroo Press, Kenthurst, NSW.

CHAPTER 2

Systems Thinking as a Perspective for the Management of Dryland Farming

J. R. Anderson and D. H. White

SYNOPSIS This chapter explains how major concepts in systems approaches can help address problems in the management of farming systems. It introduces fundamental concepts of systems analysis as applied to agricultural systems. It explains why a holistic approach is essential for studying dryland farming systems, particularly given the complex biological interrelationships within a system, the many variables that often have to be taken into account, the variability in the weather and the difficulty in extrapolating technical information to other sites and seasons.

MAJOR CONCEPTS

1 A multi-disciplinary, holistic approach is needed for the analysis and management of farming systems. This requires the definition of system objectives, boundaries and components, as well as the interrelationships, constraints and feedback mechanisms that determine the outputs.

2 Dryland farming systems are open systems in that they necessarily feature connections with an environment.

3 There are strong interactions between the components of a system whereas interactions between elements in the environment are rare.

4 A systems view of the managerial domain should clearly articulate the interrelationships among ecological as well as the more overtly commercial components of farming systems.

5 Putting systems concepts to work is much easier said than done.

6 One of the features of most dryland farming systems is the essential randomness that emerges from many environmental elements.

F arms are agricultural production systems of often remarkable complexity. Human skills are used to manipulate plant and animal growth and reproduction within a predominantly physical environment to fulfil economic and social objectives. Management decisions have to be made in the face of considerable variations between years in the weather and market prices. A multi-disciplinary holistic approach is therefore required for the analysis and management of farming systems (Chapter 1).

Some key ideas in systems thinking

Systems analysis requires a clear definition of objectives and the major outputs from the system. The boundaries of the system must be defined. The components that make up its internal structure and the interrelationships between these and the constraints under which the system is operating need to be identified. These constraints may be physical, biological or financial, along with personal prejudices, social values, lack of specific skills and even the state of health of the system manager.

The systems of concern in dryland farming systems are 'open systems' in that they necessarily feature connections with an 'environment'.

The systems boundary

A good working definition of what constitutes a boundary is a line drawn between what is under some influence of the management of the system, which defines components and features of the system, as opposed to those elements which are not under such influence and thus can be said to reside in the environment of the system. When they describe such matters economists would probably refer to such environmental elements and related variables as being exogenous to the system.

Because it is impractical to take all factors into account in analysing any system, it is essential that the boundaries of the system under study, and hence of the analysis, be defined in advance. These will depend on the objectives of the analysis and hence of managing the particular agricultural system. A comparison of different sheep production systems might ignore the application and fate of fertilizer, this being assumed common to all the sheep management strategies under study. However, when comparing different crop-livestock rotations, then the level and fate of phosphate and nitrogen in the plants and soil can become critical to their productivity, sustainability and financial viability. Changes in soil fertility must therefore be included in the boundaries of the analysis. Likewise, these boundaries would have to be extended beyond the farm gate when targeting specific markets, domestic or overseas, which are profitable for restricted periods of the year.

It is helpful to think in visual terms when considering such definitional matters. The boundary can be sketched to clarify whether designated items lie inside or outside the system as defined.

Just what properly can be regarded as a component of a system is not always clear-cut. Potential ambiguity derives from the uncertainty that may be attached to the degree of control or influence that management has over such a potential component. The most readily recognized components are physical attributes of the system and the variables that relate directly to these. For instance, sheep would be one physical resource and a counterpart variable would be the manager's stocking rate. For cropping enterprises, there is a similar ready identification of areas devoted to particular crops and to the rates of inputs that are applied to such cropping activities, such as seeding rate, crop variety, pesticide use, etc. For most such physical components and their associated variables there is also a temporal dimension which may usefully be thought of as additional components. Such components include the timing of selling operations and the strategy that is used to guide such timing, and strategies and tactics for stocking and destocking animals in response to changing environmental circumstances.

The boundary around some dryland farming systems will be cast widely indeed if it is to include adequately land uses arising from elements of society that are beyond the control of particular farm managers.

System components

Once the boundaries of a system and its analysis have been defined, the major input components and outputs of that system should be identified. For example, in a pasture-livestock system one might identify rainfall, air temperature, soil and pasture type, the initial condition of the plants and animals, animal class and the decision rules for a management strategy as the major inputs. The available soil moisture, pasture biomass, botanical composition and animal liveweights are major components. The way these vary over time and interact with one another will influence the productivity of the system, i.e. the outputs.

Interrelationships between system components

Typically, there are strong interactions between the components of a system whereas interactions between elements in the environment are rare and, if they occur at all in a manner that is significant to the managerial challenge, such interactions take place within the boundaries of the system. Indeed, to the extent that there is managerial influence over such interactions, the boundary of the system should, of course, be drawn to include them within. Some of the major environmental elements in dryland farming systems relate to climate. The climatic element that is usually of most overwhelming importance is rainfall. Rainfall itself has several dimensions or sub-elements, such as its time of arrival, intensity of fall, uniformity or otherwise over space, and so on.

Linear responses, such as the increase in the wool growth of a sheep as it eats more feed are easy to visualize. There is more difficulty in trying to visualize curvilinear relationships, such as the diminishing response of pastures to extra fertilizer, or the way that wool production per hectare varies with stocking rate (Figure 2.1). To take account of competing responses, interactions and feedbacks in, say, a pasture-livestock system is really difficult. It is only the most experienced and mentally adept that really have a firm grasp on the management and productivity of their farms and the expected cash flow, and can estimate with reasonable accuracy how these might change in response to a change in season or management.

Consider some of the interrelationships associated with crop growth in southern Australia, which depends largely on the amount of water present in the root zone of the soil. In response to favourable temperature conditions in late winter or early spring, and adequate fertilizer, a large amount of leaf material can be produced for intercepting solar radiation. However, the rate of transpiration of moisture from the plant is proportional to the leaf biomass so that the demand for water by the plant likewise also increases. In the event of an early summer, with cessation of spring rains, the moisture reserves in the soil can quickly disappear. Without water the plant is unable to translocate nutrients to fill the grain. The yield and quality of the grain in these circumstances will therefore be very low, when a good yield might otherwise have been expected given the initial vigorous growth of the crop. A crop with less leaf area and hence less potential for active growth in late winter or early spring might

have allowed more moisture to be conserved for grain formation as the plant matured.

There are many other aspects of a system's environment that impinge on a dryland farming system. Apart from the natural environment, the economic and social environments have major effects on what happens within the system. Most obvious are the effects on the prices for goods produced in and purchased for the farming system. Less obvious, but sometimes of even more profound significance, are social factors such as land tenure arrangements and their political and legislative backgrounds.

Dryland farming systems also include important ecological variables and, fortunately, these have been increasingly recognized by managers as components of systems that are worth worrying about. Some of the obligation for managers to address environmental issues such as accelerating soil degradation, prevalence of woody vegetation, maintenance of stands of palatable species, and so on, has come from heightened community awareness of the importance of such matters in the long-term custodial responsibilities of land management (Chapters 11 and 21). For the moment, the important thing to note is that a systems view of the managerial domain should clearly articulate the interrelationships among ecological as well as the more overtly commercial components of farming systems. Nowhere are these considerations more explicit than in wildlife management within the physical boundaries of dryland farming systems. The competition between, say, sheep and kangaroos, provides considerable managerial challenge and, nowadays, must be tackled within severe constraints imposed by society at large.

System feedbacks

Biological feedbacks are common in agricultural systems and will often act to reduce substantially the yield of crop or animal product that is obtained. They are usually overlooked in relatively simple analyses, leading to overestimation of the

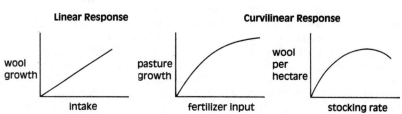

Figure 2.1 *Examples of responses to varying levels of inputs in agricultural systems.*

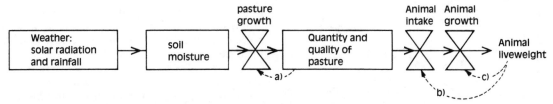

- Pasture growth is determined in part by the quantity or leaf area of standing green pasture.
- Animal intake is not only related to the quantity and quality of the pasture, but to the liveweight, age and physiological state (dry, pregnant, lactating) of the animals.
- Animal growth also depends on the maintenance requirements of the animals which are in turn related to liveweight, age, exercise, and temperature extremes.

Figure 2.2 *Examples of feedback in an agricultural system.*

benefits of applying a particular technological input or management strategy.

Electronic feedbacks are reasonably well known. For example, a microphone placed in front of a loudspeaker produces a very unpleasant shriek. Immediate feedbacks in a biological system might include the cyclical fluctuations in feeding activity of animals. Longer term feedbacks can occur when animals have, say, a higher reproductive rate in one year which, through the extra demands of lactation and stress of higher grazing pressure, results in a lower mating weight and hence lower reproductive rate in the following year. Typical feedbacks limiting plant and animal growth are shown in Figure 2.2.

In the example of crop growth, rapid biomass accumulation early in the growing season accelerated the depletion of soil moisture and reduced grain yield. Other feedbacks can accompany warm wet weather; it provides a favourable microclimate for pests and diseases within a well-grown crop, with consequent disastrous effects on crop yields.

Implementation challenges

Putting systems concepts to work is much easier said than done. Depending on the extent of implementation that is required by a manager, the challenge can range from small to great. Some of the chapters in this volume, particularly those dealing with systems modelling (Chapters 13, 15 and 16), illustrate that a fully-fledged systems analysis is anything but trivial and low-cost. At the other extreme, however, there is often considerable value in using the general concepts in varied informal and almost costless manners which still yield worthy insights. In this section, some matters

specifically connected to economic and more personal-goal aspects of management are broached.

Clarifying objectives

People inevitably have complex objectives (Chapter 4), and the managers of dryland farming systems are no exception especially when the managerial entity is not just an individual farmer but is, say, the board of directors of a large pastoral holding, a collective of individuals representing the population of a communal farming activity, or another type of managerial group. The objectives of people can seldom be boiled down to conform with the simplified assumptions that are typically made by economists. Economists all too readily presume that all that managers are interested in is seeking money profits that are as large as possible.

For those born on to a particular property, the farm is a family inheritance to be looked after and cherished, to be nurtured and improved and passed on with pride to possibly the more fortunate of one's children and grandchildren. Goals can vary throughout life, but catering for one's retirement should be one of the long-term outcomes, and may include a cottage on the property or by the sea.

Current management should equate with the long-term goals of the farm and be consistent with biological and economic principles of farm management. For example, pasture-livestock systems should not be managed to maximize animal productivity. Since costs of production increase along with stock numbers, profits will be maximized at a level of stocking that is substantially lower than that at which production is maximized (Figures 2.3 and 2.4). In fact, the selected level of stocking should be even lower, since it logically involves a

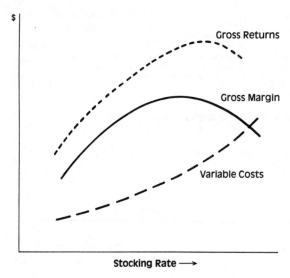

Figure 2.3 *Effect of stocking rate on gross returns, variable costs and gross margin per unit area of land* (SOURCE: White, D. H., 1987. 'Stocking rate', R. W. Snaydon (ed.), *Managed Grasslands B. Analytical Studies*, Elsevier, Amsterdam, 227–238)

compromise between profit maximization and minimizing financial risk, at the same time ensuring that the long-term stability of the soils and vegetation are assured. The selected goal or goals should be incorporated into the analysis of the agricultural system.

The dimension of time always complicates farming objectives. Quite simply, a dollar has different inherent values to a recipient depending on such things as stage in the life cycle, family and other commitments, wealth, and so on. Sometimes economists compact such matters that change over long periods of time through the 'sleight of hand' of discounting procedures. Such procedures recognize that, quite apart from more general considerations, a dollar today is worth more than a dollar tomorrow, particularly in terms of the consumption of goods and services. Thus, the discount method is often used to combine a series of profits over time into a value at some important planning moment, such as the present (net present value) or

at the end of some designated planning period (net terminal worth), or at any other convenient accounting point such as the expected end of the life of the managerial entity.

The broader environmental concerns noted above do not lend themselves readily either to cost accounting or to ready transformation into measures of profitability. In principle, methods are available for valuing goods and services such as recreational benefits of a pastoral resource or the existence value of native vegetation and animals, etc. but, in practice, such methods are not easy to use and always involve controversial weighing systems to combine 'chalk and cheese' attributes. The theoretical technology is well ahead of empirical experience in dealing with these topics.

When such complex objectives are combined with concepts of land-management custody, either for posterity or for a manager's own successors, analysis becomes very awkward indeed and it is thus important to recognize that crude indicators of economic profitability are, at best, just that. The rich complexity of individual and social objectives requires sympathetic interpretation, probably best expressed in words rather than formulas, and a recognition that simple accounting measures do not do full justice to reality.

In terms of the ideas for systems analysis noted above, all this means that those components of a system that are intended to depict the objectives held for the operation of the system will never be easy to articulate. However, it is of the utmost importance to endeavour to describe as clearly as possible what the managerial objectives are. Without making such an explicit statement, it will be impossible to judge the level of efficiency or otherwise that is attained through the practice of management.

Handling uncertainty

One of the features of nearly all dryland farming systems is the essential randomness that emerges from many environmental elements. Again, both the natural climate and the economic environment provide examples of important elements beyond the control of farm managers which determine the performance of a system in a manner that can be

Figure 2.4 *Financial returns in relation to stocking rate.*

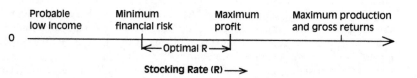

described as random. Some commentators on these topics also describe such phenomena as being intrinsically uncertain, stochastic, or probabilistic. Attempts to manage systems in dryland areas must confront the inherent uncertainty. The simplest conceptual approach is to worry only about what happens on average. Not everyone can afford to take such a position. Average performance may be acceptable to those who are sufficiently wealthy not to be trammelled by bad luck. In the dryland farming systems of industrial countries such managers may be found, especially in public enterprises. In the poor parts of the developing world, however, the managers and the beneficiaries of farming systems can seldom afford to take such a distant and risk-neutral view.

Just how uncertainty is dealt with depends on the degree of sophistication that is embodied in a systems analysis. Economists have developed elaborate methods based on the complex formalized objective functions alluded to above. Such multidimensional preference or utility functions can be combined with formal modelling of the random processes involved to yield information about how resources might best be allocated in such an uncertain world, but these are shirked here to avoid the jargon and complexities involved. These methods are, however, relatively expensive to apply and are not widely comprehended by most of the managers of dryland farming systems. So, farmers and managers deal with uncertainty more pragmatically. Any simple procedure for dealing with risk is bound to be imperfect and at best approximate.

Variability in farming systems

Many analyses of agricultural systems assume an average season. An examination of the weather records for most regions of Australia reveals that there is no such thing as the 'typical average season'. Another problem arises when one tries to define the characteristics of a poor season. For some agricultural systems and locations a lack of rain in one month would be disastrous, whereas in another month rainfall failure would have a minor impact. The sensitivity of different systems to the weather will also vary with the intensity of production. Pastures and soils supporting animals at high stocking densities will be much more vulnerable in adverse seasons. However, it is important to appreciate that pasture improvement is probably the major means of reducing soil erosion in southern Australia, yet if farmers are to get a reasonable return on their investment in seeds and

fertilizers then stocking rates will almost certainly have to be significantly increased. Interactions between seasonal climate, farming intensity and farm improvement will therefore be critical in assessing and comparing different production systems.

Australia is a 'land of droughts, fires and flooding rains'. This climatic variability greatly increases the difficulties of managing the land effectively, including ensuring that excessive land degradation does not occur through overcropping or overstocking in adverse seasons, remaining financially viable and identifying what are the most appropriate agricultural systems to be used.

It also presents difficulties in extrapolating experimental information to other sites and seasons. For example, the number of anthelmintic drenches required by weaned lambs in western Victoria can vary substantially from year to year according to the weather. This may not be apparent from a field experiment conducted over only two or three years.

Systems analysis

Farm management has long been based on partial or total budgeting. Management decisions are often based on assumed outcomes which can be costed and priced accordingly. For example, if lambing is in autumn 80 per cent lamb marking might be expected, whereas if lambing is in spring then 120 per cent marking might be expected. This is because the ovulation rate of ewes mated in autumn for a spring lambing is likely to be much higher. However, such assumptions often fail, thereby making the whole analysis invalid.

Lambing percentage can be affected by a number of factors. For example, the live weight of the ewe at mating, closely associated with ovulation rate, is itself affected by season and stocking rate because of their effects on the amount of feed available for grazing. A high lambing percentage in autumn would mean high stocking rates and heavy grazing pressure on pastures in spring. This could result in the ewes being in light condition at mating in late spring, a shortage of feed going into summer, and a lower reproductive performance. Many factors interact, often in most unexpected ways, to determine the overall productivity of a farm.

To show how complex some apparently simple decisions can be, comparison of the pros and cons of autumn and spring lambing in, say, a medium-high rainfall area of Victoria could be considered. Table 2.1 shows that there are a number of rel-

Table 2.1 *A comparison of spring and autumn lambing systems.*

	Autumn Mating Spring Lambing	Spring Mating Autumn Lambing
Ewe weight at mating	low	high
Weight change at mating	decreasing	increasing
Sunlight hours (photoperiod)	favourable	unfavourable
Weather at lambing	cold, wet and windy	warm and dry
Pasture available at lambing	high	low
Losses — pregnancy toxaemia	low	high
Neonatal lamb survival	low	high
Carrying capacity of pasture	high	low
Risk of pasture drying off whilst lambs light in weight	high	low
Supplementary feed: requirements of ewes	low	high
requirements of lambs	high	low

(SOURCE: Donald, C. M., 1981. 'Innovation in Australian Agriculture')

evant factors which have to be taken into account, and the decision is not straightforward.

The relative importance of the factors in Table 2.1 will vary with the seasons and with many other management decisions made on the farm. If lambs are to be sold direct to market then one must also take the variation in lamb price per kilogram with time of year into account. If a substantial part of the farm is in cereal crop each year, the amount of feed available to the ewes at different times of the year is greatly affected.

A scientist can conduct an experiment for, say, three years to determine the optimal date of lambing for a particular district. But given the substantial variation between years in most farming regions of Australia there is a reasonable chance that the results may represent atypical seasons. If many of the local farms are in crop or stocked at much lower rates, with pastures that are much less improved, then how can the results be extrapolated to these, let alone those farms that are some distance away? Furthermore, the weather conditions experienced by different generations living in the one location are often remarkably different, so that appropriate management strategies for a particular property must change over time.

Fortunately new tools have recently been developed that are ideally suited to sorting out what is likely to happen on individual farms over a wide range of seasons, in different environments and in response to changes in management strategies or technological inputs. These are computerized simulation models or mathematical programs which can help us manage our farms (Chapter 13).

The role of research and development in expanding systems

An important role for managers is to consider adding new or different techniques, methods, enterprises, etc. to the components of a managed system. It is in this context that both formal and informal research and development activities have a potentially important role to play. Better ways of accomplishing activities or completely new activities, such as growing non-traditional dryland-adapted crops, can add to the portfolio of activities in the system and possibly exploit important useful positive interactions between new and existing components. Such measures may, in the light of the simple risk-adjustment process noted, also effectively reduce the overall riskiness of the enterprise in a worthwhile way.

Thus it is that describing more fully the environment of dryland farming systems might usefully highlight the potential for innovations from either formal public-sector research systems or innovations emerging from the experience of other dryland farming systems that are variously transmitted through extension programs or by other means of communication. There may be a link back to some of these services through both traditional and potentially novel ways of paying for such innovative knowledge. Farm managers themselves can engage in innovative activities which would place such research and development activities within the boundaries of a system. The extent to which a manager should invest in such activities then becomes yet another risky decision to be faced.

Conclusions

Whether or not systems thinking turns out to be useful in any particular case naturally depends on many individual factors. For those who have been exposed to systems concepts, it is often found that merely sketching symbolic charts of systems and depicting key interrelationships can be a useful activity in helping to conceptualize a management problem. This can be so even without any overt attempt at quantification of what are often quite complex things to measure. As with all artistic endeavour, its beauty lies in the eye of the beholder, its satisfaction lies in the mind of the creator, and the quality and insight of the representation directly reflect the experience and wisdom of the system perceiver/analyst.

Further reading

Snaydon, R. W. (ed.)(1987). *Managed Grasslands B. Analytical Studies*, Elsevier, Amsterdam.

Wilson, J. (1988). *Changing Agriculture: An Introduction to Systems Thinking*. Kangaroo Press, Kenthurst, NSW.

CHAPTER 3

Factors in the Development and Classification of Dryland Farming Systems

P. G. Tow

SYNOPSIS This chapter discusses the major factors determining the nature of Australian farming systems and how they relate to a general world classification. It defines the characteristics of Australian farming systems which are important for their further classification, comparison and design.

MAJOR CONCEPTS

1 Classification of farming systems is useful in order to compare them and to choose and plan them for particular purposes.

2 The development of farming systems is a response to environmental, technological, economic, political and sociological opportunities and constraints.

3 The main determinant of Australian dryland farming systems is the seasonal distribution and level of rainfall.

4 Australian dryland farming systems show some similarities with systems of the Mediterranean region, western Europe and North America. These are based on climate, tradition and land development patterns.

5 Australian farming systems have developed their own distinctive characteristics. These are due to the importance of livestock (especially sheep) products and wheat as export commodities, the availability of pasture legumes for incorporation into crop rotations and the need for efficiency imposed by farm size.

6 Australian dryland farming systems can be defined in structural and functional terms which farmers can relate to goals, environmental constraints and economic assessment.

I n any consideration of farming systems, we need to be able to describe and compare them (Chapter 6). Better still, we want to be able to analyse and evaluate them (Chapter 13) and to know how to get the best out of them. Such procedures are followed for other types of systems. For instance, good descriptions and assessment of machines enable us to compare them, choose one suited to our purpose and use it to best advantage. Well-known types are given names and when a wide range is available, they are classified according to important characteristics. The same may soon apply to dryland farming systems. This is because of their wide diversity of composition and operation and the increasing need for them to be not only technically efficient, economically profitable and productively stable, but also managerially flexible and ecologically sustainable.

Here we need to distinguish between 'farm systems' and 'farming' systems. A particular design of agricultural system which is well defined and distinguishable from others is generally known as a 'farming system'. When dealing with individual operational units, it is more useful to use the term 'farm system'.

Changes at the level of the farm system may be induced by changes occurring at lower or higher hierarchical levels. An example of the former would be the introduction of new varieties of crop or pasture, new species of weed or new strains of pathogens. At the higher level, the farm system may be changed in response to human population increases, changes in markets for old products, demand for new products, introduction of subsidies and so on. Farm systems tend to become more intensive (higher levels of inputs and outputs) and more specialized, in response to increasing per capita income in the population that comprises the market.

The components of farming systems in a particular region are determined to a large extent by climatic factors. In the past, this relationship was learnt by trial and error. Now it is more predictable, although former predictions are becoming less reliable because of climatic change and the success of plant breeders in extending the climatic range of economic species.

Other features of our farming systems have been determined by particular developments in our history. Many of these have left a permanent impression and an influence on system evolution. Some of the developments, e.g. technological advances, have had parallels in other countries.

The development of particular farm systems is a response to a very wide range of factors — environmental, technological, economic, political and sociological opportunities and constraints, as well as to personal and family goals. Some of these determinants are internal to the farm system, others external. Farm systems are goal-oriented (Chapter 4) and change as farmers adapt to changes in internal and external relationships.

To show how farming systems develop and evolve, it could be useful to classify them according to differences in structure and management that occur in response to changing relationships. This, however, has not yet been done. A classification of farming systems would be important to different people for different reasons. (Its design would also be different.) It is important to policy makers if it helps define, concisely, the types of farming systems most suited to particular regions and the inputs of economic support needed (infrastructure, incentives, advisory services, research, etc.). Researchers need to define systems (e.g. by the way they function), in order to determine their capabilities and limitations and ways to improve them. At the farm level, a classification is useful if it helps the farmer answer the questions: 'What sort of system best suits my goals?'; 'How can I compare systems to choose between them?'; 'How can I operate such a system efficiently and profitably while conserving the resources of the land for the future?'.

Climatic determinants of farming systems

At the most general level, climates relevant to dryland farming in Australia are divided on the basis of rainfall distribution and amount available for use by crops and pastures. There are three main categories of distribution: winter-dominant, summer-dominant and transitional between summer and winter dominance (uniform). These determine whether summer-growing crops or winter-growing crops or a mixture of the two can be grown (Chapter 6).

Within each of these categories the amount of rainfall also determines the type of system and its components. At the dry end of the spectrum, the risk of crop failure due to inadequate rainfall is high and extensive grazing is the most appropriate system. At the wet end of the spectrum, there may be adequate rainfall and in fact, too much. Too much rain may bring high risks of disease and interference with land preparation, sowing and harvesting. This environment is therefore most suited to intensive grazing enterprises. In between these rather extreme situations, land with suitable soil and topography (and not set aside for urban use) is now used for dryland agriculture. 'Dryland' refers to a climatic situation where, in the absence of irrigation, rainfall is sub-optimal to a greater or lesser extent, i.e. inadequate to achieve the potential set by solar radiation input. Thus soil moisture is normally an important limiting factor in dryland agriculture and the degree of deficiency a means of categorizing regions.

Total annual or seasonal rainfall is the simplest quantity used to characterize the moisture regime of a region. With experience, total rainfall in dryland areas can be a useful local indication of production potential. However, the amount of rain needed to produce a unit yield of crop or pasture varies from region to region. Thus in South Australia, 400 mm may be regarded as a reasonable annual rainfall while in Queensland it may be 700 mm. Such differences are due largely to the decreasing 'effectiveness' of rainfall for maintaining plant growth because of increasing evaporation. A long used measure of rainfall effectiveness is the ratio P/E where P = precipitation and E = potential evapotranspiration or an index of it such as pan evaporation. A moisture regime with P/E = 1 is the ideal situation for agriculture and has been called the hydroneutral regime (neither too little nor too much moisture). Some Australian dryland farming areas have a moisture regime which approaches hydroneutrality. An example of this is north-eastern Victoria (parts of the Goulburn and north-eastern Divisions).

The closer the moisture regime is to hydroneutrality, the more flexibility there is in farming systems. Examples of flexibility include:

• Wider choice of crop and pasture species which can be grown.
• Wider choice of rotations or mixes of enterprises.
• Higher likelihood of achieving economic responses from fertilizer nitrogen and pesticides.

Thus farmers in the hydroneutral zone are readily able to take advantage of variations in demand and price for crop and livestock products by varying the composition of their farming systems, introducing new crops or increasing material inputs. Farmers in the drier zone of more extensive cropping or crop–livestock systems have much less flexibility and are thus open to greater risk, not only from drought but also from low prices for their limited range of products. Their survival depends on making highly efficient use of available moisture and minimizing material inputs.

In southern Australia the concepts of 'effective rainfall' and 'length of growing season' have been useful in comparing the moisture regimes of different areas. They are particularly relevant to those areas which have a Mediterranean-type climate — winter-dominant rainfall and a clear start ('break') and end to the rainfall growing season.

The break of season is said to have occurred when the amount of rainfall received equals or exceeds the effective rainfall. 'Effective rainfall' is defined as the amount of rain needed to establish plants from seeds and maintain them above wilting. This amount equals $1.21E^{0.75}$, where E is the monthly evaporation (in millimetres) from a standard Australian tank (about 0.86 evaporation from a Class A pan with a bird guard) (Figure 3.1).

The mean length of growing season for a particular place is the number of consecutive months over which mean rainfall is at least equal to mean effective rainfall. Rainfall received outside this period (approximately April to October) is ineffective for plant growth. The dryland cereal areas of southern Australia have a mean growing season length varying from 5 to 7 months. The method of defining the length of the growing season is illustrated in Figure 3.2. This method does not take into account excess moisture stored in the soil during winter and used by the crop in spring, thus increasing the length of the growing season. Length of growing season can be calculated more accurately by a water budgeting procedure, using rainfall as input and evapotranspiration as output. The simplest way to estimate evapotranspiration is to replace the value 1.21 with various other constants corresponding to different levels of crop growth and water use. Figure 3.2 indicates that, in a typical southern Australian cereal region, the moisture stored in the soil in winter makes only a small difference to the estimated length of the growing season.

In cereal-growing areas of NSW and Queensland, effective rainfall may occur at most times of the year, but not necessarily in a block of five or more consecutive months. Gaps in rainfall input are bridged by accumulation of moisture in fallowed areas prior to sowing a crop. Even then, high rainfall variability makes the concept of 'length of growing season' less useful than in southern Australia. Instead, processes of water budgeting and data analysis have been devised to estimate the probability of receiving various levels of moisture for either summer or winter crops. When this is linked to crop modelling procedures, the probabilities of obtaining different levels of crop yield can be estimated using historical climatic data (Chapter 16). This is a very useful way of evaluating the climatic regime of a region and also to compare regions.

Thus, throughout the Australian cereal zone different regions can be categorized according to seasonal distribution of rainfall, level of moisture availability for crop production and to the yields associated with such levels.

Figure 3.1 *Class A Evaporation pan with bird guard with standard rain guage, automatic rain guage, and anemometer.* (PHOTO: D. Belton)

Figure 3.2 *Mean monthly temperatures and rainfall for Roseworthy, South Australia, together with estimates of growing season length using effective rainfall and water budgeting methods.*

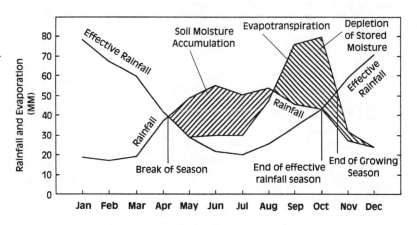

General classification and comparisons of farming systems

On a world scale, classification of farming systems has been based not only on the farms' climate (as discussed above) but also on their:

- Purpose (type of outputs and their use).
- Structure (types of enterprises, rotations, equipment, farm size and tenure, etc.).
- Types and intensity of inputs (labour, machinery, chemicals, management etc.).

Australian Dryland Farming Systems do not fit neatly into existing classifications, but they have similarities to three major farming systems:

1 Mediterranean agriculture.
2 Mixed farming of western Europe and eastern North America.
3 Large-scale grain production.

Mediterranean agriculture

This system has had a large influence on the agriculture of southern Australia, because of the Mediterranean-type climate there and because of the importation of crop, pasture, weed and livestock species from the Mediterranean basin. Australia has had its own distinctive influence on Mediterranean agriculture and has sought to re-export its own system back to the countries of the Mediterranean basin (Chapter 19).

Dryland farming zones in the Mediterranean basin are flanked by zones of higher and lower rainfall. These zones are related through transfer of livestock and livestock feeds and complementary agricultural enterprises. This also applies to dryland farming in Australia. Opportunities also exist for complementary farm investments and seasonal exchange of labour and other resources.

Mixed farming of north-west Europe and eastern North America

Because of the lack of temperate, summer rainfall areas in Australia, the influence of this system on Australian agriculture, while important, has been largely indirect.

Features common to both Australian and European systems include:

- High degree of commercialization (little subsistence farming); dependence on urbanized markets and industrial inputs.
- Farms mainly owned and run by families; declining labour force.
- Farms produce both crops and livestock and the two enterprises are integrated. Sown grassland occupies a substantial proportion of arable land and is carefully managed.
- A range of crops is grown in rotations.

Ways in which the European systems differ from those of Australia include: higher levels of inputs in keeping with higher rainfall; smaller farm size; greater use of crops for livestock feeding, and higher levels of government subsidies to farmers (also related to small farm size).

Mixed farming areas of the USA have had the features of large farm size and dependence on mechanization in common with Australian cereal producing farms. This has been due to the initial abundant supply of new (fertile) land and the rela-

tive scarcity of labour. Although animals (e.g. beef and pigs) are important in the US system, they are fed on maize and other cereal grains rather than pasture. Nitrogen is supplied by fertilizers rather than by legumes. A rich national economy has been able to support this degree of intensification.

Large-scale cereal grain production

Large-scale cereal production has been conducted in parts of the USA, Canada, Argentina, the USSR and Australia. It is commonly a feature of the exploitation of large areas of land newly brought under cultivation. Australian examples are the black, grey and brown clays (mainly vertisols) of northern NSW, southern Queensland and central Queensland.

Because of initial high fertility, inputs of fertilizers are low for many years. The main aim is to maximize output per man, since labour is expensive and land relatively cheap. The farms are thus highly mechanized (Figure 3.3). Rainfall in such areas is generally variable, making average yields moderate to low. Because of this and high machinery costs, the farmer is particularly vulnerable to fluctuating grain prices and to high costs of transport. Livestock are generally not integrated into this system, partly because of lack of farmer interest in livestock and partly because of the lack of suitable pasture legumes.

Because of the lack of flexibility, risk of soil erosion, build-up of disease organisms and periodic low prices, this system is not sustainable. In recent years, some success has been achieved in finding alternative crop and pasture species and to overcome erosion problems by conservation farming (maintaining crop residues on the soil surface). (Chapters 6 and 7).

Distinctive features of dryland farming systems developed in Australia

Sheep (especially) and cattle have been important components of Australia's rural systems since the early years. Success of the small colony depended on the export of materials produced and transported at low cost and in high demand. Wool was the only commodity which fitted into these criteria in the early decades of settlement.

Graziers with their flocks moved westwards, northwards and southwards from Sydney, achieving economic success by:

Figure 3.3 *Grain production is highly mechanized. Harvesting equipment is a major investment.* (PHOTO: Queensland Department of Primary Industries)

- Using land without paying for it.
- Concentrating on areas served by year-round flowing streams.
- Using cheap labour of convicts and ex-convicts, and other cost-saving devices.

By 1850 a huge area of south-eastern Australia, stretching for some 800 km west of the mountains and from the Darling Downs in southern Queensland to Adelaide in South Australia, was scattered with large sheep runs occupied by 'squatters'. A similar movement in Western Australia occurred more slowly. Much of this area is now occupied by the 'cereal belt'. After 1860, to solve the problem of unemployment at the end of the gold rush much of the squatters' land was subdivided for cropping. As in all periods of Australian history, subdivision was planned to provide sufficient area for a family to make a reasonable living. In most cases until recent years, the size has been inadequate. Over-optimism about profitability has been partly due to disregard for the periodic occurrence of drought and low export prices. Poverty or complete failure have often resulted.

A situation of inadequate land reduces the capacity of farmers to adapt to new constraints and opportunities. Yet adaptations have occurred repeatedly over more than a century to make Australian dryland farming efficient and profitable.

One of the most notable features was that, after the subdivision of grazing areas, sheep remained a component of the new farming system. In the 1860s because of the limited local market for wheat and high costs of transport and labour, many selectors had to rely on sheep and cattle grazing for a living. Wheat became more profitable than sheep after 1870, but sheep were still retained and the two enterprises were integrated into one system.

Wheat became a profitable component of Australian dryland farming systems because of innovations which increased production per man as well as per hectare and reduced the costs of production and transport. These included:

- Invention of the stripper harvester by Bull & Ridley in 1843. By 1870, 90 per cent of the South Australian wheat crop was harvested by strippers.
- Development of 3, 4 and 5 furrow ploughs.
- Invention of the stump-jump plough for use in newly cleared land by Smith in 1876.
- Invention of the stripper-winnower which delivered grain ready for bagging.
- Increasing use of superphosphate from the 1880s.
- Building of railways to inland farming areas from the 1850s onwards.
- The availability of cheap transport to Britain in clipper ships.

Cereals and sheep were integrated into the

farming system in its early years because they tended to be complementary rather than competitive. For instance, the grazing of wheat and oats in winter provided sheep with feed and also improved the tillering of the crop, preventing it from lodging in spring. Only an upright crop could be harvested with strippers.

It was also the practice to rest wheat land regularly and use it for grazing. The plants in such a 'weedy fallow' may have been annual grasses, legumes and forbs of Mediterranean origin. Certainly, by the 1880s annual species of *Medicago* and *Triofolium* had been noted growing wild in parts of southern Australia. These provided good livestock feed and improved the soil nitrogen content for the following crop. About this time also, a bare fallow was introduced into the farming system to conserve moisture in one season for use in the following one (Chapter 7). It also helped control weeds and diseases and accumulate nitrate nitrogen. Cereal yields were stabilized, although only one crop could be harvested every two years. In the early 1900s in South Australia a rotation of fallow, wheat and pasture was used and the common burr medic was obvious because the spiny pods had to be removed from the wool. However, another few decades were to pass before the unsustainability of the fallow system and the potential value of the annual pasture legumes in the farming system were to be fully appreciated.

The rotation of cereals and annual pasture legumes became known as the southern Australian ley farming system. The complementarity of cereal and sheep is an important feature. The main integrating factor is the legume pasture which provides high quality feed for sheep and extra soil nitrogen for the following crop. Another integrating feature is the grazing of crop and pasture residues. The sheep obtain feed in the dry half of the year and reduce the volume of the residues and also weed growth to assist in land preparation. The annual legumes are self-regenerating from seed set each year and deposited in the soil. A high proportion of the seeds (especially those of annual *Medigaco* species (medics) remain impermeable ('hard') for varying numbers of years before germinating. Thus self-regeneration readily occurs after an intervening cereal crop.

A package of management procedures is practised to ensure the continued presence of the pasture species in the rotation, e.g. shallow tillage to avoid deep burial of the legume seed, herbicides for weed control in the cereal phase, phosphatic fertilizer application to both crop and pasture and management to avoid both overgrazing at critical development stages (flowering and early seed-set) and also consumption of pods on the soil surface.

Since the introduction of the ley farming system in southern Australia, large tracts of land have been opened up for crop production, particularly in Queensland, New South Wales and Western Australia. The farming systems have been characterized by relatively large farm size, the use of large machinery for timeliness of operations, and the aim of exploiting the initial high fertility of many of the soils. Unfortunately this exploitive phase has often been accompanied by soil degradation (particularly erosion and salinization). Thus attempts are now being made to include 'conservation farming' practices in the farming system (Chapters 6, 7 and 11). As yet however, few farms of the northern cereal zone have adopted a ley farming system. In most cases this is due to the lack of suitable pasture legumes and/or the higher profitability of crops than of livestock products.

Alternatives to cereal pasture rotations have been tested throughout the Australian cereal belt and have included cereal-grain legume rotations. The diversity of farming systems has increased markedly and it would be useful to compare them for productivity, profitability, stability, flexibility and sustainability. This is probably only partly within our reach.

However, farming systems can be defined in structural and functional terms which farmers can relate to their goals, to the environmental constraints and risks imposed on them, and to the inputs and outputs which will require economic assessment. A list of such terms is given below, based on concepts discussed in this chapter and other chapters of the book.

Terms for defining Australian dryland farming systems

- Season of Cropping.
  — Summer; summer and winter; winter. This is determined by rainfall distribution and the capacity of soil to store water from one season to the next.
- Enterprise Orientation.
  — Mainly cropping; integrated crop-livestock; mainly livestock. This is determined by farmer preference, tradition and the value placed on the complementarity and stability provided by combining crops and livestock.
- Degree of Diversification of Enterprises.
  This is determined by the average level of avail-

able moisture and the availability of suitable crop and pasture cultivars and livestock breeds.

- Degree of Flexibility in Choice of Rotations and Enterprise Mixes.
  — Fixed rotation imposed by constraints of disease carry-over, available range of plant cultivars, risk of soil degradation or management requirements of specialized enterprises;
  — Opportunistic and flexible system to make best use of rainfall and market opportunities.
- Degree of Integration with Surrounding Regions (with respect to transfer of livestock and livestock feeds; labour and investment).
  Such integration can be a source of complementarity and stability.
- Degree of Reliance on External Support and Investments (subsidies, incentives, market stabilization, alternative employment and investments) for sustaining the system.
  This is related to the need for special measures to cope with high levels of climatic, market and soil erosion risks.
- Source of Nitrogen for Crops.
  Legume nitrogen; combination of legume and fertilizer (inorganic) nitrogen; inorganic nitrogen only.
- Methods of Soil Management.
  — Tillage, soil protection and fertility main-tenance/improvement methods. These relate to the efficiency of soil water and nutrient use and the sustainability of the system.
- Relative Reliance on Chemical and Ecological Methods for Control of Weeds, Insect Pests and Diseases.
  — Relates to sustainability of the system.
- Level of Available (or Required) Management Skill, Economic Advice and Research Support.
  — Relates to stability and sustainability of the system.

Conclusions

With the historical and geographic perspectives provided, it is clear that farming systems change at varying rates over time in response to a wide range of factors related to cost of land, labour and materials and the variability of climate and market prices. In a situation of generally sub-optimal rainfall, future changes in farming systems may lean less towards intensification of material inputs than towards greater biological efficiency, scientific innovation and technical, managerial skill, together with conservation of soil and other resources. These characteristics could be linked with types of inputs, outputs and management to produce a useful classification of Australian dryland farming systems.

Further reading

Davidson, B. R. (1981). *European Farming in Australia — An Economic History of Australian Farming*, Elsevier, Amsterdam.

Duckham, A. N. and Masefield, G. B. (1970). *Farming Systems of the World*, Chatto and Windus, London.

Fresco, L. O. and Westphal, E. (1988). 'A hierarchical classification of farm systems', *Experimental Agriculture 24*, 399–419.

Grigg, D. B. (1977). *The Agricultural Systems of the World. An Evolutionary Approach*, Cambridge University Press, Melbourne.

Prescott, J. A. and Thomas, J. A. (1948). 'The length of the growing season in Australia as determined by the effectiveness of the rainfall — A revision', *Proc. Royal Geographical Soc. of Australasia*, South Australian Branch, *50*, 42–6.

Ruthenberg, H. (1980). *Farming Systems in the Tropics*, 3rd edn, Clarendon Press, Oxford.

Spedding, C. R. W. (1988). *An Introduction to Agricultural Systems*, 2nd edn, Applied Science Publishers, London.

CHAPTER 4

Goals of Farming Systems

M. Bartholomaeus

SYNOPSIS The concepts of mission statements, goals, objectives and management strategies are introduced within the context of a landcare ethic and sustainable farming systems. These concepts are then carried through a discussion about economic pressures on farming, and the resulting implications for land use. To assist in the dilemma of economic pressure conflicting with sustainable land use, the concepts of whole farm planning and farming systems are introduced to illustrate how sustainable dryland farming and profitable farm businesses can evolve around a basis of goals and objectives focusing on a landcare ethic.

MAJOR CONCEPTS

1 A mission statement is a one- or two-sentence statement which says what the organization is, what it does, and captures the underlying philosophy of the business.

2 A goal is a statement which is consistent with the mission statement which will give the business specific direction.

3 Objectives are more specific than goals and often say exactly what is to be done, and when it should be completed.

4 Strategies are management actions set in place to initiate work towards stated goals and objectives.

5 The economic system in developed countries generates economic pressures for agriculture which manifest themselves as a cost-price squeeze for farmers.

6 Farmers add to those pressures by improving productivity, increasing output and driving prices for farm products down.

7 Whole farm planning involves assessing resources available and planning for repair of land degradation, and planning for sustainable use of different land classes.

8 The key to sustainable land use is the farming system put into action on each land class.

9 Farming system design revolves around product selection, rotation of land use, tillage, pest control, and crop and pasture nutrition.

Farming is a business, and like any other business a farm should have a business plan with a mission statement, clearly stated goals, and well-defined objectives. When it comes to achieving these stated goals and objectives there will be a set of strategies. The implementing of field-oriented strategies is really the function of using land for farming in ways consistent with the goals and objectives of the business.

Mission statements

A well-written mission statement for a business should not need revision unless the business is undergoing a major change of direction. Even then the mission statement would not normally change as it should remain in place to guide such changes in direction.

In southern Australia, for example, where agriculture is based on dryland farming, a mission statement should:

- State the type of farming on which the business is based (i.e. what the business is).
- State the context in which goals and objectives are set (i.e. state the basis for decision making).
- Make a comment about resource use and preservation (i.e. make a comment about long-term survival of the business).

For example: Callum Downs is a dryland farm in the Australian wheat/sheep zone, where production is based on the growing of dryland winter crops and pastures. The land available is to be classified according to its capability and each land class is to be used in its most productive and profitable way from all the sustainable alternatives.

The key components of a mission statement are in the example of Callum Downs:

1 What type of farm business is it? The description reveals that the farming system will be rainfed, is located in the winter-rainfall dominant part of the Australian agricultural region where production revolves around winter cereals and animal products derived from broadacre grazing of animals.

2 On what basis will decisions be made? The statement on land-class use suggests that the business is to be centred around using a land resource and that the business must be profitable with productivity of the land being the key to achieving profitability.

3 What is the long-term view for the business? The statement that land will be classified and used according to its capability is an expressed desire not to exploit any land and leave it open to degradation. This is further strengthened by the statement that only sustainable alternatives will be considered, indicating a long-term desire to produce from the land resource. It expressly excludes profitable uses of land which are not sustainable.

The key to the mission statement will revolve around the manager's interpretation and understanding of the term 'sustainable'. This should

become apparent as the goals and objectives of the business are specified.

Goals

Goals are not normally ends in themselves and so may never be achieved. There is no need, therefore, to be able to measure how close we are to achieving our goals, but we do need to be able to monitor whether we are still moving in the direction of our goals or not.

A set of goals for the Callum Downs example may be:

1 To classify all land according to its capability.

2 To use all land according to its capability.

3 To reclaim degraded land and make it productive and prevent further encroachment of degradation on to the productive land.

4 In the event of land degradation problems arising, implement control and repair strategies immediately.

5 Implement strategies to improve the productivity of the different land classes.

6 To have a mix of crop and livestock products which is profitable and provides a stable income.

7 To have annual growth of income to allow living standards, relative to the rest of society, to be maintained.

The first two goals are obviously direct statements from the mission statement, while the next two goals get their direction from the mission statement and are consistent with the mission statement.

The mission statement requires that each land class be used in its most profitable and productive way. The fifth goal addresses this by having, as a direction in planning, a clear intention to improve productivity if possible.

The sixth acknowledges that the business is operating in an environment suited to both crop and animal production, as stated in the mission statement. It also gives the direction that the balance between livestock and cropping must be a profitable one (as stated in the mission statement), and that the livestock cropping mix be used to stabilize income — a notion consistent with the sustainability thrust of the mission statement, if it is acknowledged that long-term business survival is as much a part of sustainable agriculture as is the preservation of the soil, water and genetic resources used in agriculture.

The last goal is also directed at sustainability of

the business. The best farming system in the world will not be sustainable if it does not allow the farm business to grow at a rate comparable to the economic growth elsewhere in the economy. However, while such a goal gives direction to the planning process for the farm business, such growth will have to be consistent with the other stated goals and the mission statement. In this way business growth can only occur within the limits of good land management practices.

Strategies

In the process of setting the management functions to progress in the directions set by the stated goals, it is common for a number of strategies to be set in place. Strategies outline what is intended to be done in working towards one's goals. For example, the following set of strategies might apply to the goals stated by the manager of Callum Downs:
1 Prepare a whole farm plan for the entire property.
2 Re-fence the property along land-class boundaries where needed, and where practical.
3 Develop farming systems for each land capability class based on crop/pasture rotations for those areas capable of being cropped, and on improved pastures on other areas, with appropriate livestock management systems.
4 Commence reclamation work on areas where degradation has occurred.
5 Put in place regular soil testing programmes to enable organic carbon levels, soil aggregates, and nutrient levels to be monitored as a measure of land-use sustainability.
6 Implement suitable financial recording and analysis systems to allow the progress of the farm business to be monitored.
7 Generate initial growth in farm income by realizing more of the current property's productive potential.
8 Plan for future property expansion to give further growth in farm income.
The above strategies are consistent with the goals of the business. Which of these strategies should be followed first, and how many resources should be devoted to it can be better ascertained by specifying objectives.

Objectives

Objectives are measurable and one will know when an objective has been met. At this point new objectives will be needed. Strategies may accompany objectives to set out how tasks will be done.

The manager of Callum Downs may set the following objectives initially:
1 Complete a whole farm plan during autumn next year.
Strategy: enrol in a Technical and Further Education course which will enable the manager to do his own plan.
2 Re-fence the most recently purchased land according to the whole farm plan, to be completed in time for the next year's cropping programme.
Strategy: employ a fencing contractor to drive all posts and use own farm labour to install all wire.
3 Implement a sustainable rotation on all cropping paddocks so that within three years all cropping areas are incorporated into the new sequence.
Strategy: use grain legumes and sub-clover pastures to provide disease breaks for the cereal crops and input nitrogen into the system.
4 Lift all cereal yields to within 70 per cent of potential (as determined by growing season rainfall) within three years.
Strategy: implement the new rotation, fence out any areas previously cropped but now classified as not suitable for cropping and establish permanent pasture on those areas, match fertilizer inputs more closely to what is being removed by crops and by-products.
5 Stop all burning of crop residues within two years.
Strategy: replace existing tillage equipment with trash handling equipment and purchase a set of rubber-tyred rollers to assist with stubble breakdown.
6 Have at least two soil test results for each paddock within six years.
Strategy: begin testing each paddock at the beginning of each cropping sequence and at the midpoint of the rotation for each paddock.
7 Prepare an annual profit and loss statement for the business to monitor business progress.
Strategy: employ a farm secretary to keep computerized financial records to allow management accounting statements to be drawn up as well as taxation accounts.

Such objectives and accompanying strategies are consistent with the business goals and the mission statement. For example, the objective to lift cereal yields to 70 per cent of potential and the strategies for doing this are consistent with the goal of improving the productivity of the different land classes. This is consistent with the mission statement which asks that each land class be used in its most productive and profitable way. The other goals and objectives stated will ensure that this increase in yields to 70 per cent of potential is done in a manner which is sustainable for the resources being exploited, and is sustainable for the business (i.e. that the way in which the yield increases are achieved also increases business profit).

The formal stating of a mission statement, as well as goals, objectives, and strategies for achieving goals and objectives are extremely important if a long-term view of the business is to be taken. In the social environment of the late 20th century and early 21st century, it is going to be important that farm businesses are operating in the context of a landcare ethic which will improve the state of the land resources being used, maintain those resources in their improved state, while still allowing those businesses to be profitable with reasonable prospects for growth of profit over time. Mission statements, goals, and objectives must be written in this context.

Economic pressures on farming

While it is important for farm businesses to be operating in the context of a landcare ethic, the need for those businesses to also be profitable means that the economic pressures facing agriculture must be clearly understood if sensible goals and objectives are to be set. Economic pressures on farm businesses come from outside the sector

from the pressures of economic growth, and from within the sector from the pressures generated by adoption of new technology. The combined effect of these economic pressures will have implications for the pressure on land use.

Economic pressures from outside agriculture

The terms of trade of the rural sector is one of the most important economic indicators relating to agriculture, and embodies the external economic pressures which come to bear on the agricultural sector.

Farmers' terms of trade are the ratio of prices received (for outputs like wool and wheat) to prices paid (for purchased inputs). Historically the terms of trade have been declining in developed economies, giving rise to the so-called cost-price squeeze on agriculture. In other words, the prices paid for inputs have been rising faster than the prices being received for farm products (Figure 4.1).

The impact of a cost-price squeeze on farm businesses is relatively easy to identify. If profits are to be maintained then there must be offsetting productivity increases. The real problem is that there is unlikely to be any cessation of the declining terms of trade, and this must be taken into account when formulating goals and objectives for sustainable farm businesses.

The reasons for an expected continuation of the declining terms of trade are related to the goal of economic growth embodied within capitalistic economic systems. Economic growth leads to increased real incomes and higher standards of living. However, when real incomes rise consumers

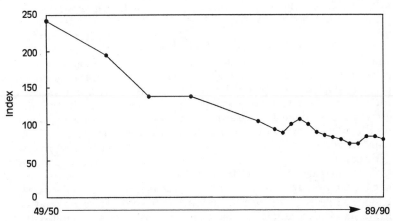

Figure 4.1 *Australian farmers' terms of trade, 1949–50 to 1989–90.*

usually do not increase their expenditure on food. Most people in developed countries have sufficient food to eat and so any additional disposable income is spent on other things, such as:

- More processing of food (cooked, canned, frozen, fast food).
- Better foods (fillet steak instead of feral rabbit).
- Consumer items (television sets, cars, houses).
- Services and recreation (automatic account paying, tourism).

The net result is that extra income accruing within the economy is spent almost entirely within the non-agricultural sector.

Economists refer to basic raw food products as being *income inelastic*. An item is income inelastic if a 1 per cent increase in disposable income leads to less than a 1 per cent increase in the quantity of the good being demanded. As a consequence, non-agricultural industries grow, make good profits and require more land, labour and capital to supply the increase in consumer demand. With their increased profits such firms bid against each other for labour (pushing up wages), land (pushing land prices above their agricultural capacity, particularly close to towns and cities) and capital (pushing up interest rates). In a deregulated capitalistic economic system the farm sector also pays these higher costs even though the prices it receives for basic unprocessed food items have not gone up.

The direct result of general economic growth within a developed market economy is a cost-price squeeze for farmers as prices for inputs get pushed up, but demand for basic farm products remains unchanged. Exceptions to this generalization are products like wool and cotton. These raw products are used in apparel production, which does increase as living standards improve. The cost-price squeeze may be lessened for such industries. In their planning for a sustainable farming system and a sustainable farm business, farm managers may be able to exploit this situation by including products like wool and cotton in their production mix.

Demand for food increases as a country's population grows, but in most developed countries population growth is slowing (Japan is a prime example), and demand for consumer items from the non-farm sector also grows as population growth occurs. Population growth within a developed country is therefore unlikely to provide any relief from the general economic pressures facing agriculture.

There may be some relief for farmers in developed countries by way of increased exports of basic food items, as developing countries improve living standards and their people respond by consuming more food per capita, and by consuming different foods derived from wheat- and meat-based food stuffs, which are the common products from the dryland farming regions of the world. However, while it may allow increased volumes of agricultural products to be traded, it is unlikely that this will result in significant price rises because of the political interference in the world food markets. There will still be economic pressure on farmers to chase productivity increases to retain an acceptable profit margin between cost of production and prices received.

Economic pressures from within agriculture

Farmers must respond to the cost-price squeeze imposed on them by increasing productivity. Genetic improvement, and use of herbicides, fertilizers, and labour-saving devices are examples of new technologies being constantly adopted in a bid to increase output per unit of input, and maintain profit levels. While this is a response from an outside economic pressure, such action within the agricultural sector generates further economic pressure.

Many of the changes made at farm level to offset the effects of the cost-price squeeze result in an expansion of farm output without an expansion in demand for many of those products, particularly while population growth is stagnating. As a result, food prices may actually fall (in real terms as well as nominal) without an increase in consumption if people already have sufficient to eat.

In economic terms this process is shown in Figure 4.2. Initially the market equilibrium is at 20 units of output at a price of $80 per unit. With the adoption of new technology, the supply curve moves to the right and a new equilibrium of 30 units (a 50 per cent increase), but at a price of only $30 per unit (a decrease of 62.5 per cent).

The demand for such basic food products is said to be *price inelastic*. An item is price inelastic if a 1 per cent change in its price results in less than 1 per cent change in the quantity demanded. The changes in technology which allow output to increase will shift the supply curve to the right. This reflects the lowered cost of production (or increase in productivity) because it displays an increase in output for the same cost per unit. The net result of a small expansion in output of a price inelastic product, is a dramatic fall in the price. It is a vicious circle, as this will, in turn, exacerbate

Figure 4.2 *Supply and demand: inelastic farm product.*

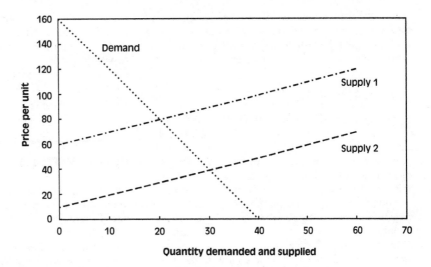

the cost-price squeeze, which was the trigger for the farmers to increase output in the first place.

Again, products like wool and cotton are exceptions to this scenario. As the price of clothing products fall, consumers will buy more, meaning that farm products such as raw wool and cotton are *price elastic*. It is therefore sensible for such products to be included in the product mix of a sustainable farming system if technically feasible.

Implications for land use

Few farmers have any choice but to react to the economic pressures by increasing production and productivity. This means that farm businesses will be in a constant state of change. It is therefore important to have a properly documented mission statement and goals and objectives. If these are written within the context of a landcare ethic, then it is likely that the farm business will develop in a way consistent with resource preservation and sustainable farming systems.

The best farm managers will not shy away from the need to pursue productivity gains. Rather, they will be the innovators, because the first people to get productivity gains will do so before widescale production increases begin to drive prices down, as depicted in Figure 4.2. Also, the best farm managers will try and achieve productivity gains without relying on increased use of purchased inputs. In southern Australian cropping areas rotations can be used to control diseases and supply nitrogen to the system, and attention to detail with time of sowing and depth of sowing can lead to dramatic yield increases without the need for a system driven by more purchased inputs.

Whole farm planning

An important component of preparing a business plan for a sustainable farm business is to have a whole farm plan for the property involved.

Whole farm planning involves:
- Assessing the resource base available.
- Recognizing past land degradation.
- Preparing strategies to halt and reverse degradation.
- Identifying land capability classes.
- Preparing land uses for different areas of the property based on the aspects of degradation and land capability, within the context of retaining a profitable farm business in the long term.

Assessing the resource base

The land is the most important physical resource of any farm. Topography and soil type are the major components, with rainfall and vegetation also being important. A complete soil survey of the property should be undertaken and appropriate maps prepared.

Other resources available to the farm business are capital — usually tied up in the form of machinery and land improvements (e.g. fencing, watering facilities), the ability to get access to more capital (ability to borrow, or financial reserves), and labour, in terms of quantity and capability. While these resources may not normally be associated with a resource survey in the context of a landcare programme, they are essential ingredients when the business component of farming is imposed on the preservation and use of the land-based resources of a farm. Labour and capital resources of individual farm businesses will be fundamental in dictating sustainable land use for different land capability classes.

Land degradation

A part of the soil mapping process of whole farm planning is to identify where land degradation has occurred, or is occurring. The types of land degradation most common are: forms of soil water erosion; forms of soil wind erosion; soil salinization; soil acidification; loss of soil structure; depletion of soil organic matter and nutrients; and loss of vegetation species from an environment. For sustainable, profitable farming, land degradation must at least be halted, and if possible reversed, and the land should not be left vulnerable to further risk of degradation, although it will be impossible to completely remove all such risk. An integral part of whole farm planning is to plan for those works to take place within the financial capability of the farm business.

Land capability classes

Long-term productivity of land for sustainable agriculture is determined by knowing the limitations of each parcel of land, and the management inputs required for long-term stability. Classifying all the land available to a farm business on the basis of that land's capability will enable the correct decisions to be made about the underlying farming system.

Land capability classification is the process of grouping land with similar capability so that various cropping and management strategies can be implemented. Most properties in the dryland areas of Australia can be divided into several land-use capability classes based on soil characteristics and topography.

Examples of different land capability classes in dryland farming areas are:
- *Class I* — land capable of being used permanently for cultivation with very low risk of wind and water erosion.
- *Class IV* — land with severe wind or water erosion hazard which cannot be controlled for regular cropping. Such land can be used for improved pastures, occasional crops or for perennial crops.
- *Class VII* — land with extreme limitations which requires protection with perennial vegetation and can be used for limited grazing only.

By mapping a property into its land capability classes, it is often possible to re-fence areas to combine like areas for efficiency of management and improved productivity. It allows areas of like capability to be used to their full potential, without being restricted by areas of lesser capability which require different management. Land use of a property should then be planned.

Farm systems

The key to successful dryland farming is to implement a farm system, or systems, which are compatible with the stated aims, goals and objectives of the business, and as such, should be sensitive to the economic pressures on agriculture, and be consistent with the capability of the land involved and the need to prevent degradation of the land resources.

There should be a defined farm system for each land capability class on a property. The key parameters involved in defining such systems for dryland agriculture are: identifying a product range; defining a rotation of land use if necessary; tillage practices; weed control; insect control; crop and pasture nutrition; flock and herd structures for animal enterprises, and animal husbandry practices.

Identifying a product range

The range of products suitable for production on a particular dryland farm will be determined by such things as climate, land capability classes represented on the property, and the location of the property relative to markets.

Many of Australia's dryland farms are in the wheat/sheep zone where cereal crops (wheat, barley, oats, triticale, and cereal rye), grain legumes (peas, beans, lupins, and vetches), other legumes (medics, subterranean clover, lucerne etc.) can be grown, and animal products like wool, mohair, cashmere, and meat can be produced. In the main it will be environmental considerations like rainfall, mean daily temperatures, and soil type and condition which will influence the final range of possibilities.

The final choice of actual products produced on a farm is determined largely by the overall goals and objectives of the managers, and the strategies they employ to meet those goals and objectives. For example, if there is a goal to achieve stability of income and growth of income over time, then it is likely that the farm system will include some livestock products like wool, some meat production, and a range of cereal and legume crops.

The inclusion in the system of a product like wool allows the business to break out of the production of goods which are income and price inelastic, thus giving some potential for income growth as general living standards rise. Meat, as a food item, falls in the same category, as developing countries demand more meat as their living standards rise.

Cereal crops are often quite profitable in their own right and so deserve a place in the system even though such products, in the future, may be more exposed to the effects of the cost-price squeeze. It is also for this reason that a farm system based entirely on cereal production is not likely to achieve long-term income growth for the business.

Legume crops are included in the product mix to help meet the stated objective because they can be used for both animal and human consumption. Their market value is therefore linked to rising living standards via the animal industries, which will offset the negative aspect of potential falling demand for human consumption.

Overall, a reasonably large range of products within a farm system will also help stabilize income from year to year. It is unlikely that poor seasons will affect all products in the same way, and it is unlikely that a downturn in market conditions will hit all products at the same time if the range of products is broad enough.

Other goals and objectives can also influence the range of products grown. A goal to minimize the use of chemicals for pest control may lead a manager to include resistant crops in the product range. Similarly, a goal to maintain soil structure and organic matter levels may lead to the inclusion of annual pastures, and therefore some grazing animals, in the system.

Rotations

Overall, the range of products chosen to help meet various goals and objectives will often lead to the formation of rotations, or sequencing of production, on a particular piece of land. On Class V, VI or VII land, where there is only one land-use option, namely grazing, the issue of a rotation does not arise. However, on other land classes the rotation becomes the central parameter of the farming system employed.

In establishing a rotation, the most profitable crops or livestock activities should be determined, and then a rotation built to maximize the production of those enterprises. For example, it may be assessed that wheat production is likely to be the most profitable product in the medium term. A rotation would then be implemented to maximize the production of wheat, while still working within other overriding goals and objectives aimed at long-term sustainability. For example, Figure 4.3 shows a rotation that may be adopted.

Such a crop sequence allows wheat to be grown three years out of eight, but more importantly, allows it to be grown in situations where soil borne diseases of wheat are likely to be at low levels, thus maximizing the potential yield of each wheat crop. This is achieved by incorporating lupins and legume-based pastures as breaks between wheat crops. Crop rotation should be at the heart of a sustainable farming system (Chapter 6).

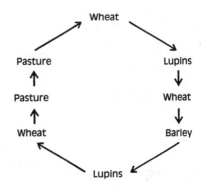

Figure 4.3 *Crop rotation should be at the heart of a sustainable system.*

Tillage

Another important part of the farming system relates to the husbandry practices followed within the rotation. Tillage is important in preserving the soil. In sustainable farming systems the objective must be to minimize soil disturbance and retain as much cover on the soil as possible in order to preserve soil structure and protect the soil from erosion. Minimum tillage, direct drilling and stubble mulching are examples of tillage practices aimed at retaining soil structure, preserving soil organic matter levels, and protecting the fragile soil surface from wind and water erosion. As well as preserving a non-renewable resource, such techniques are also likely to help maintain crop yields in a balanced system (Chapter 7).

Weed control

At some stage a decision needs to be made about the role of cultivation in weed control. Obviously a move to minimum tillage or direct drilling is going to mean a shift to chemical weed control. Here there is a danger of resistance to chemicals emerging if chemical use is not properly planned. The rotation can play an important role here. By having a variety of crops being grown over time, a variety of chemicals can also be used to control weeds. Also, if pastures are included in the farming system, grazing animals can be used to control some weeds.

Crop and pasture nutrition

In most dryland situations in Australia, phosphorous and nitrogen will be major elements in short supply. Other trace elements like zinc and copper may also be deficient. All of these elements can be supplied in various artificial and natural forms of fertilizer. In the case of nitrogen it is possible to inject biological nitrogen into the system via legumes, which have the ability to fix their own nitrogen requirements from atmospheric nitrogen. Legumes in the form of legume pasture species and grain legumes are therefore very important in an efficient dryland farming system.

Flock and herd structures

For flock or herd grazing pastures and crop residues in a farming system, the flock or herd structure is the basis of the production system. Some systems revolve around dry stock, which obviously maintains itself with a regular system of buying and selling stock. Other systems involve breeding and need to be carefully structured to maximize genetic progress or to maximize the benefits of a cross breeding programme.

It is important to understand the way in which livestock integrate into the system through grazing habits and overall demand for fodder. Sheep, goats and cattle are the major grazing animals in dryland systems, and all have different grazing patterns. It is often possible to run a few cattle with sheep without affecting the total number of sheep being grazed. Sheep and goats can also graze together because of their different grazing habits, although there may be other problems associated with health and management.

The decision to run breeding animals rather than dry animals will also affect feed requirement patterns. A lactating female will have a much higher feed requirement than a dry animal, and this peak feed time will tend to occur at the same time each year. It is then important to ensure that the feed production system, often dictated by the crop rotation, is consistent with the peak requirements.

Conclusions

Dryland farming systems are complex, being made up of many small but integrated components. The rotation may be the basic foundation, but out of that comes implications for tillage, weed control, crop nutrition, disease control, and the place for livestock. The system must also be consistent with overall business goals and objectives which, while normally including aspects of a landcare ethic and sustainable production systems, must also take account of business profitability and long-term economic survival within economic systems where rising living standards for all citizens is the normal expectation.

CHAPTER 5

Farm Planning and Land Management

I. S. Ferguson, A. D. Wilson and C. A. Campbell

SYNOPSIS This chapter defines the critical characteristics of joint production involving agriculture and forestry and describes some of the properties and interrelationships of the major dryland farming systems that integrate agriculture and forestry. The role of whole farm planning is outlined.

MAJOR CONCEPTS

1 Whole farm planning is needed to determine the most efficient and effective methods of the design of shelterbelts and wood lots and the role and stocking of trees throughout permanent pasture. Associated changes in the design of paddocks, access and fencing frequently offer substantial savings in the cost of establishment of trees or shelterbelts, as well as savings in the labour involved in livestock management.

2 The integration of agriculture and forestry in dryland farming is an example of joint production.

3 The essential characteristics of joint production are the nature of the interrelationships between the products and the ratios of the prices of the products.

4 The interrelationships may be characterized as complementary or competitive. Most involving dryland farming are competitive.

5 Because many of the products or services are non-market or unpriced in character, it is often difficult to quantify the appropriate combina-

tion of production of the respective goods or services.

6 Crop-shelterbelt systems are severely competitive in character. Nevertheless, they offer potentially significant benefits through the reduction of soil erosion, the provision of fuel wood and fencing, and the improvement of aesthetic values of the property.

7 Pasture-shelterbelt systems are similarly severely competitive and offer similar prospects to crop-shelterbelt systems. They may also assist in the amelioration of salting, provided the revegetation is focused on the major recharge zones.

Because salting may affect downstream neighbours more than the property concerned, more information, more incentives and perhaps more regulatory measures may be needed to encourage the changes needed.

8 Pasture-tree systems involve an intermixture of trees or shrubs and pasture and a moderately competitive relationship under normal conditions. The long-term effects of trees in sustaining pasture production by improving soil

structure and fertility are not well understood but some research suggests that the gains could be substantial. The benefits from shade are better known and can be important as can those from timber production and amenity and other services.

F arm planning has been given a new impetus in the past few years. Many farm plans involve tree planting for a mix of reasons. The integration of agriculture and forestry in dryland farming systems is of growing interest to farmers for several reasons. Awareness of the environmental problems associated with the continuing removal and decline of residual woody vegetation on farms has increased markedly. The aesthetics of the farm and the preservation of native flora or fauna are now a matter for serious consideration by many farmers. The negative impact on farm income of having to purchase additional energy for their household heating or additional timber for fencing and sheds has also become apparent to many. The joint benefits of cropping and/or livestock management together with that of timber or other tree crops are another motivation. The protection that wood vegetation offers to adjacent crops and to animals, and the effects on salinity, offer further reasons.

The level of management expertise required to successfully integrate agriculture and forestry within a dryland farming system is significantly higher than that required to carry out either enterprise separately, which highlights the need for a carefully planned approach over the whole farm.

Whole farm planning

Whole farm planning refers to a process of planning property design and management based on a consideration and integration of ecological, economic and social factors. Thus whole farm planning implies planning of the physical layout of a property, as well as the financial budgeting and management, and the provision of nature conservation, amenity and protection values; not an exclusive focus on any one of them. In this process, particular account needs to be taken of the long-term impact of production systems on the productive capacity of the land and thus of the sustainability and profitability of the farming system.

Briefly, the steps in the whole farm planning process are as follows:

- Objectives. The farm owners need to review their reasons for farming and long-term goals in terms of farm, business, lifestyle, family and personal development.
- Inventory. The first step in an inventory is to prepare a map showing: present layout, including land tenure boundaries and the location of residences, sheds and other structures; fence lines, access tracks, dams, drainage lines, soil types and water courses; existing areas of trees, their species and condition; and problem areas susceptible to or showing erosion, salting, rock, swamps or unattractive areas. Because of their critically important role in the natural ecology, opportunities to restore or preserve the ecological integrity of natural watercourses and wetlands need to be identified and noted. Finally, the base map needs to delineate natural land units and their land capabilities.

Natural land units are broad-scale but relatively homogeneous land types, based on soil types, slope, aspect, drainage, elevation, pasture composition and remnant vegetation. The land capability of each of the land types is then identified in terms of the present and potential uses.

This process can be most readily achieved by using aerial photographs of the property enlarged to an appropriate scale. The information on the aerial photographs can then be transferred on to a topographic map of the same scale, using clear plastic overlays. This preliminary map needs to be further annotated and checked from field inspection to finalize a base map. Subsequent design work can then be pursued on plastic overlays.

- Design. Design requires some graphical skills, a capacity to visualize the results, an appreciation of the possible impacts on production and management, and above all the perseverance to map by trial and error a wide range of possible alternatives to the existing design that will better meet the objectives of the owner and family. Consultants and government advisers can assist in this process but the owner must be heavily involved and prepared to contribute.

Experience suggests that one of the greatest limitations in present-day farming is that the design of paddocks reflects history rather than the present realities of management with a minimum of labour. Changes in design offer opportunities to develop shelterbelts or wood lots associated with new fence lines and access tracks, as well as improving the efficiency of management. However, replacement of fencing is expensive and can seldom be done in one sweep. Thus a new design needs to identify the long-term layout it is designed to achieve and to break that into a series of small stages that can be accomplished over time.

Once the location of shelterbelts and other areas to be planted with trees has been determined, appropriate species and sources of planting stock or seed need to be selected. Tree growth is a relatively slow process compared with that for annuals and watering, cultivation, weedicide treatment, fertilizing, and protection from stock are often essential elements over several years following establishment. Failure to do so is a sure recipe for a wasted and uneconomic investment.

In any event, integration of trees and other revegetation measures into the system would thus take place after a basic layout has been defined and the needs for trees identified, ensuring that trees are likely to be established in the right place for the right reason.

Implementation of whole farm plan

Having developed a plan that involves workable stages in the pattern of development, care must be taken to implement that plan in a timely manner with respect to season and resources. Experience in implementation will highlight changes that need to be made in the further stages, and so the refinement of the design continues. Advice from consultants, government advisers or neighbours that are further advanced in similar work should be sought where possible. Where salting and gully erosion are involved, co-operative action at a catchment level may be needed (Figure 5.1). In any event, participation in landcare groups provides a valuable source of information and communication.

Working through this process encourages farmers to further their knowledge of the land and gives them ownership of the decision-making process, ensuring that plans are more likely to be implemented. Whole farm plans provide a useful framework for co-ordinating advice from a diverse range of sources including government agencies. The process of whole farm planning quickly exposes farmers to ecological issues which do not respect farm boundaries, such as rising water tables, stream water quality or wildlife habitat, and consequently it can be used to provide linkages between farm, catchment and regional

Figure 5.1 *Soil erosion is significantly reduced in catchments protected by well-maintained contour bank layouts and sound land management practices. Whole farm planning is best done where there is co-operation at the catchment level* (PHOTO: Queensland Department of Primary Industries)

land-use plans. The process also encourages group activity, by exposing the interdependence of farms and providing a mechanism to assist groups of farmers to co-operate in tackling common problems.

Integration of agriculture and forestry

The process of integration of agriculture and forestry is best defined in terms of the joint production of two or more products or services. The emphasis on joint production is appropriate because there are important interactions between the two components of agriculture and forestry, if for no other reason that an area occupied by a tree or shrub could potentially be used for some other purpose. Generally, however, stronger sources of interaction exist and these form a major consideration in this chapter.

For descriptive purposes, the major dryland farming systems in which joint production is practised fall into three broad systems: crop-shelterbelt, pasture-shelterbelt and pasture-tree systems. Each has distinctive properties with respect to the relationships between the products or services concerned, although there is clearly much overlap between crop-shelterbelt and pasture-shelterbelt systems in the areas where integrated crop and livestock production is practised.

Crop-shelterbelt systems

In the drier cropping areas, such as the Mallee in Victoria, the retention of dispersed shelter interferes with the large machines used for cultivation and harvesting. Wind erosion is usually kept in check in normal years by improved farming procedures such as stubble retention. However, in drought years crops may not be planted or may fail after planting, leaving the soil vulnerable to wind erosion. In such a situation, winds of 10 m/s have resulted in the removal of 27 t/ha of topsoil and the sandblasting of a nearby crop.

Shelterbelts can be created by the retention or planting of strategic belts or clumps of trees to provide protection from wind and to ameliorate dryland salinity through their role as water pumps,

as well as providing amenity and other values (Figure 5.2). Shelterbelts are severely competitive with crop production in that the land is taken out or reserved from crop production. The issue is then whether the net benefits derived are greater than those from cropping and this depends on the prices or imputed prices involved, with all the obvious difficulties that this poses with regard to non-market services such as protection from wind or salting, preservation of flora and fauna or enhancement of aesthetic values.

Overseas research indicates that the height, porosity and profile of shelterbelts are important determinants of the extent of protection from wind. In general, the higher the shelterbelt, the wider the band of protection afforded from wind on the lee side. However, this effect is complicated by interactions between height, porosity and profile. Shelterbelts with very dense foliage obviously present a much greater resistance to wind. In doing so, however, they may create turbulent flows such that while windspeeds on the immediate lee side of the break are reduced, those further out are actually increased for a significant distance. Shelterbelts with more porous foliage do not generate turbulent flows to the same extent. The profile or cross-sectional shape of the shelterbelt also has a potential impact on the extent and magnitude of turbulent flow. From the windward side in, successively taller rows of species are

Figure 5.2 *Clumps of trees provide protection, amenity value and can reduce the loss of soil through runoff.* (PHOTO: Queensland Department of Primary Industries)

likely to reduce the prospect of significant turbulent flows.

The effects of shelterbelts of radiata pine (*Pinus radiata*) and mallee (*Eucalyptus polybracta* or other mallees), two species with markedly different heights and porosities, are illustrated in Figure 5.3. While the data for radiata pine are only germane to the wettest fringe of dryland farming areas, with an annual rainfall in excess of 500 mm/year, the contrast in heights is useful.

The data for the pine shelterbelt at Mt Compass illustrate the possible effects of turbulent flow, in that some twenty times the tree height from the belt, wind speed is higher than that on the windward side of the belt. All of the data show the gains to be made in the reduction of relative windspeed.

One example of the effects of a shelterbelt on the yield of barley in South Australia is shown in Figure 5.4. The shelterbelt comprises two lines of trees on either side of a 20 m road, the windward row being the taller mallet (*E. astringens*) and the lee row being the lower wandoo (*E. wandoo*). The prevailing wind is persistent and strong.

The data show a substantial increase in yields within a band of about twenty times tree height on the lee side.

A further option is that of alley cropping. This

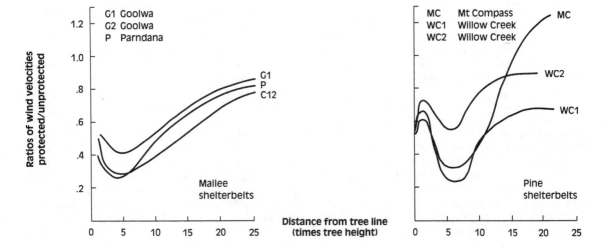

Figure 5.3 *Effects of shelterbelts on wind velocities.* (SOURCE: SA Bushfire Research Committee [unpub])

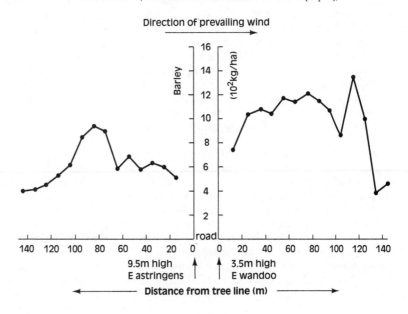

Figure 5.4 *Effects of shelterbelts on crop yields.* (SOURCE: van der Sommen, 1984. Workshop on the benefits of trees on farms, Department of Home Affairs and Environment)

involves arable cropping between rows of a leguminous tree. This system has been used in tropical climates (e.g. maize and leucaena (*Leucaena leucocephala*) in Nigeria). The trees are pruned prior to crop planting and the prunings used to mulch the crop land. The system requires the selection of special tree species that are deep rooted, coppice after cutting, and have a use such as forage or fuel wood. The advantages are mainly in terms of the protection from erosion and improvement in fertility of the soil, together with the growing of another product. It could be adapted to dryland cropping regions with species such as tagasaste and saltbush if the logistic problems of planting and grazing could be solved.

There is a zone close to the shelterbelt in which the potentially beneficial effects of shelter from wind may be offset or even reversed by shading or competition for soil moisture between the trees and the crop. The extent of this zone depends on the species and climates involved but may be up to twice tree height.

In cropping areas, shading is not generally of critical importance because light is seldom limiting. Indeed, the shading effects may even be beneficial where extremes of temperature occur in the growing season. However, competition for soil moisture and nutrients is important and much depends on the rooting habits of the two potential competitors. The greater the tendency to deeper rooting and less dense surface rooting in the tree species the better, because most crop species are shallow rooted.

Although many studies are currently in progress, little is known about the best species or combinations of species for Australian conditions. The broad principles enumerated above provide the principal guidelines for establishment of shelterbelts. However, shelterbelts need to be located at intervals of about twenty-five times their height and placed across the direction of the strongest winds (e.g. generally running north-south).

Whatever the nature of the shelterbelt, the frequency and consistency of direction of wind at levels that would otherwise curtail transpiration and photosynthesis during the growing season are of prime importance. Shelterbelts for protection from wind are most effective where wind of significant speed from a particular direction is a dominant and persistent element in the climate, especially during the growing season.

Many dryland cropping areas face difficulties with rising water tables and associated dryland salinity, arising from changes in the natural hydrological balance. The original tree and shrub canopy had deeper roots and a longer growing season than the annual cereal crops (and pasture, in mixed farming areas) that replaced them, so that more of the rainfall now percolates through to the water table. Since the deeper soil profile contains large amounts of stored salt, the rise of the water table brings salt to the surface at the lower parts of the landscape. As a consequence, regulations have now been introduced to prevent further clearing of Mallee vegetation in the Murray geological basin and elsewhere. However, over the wider part of the landscape, clearing was completed many years ago and dryland salting has become an increasingly serious problem in the amount of land taken out of production or showing reduced yields.

In dune-swale mallee systems it is likely that most recharge of the water table occurs on the lighter soils of the dunes and it is these that should have priority for tree retention or revegetation.

Shelterbelts can provide a variety of other values besides protection from wind and as water pumps. They can provide an opportunity to preserve native flora and fauna and so contribute to the breadth and frequency of representation of endangered ecosystems. While such endeavours may contribute to the farmer's personal satisfaction or utility, there may be additional benefits to farm production through the effects of associated populations of birds on insect pests. However, there is little hard evidence about these effects in this farming system. It may also be that there are additional costs through damage to crops by bird and insect populations.

The aesthetic benefits of shelterbelts on crop lands are well established through the prices paid for properties. A landscape without trees is not only particularly vulnerable to erosion in drought years: it suffers a loss of capital value because of visual monotony.

Because of the low rainfall in the cropping areas, the opportunities for producing commercial products in shelterbelts are limited. Reference has already been made to the possibilities for fodder in alley cropping. Fuel wood and light fencing and poles are the other main commercial products. Because these are primarily consumed by the farmer, little is known of the price ratios and thus of the significance of their contribution to the economics of the enterprise.

One other product that could be grown commercially is broombrush (*Melaleuca* spp). Now that commercial harvesting of broombrush on public lands has been curtailed and may cease, opportunities for the development of broombrush shelterbelts may become attractive.

Pasture-shelterbelt systems

Because cropping and livestock are integrated over very large areas of intermediate rainfall, many of the preceding principles hold for pasture-shelterbelt systems. The presumption is that in part we are dealing with those areas on which the retention of dispersed shelter interferes with the large machines used for cultivation and harvesting of crops. However, many of these areas are also used for pasture production on a rotational basis or as economic conditions dictate.

The effect of wind on livestock, and hence the value of shelter, can be evaluated from known physiological response of sheep. Heat is lost from the body according to multiplicative effects of air temperature and wind speed with an additional cooling effect of rain. These factors have been combined into a chill index which may be used as a measure of the effect of the total environment on the animal and becomes a dependent variable in studies on lamb and shorn sheep survival in inclement weather. Whilst several factors are involved in cold stress, it is only wind that is subject to management control. Windbreaks can halve the lamb mortality in cold climates and wood lots can eliminate losses of shorn sheep.

Exposure to cold also increases the amount of energy diverted to maintaining the animal and hence decreases the energy available for growth. At Armidale in NSW, improvements of 18 per cent in pasture growth and up to 31 per cent in annual wool production have been recorded from the use of artificial windbreaks.

In general, the same principles prevail for the establishment of shelterbelts in pasture as in crop production. The effects of competition between shelterbelts and pasture are not well known and warrant more research. For example, Monterey cypress (*Cupressus macrocarpa*), a species widely used for windbreaks in wetter pastoral areas, is renowned for the exceedingly dense mat of relatively shallow roots around the tree, with obvious possible consequences for pasture growth in the immediate vicinity, but there are no data available on the extent and magnitude of that competition.

Despite the lack of information, the choice of tree species is wider in this system and the possibilities of deriving other commercial products is better. However, most tree planting in dryland areas, both cropping and pastoral, has used the traditional method of planting seedlings raised in nurseries. Effective weed control to conserve soil moisture, with a residual herbicide, or by grading or trenching, is essential. Watering is required in the first year and this greatly reduces the number of trees that can be planted. Direct seeding of local species of she-oak (*Allocasuarina* spp), wattle (*Acacia* spp) and paperbark (*Melaleuca* spp) species has also been successful, although the results may vary with the seasonal rainfall in the year of seeding. Once again weed control with a residual herbicide is essential for success. The development of such low-cost techniques is essential if the benefits of replanting on a wider scale are to be realized.

While shelterbelts can play a useful role in providing shade for stock, they are much less desirable than scattered trees because they tend to encourage camping by stock in the major areas of shade, with consequent trampling and overgrazing to the associated pasture.

The use of shelterbelts or wood lots to reduce the input of water into the recharge zones and so reduce the extent or spread of salting seems likely to be of more significance in this system, given the levels of rainfall and inland drainage patterns common to it. In general, the most significant recharge areas are often on stony or sandy ridges. This poses a particular problem for fine-wool producers because these areas are the most important source of fine-wool production and their removal from wool production into shelterbelts or wood lots may have significant financial repercussions. In the longer term, grazing may be resumed in these shelterbelts or wood lots once the trees are large enough and given suitable periodic thinnings. But there is little reliable information on the actual efficacy of tree-planting for this purpose, nor on the thinning and management regimes needed to sustain some grazing.

Given these uncertainties, it is unlikely that farmers will entertain the establishment of shelterbelts or wood lots for the amelioration of salting, especially as those farmers with the recharge areas may suffer much less from salting than those at lower elevations in the catchment. More information, more incentives or even more stringent measures may be needed before substantial progress can be made in this regard.

As in the case of crop-shelterbelt systems, the preservation of native flora and fauna may contribute to the farmer's personal satisfaction or utility. There is an increasing awareness of these values, especially in relation to encouraging populations of native birds.

Additional benefits to farm production may occur through the effects of associated populations of birds and insect populations on insect pests. For example, scolitid wasps are said to control scarab beetles that otherwise sometimes spoil pasture growth in temperate regions. These wasps are reliant on sources of nectar from native species at particular times of year for their survival and will thus be maintained better where shelterbelts or wood lots or reserves of native flora are maintained or established. There is much anecdotal evidence on the roles of various birds, mammals, reptiles and insects in eating or controlling various insect pests but very few hard data.

In some cases, small amounts of income in the higher rainfall areas can be generated from leasing rights for apiary sites where special species such as the summer flowering red ironbark (*E. sideroxylon*) are available.

The opportunities for growing nut and fruit-bearing trees are somewhat limited in the dryland farming areas, because the rainfall is generally too low to enable the best-known nut species such as almond (*Prunus dulcis*), chestnut (*Castanea sativa*) and walnut (*Juglans regia*) to be grown unless irrigation is available. Furthermore, these species often require much more intensive cultivation and tending in the early years to enable successful establishment. Some of the stone pines (*Pinus pinea, P. cembra, P. cembroides*) can be grown successfully in higher rainfall areas and yield edible pine nuts but there has been little commercial development of these species in Australia; all of the commercial pine nuts sold being imported. Fruit trees are even more restricted in scope, for similar reasons. Some promise has been reported with quandong (*Santalum acuminatum*), dates (*Phoenix dactylisera*) and the capricornium plum (*Terminalia ferdinandiana*) but principally in the context of plantings around or adjacent to the homestead.

With some exceptions, the prospects for growing timber for sawmilling are generally not strong in dryland pastoral areas. Rates of growth are slow. The volumes are also insufficient to attract any but mobile or small-scale local sawmills and prices received therefore tend to be irregular and low.

The most notable exceptions of tree crops in dryland pastoral areas with potentially high yields of sawlogs, round timbers and/or fuel wood are those grown on land irrigated with sewerage effluent or other waste water. Spectacular successes have been achieved at several locations, notably with river red gum (*E. camaldulensis*) at Alice Springs and flooded gum (*E. grandis*) in northeastern Victoria.

Nevertheless, as noted for crop-shelterbelt systems, there are good opportunities for growing fuel wood, fencing timber and poles to be consumed on the property, and these represent a potentially significant contribution.

Pasture-tree systems

Pasture-tree systems refer to the situations in which the trees (or shrubs) are intermixed with the pasture. Here the relationships are moderately competitive to some degree because of the mutual competition between pasture and trees for soil moisture, nutrients and light. Assuming that the prices pertaining to the services of trees are positive, and this seems clear, the issue is one of determining the best stocking of trees rather than whether there should be any.

The role of trees when intermixed with pasture for amenity and other related values is little different to that described for pasture-shelterbelt systems and need not be repeated here. However, there are a number of other distinctive features of pasture-tree systems.

The complete clearing of trees to facilitate cropping or grazing is a widespread practice (Figure 5.5). The trees may provide no direct economic

Figure 5.5 *Clearing of trees to facilitate sowing of improved pastures was a common practice in the past.* (PHOTO: N. Bonney, Greening Australia)

return, obstruct machines, and compete with grasses for water and minerals. The savannah woodlands of semi-arid and subtropical Australia are an example. Clearing all the trees gives a substantial short-term increase in herbage production because more water is then available for the grasses.

Although complete clearing has been widely recommended and practised, the advantages in the long-term are not so clear. In time, the salinity of the surface soil increases markedly wherever the soils are sodic and soil structure may deteriorate from greater exposure to raindrop action and trampling. Also, trees draw nutrients and water from depth and have a role in these ecosystems in maintaining the infiltration and fertility of the soil surface. The same effect has been observed in southern Australia. Nutrient return from rainfall drip and litter can be substantial in nitrogen, phosphorus, potassium, calcium and magnesium and contribute to better pasture growth under the canopies of scattered yellow box (*E. melliodora*) and river red gum (*E. camaldulensis*) than on the intervening areas.

One study has demonstrated a decline in production with time after clearing in the subtropics. On the basis of a combined production and financial analysis, the best option for that region was to improve the pasture beneath the trees by introducing a legume and fertilizing with phosphorus and not to remove the trees at all. The trees are costly to remove and are an insurance of long-term sustainability.

This conclusion is not necessarily transferable to other regions — it will depend on the soil type, tree species, density and production responses. For example, a short-term study in Western Australia suggested that scattered marri (*E. calophylla*), jarrah (*E. marginata*) or wandoo (*E. wandoo*) contributed to a small decline in both dry-matter production and clover content of pasture. However, the long-term effects need to be considered, as well as the immediate production changes.

Comparative studies of soil organic matter elsewhere show that continuous cultivation resulted in a marked decrease in soil organic matter compared with the original woodland. Pasture production over three to ten years resulted in a major but lesser reduction. Thus some reduction in tree and shrub cover may be necessary to improve forage production, but a degree of tree cover (5–10 per cent) may have an important role in providing long-term stability to the ecosystem, as well as other possible benefits discussed below.

Livestock will seek shade when the temperature exceeds about 30°C, but production responses are not evident until the heat load is sufficient to produce physiological stress. Sheep are insulated from the radiant heat loads by their wool and heat stress of sufficient magnitude to affect production is unlikely to occur in southern Australia. Shade is nonetheless commonly provided as a welfare measure on most farms (Figure 5.6).

Heat stress is more common in northern Australia where average temperatures are higher

Figure 5.6 *Shade for livestock is an important protection from the hot sun.* (PHOTO: V. Squires)

and sheep may be lambed in summer to take advantage of the summer growing season. As an example, the provision of shade on the Mitchell grass plains can increase lambing percentages by 10–15 per cent. Heat stress at joining may reduce oestrous and increase early embryonic mortality, while stress at lambing reduces birth weight and lamb survival. In this region shade may be provided by planting Athel pines (*Tamarix articulata*). Shade is also needed to protect newly born calves from heat stress.

Trees and shrubs have the potential for an important place in dryland farming systems in providing a source of reserve forage or a supplement of protein at a time of the year when grasses are either scarce or of low quality.

Tree fodders are not necessarily of the best quality, indeed many are quite fibrous and hence are of low energy availability. The digestibility of dry matter is commonly in the range of 40–50 per cent, which is sufficient only for maintenance of body weight. However, a few species are of much higher value and their particular advantage lies in having a low seasonal variation in quality, and by virtue of their deep rooting, maintaining green leaf into the dry season. An example of an Australian species is the wilga tree (*Geijera parviflora*), which contains more than 2.5 per cent nitrogen in its leaf at all times of the year. Another characteristic is that the leaves of trees often contain anti-nutritional substances, such as oils and tannins, that render them unpalatable or restrict their intake. This is particularly true of Acacia species, which contain substantial tannins, and hence are of little value as forages. The tannin binds the protein and makes it indigestible to domestic ruminants unless mixed with costly additives such as polyethylene glycol. Some browsing animals such as deer, and to some limited extent, goats, produce a salivary mucoprotein which neutralizes the tannins, leaving the plant protein available for digestion. Hence the value of the fodder is partly dependent on the animal system in which it is to be used.

A major consequence of the presence of these anti-nutritional substances is that quality cannot be determined by the normal laboratory tests used to assess forages. Food intake and weight gain may simply not be related at all to nitrogen content and *in vitro* digestibility (the normal measure of energy availability). The widespread use of such tests alone is a serious impediment to the evaluation of fodder trees and shrubs. They can only be tested in feeding trials.

In view of their better relative value in the dry season and the prevalence of anti-nutritional substances, fodder trees should always be viewed as supplements rather than as sole forages. If eaten in small amounts they may rectify a protein deficiency without giving rise to the metabolic problems that arise when eaten in larger quantities. They may also be used as drought reserves because they accumulate leaves, often with quite high moisture contents, from wet to dry season and to some extent from year to year. A useful example is the tree-legume leucaena (*Leucaena leucocephala*), which is valuable in the dry tropics as a dry season forage for cattle. It has a high protein content. Where one-quarter of speargrass country has been planted to leucaena, cattle weights are more than 100 kg heavier at 33 months of age. Stocking rates may also be increased by 25 per cent. Leucaena contains the toxic substance mimosine, but rumen bacteria have been introduced from cattle in Hawaii to detoxify this substance.

In Mediterranean climates a similar place may be served by the shrub tagasaste (*Chamaecytisus palmensis*), although the advantage is probably less simply because the normal grass forages in winter-growing climates are of better quality than their tropical counterparts. Tagasaste also accumulates large amounts of fodder — 10–15 kg per tree — which is available for cutting or grazing in the dry season. Nitrogen is high but phosphorus, potassium and sulphur contents are low and may need supplementing. Liveweight gains are less than expected, which reinforces the view that it is a plant for specialist dry season niches rather than as a wonder fodder — an accolade sometimes awarded to it by its advocates. It also has specific disadvantages of requiring harvesting rather than grazing, because high branches are not eaten and plants may be damaged by bark chewing. It is commonly grown in rows along fence lines and access roads to facilitate harvesting.

Another shrub of specialist value is the chenopod *Atriplex nummularia*. This shrub grows well in semi-arid climates and has a high nitrogen content. However, its leaves also contain more than 25 per cent mineral matter, mainly in the form of sodium chloride, and it is relatively unpalatable to livestock. Once again, food intake is much lower than expected from laboratory analyses and its value is greater as a supplement than as a sole fodder. It was widely advertised in the Australian popular press of 1988–9 as a plant that could drought-proof a grazing property when planted in closely grazed plantations. It does have some valuable properties in this regard, but remains quite uneconomic because of establishment costs which may exceed $600/ha. It will remain so until technology is developed that allows successful direct seeding.

Although fodder trees and pasture are essentially moderately competitive, the extent of the competition is dependent on rainfall. In a variable rainfall environment, competition is greatest in years of moderate rainfall. It is least in years of high rainfall when moisture is abundant and in droughts when the pasture may not grow at all. Given that the competitive effects do not appear to be dramatic even under conditions of moderate rainfall, the value of this system is that it provides a low-cost buffer against possible disasters due to drought. The relative net benefit of the nutritional value versus that of pasture is therefore of little moment because the alternatives during drought are worse.

The retention of trees in pasture can also provide a source of timber for domestic use or sale. Durable species such as red gum, boxes and ironbarks can supply strainer posts and other materials for fencing. Their open-grown form is often sufficiently straight and free of branches to provide adequate round or split timber for the purpose. In drier regions, cypress pine (*Callitris bugelii*), mulga (*Acacia aneura*), wandoo (*E. wandoo*) or raspberry jam (*A. acuminanata*) can also provide useful fencing and post timbers. Durable and non-durable species can also supply fuel wood, although the denser species such as the boxes or wandoo (*E. microcarpa* and others, *E. wandoo*) are generally to be preferred.

Cypress pine (*Callitris bugelii*), occurring in the mid-west of NSW and Queensland, is an example of a tree species that produces valuable sawlogs, albeit slowly grown under the relatively low rainfall of that region. Cypress pine regenerates freely after fire provided livestock or rabbits do not decimate the seedlings. Although relatively slowly grown, cypress pine can be sawn into timber at quite small sizes by comparison with eucalypts. The timber is termite resistant and the multitude of knots are often considered to provide a decorative appearance. Much of the public land carrying cypress pine is leased for grazing and subject to limitations on clearing and on the payments to lessees with respect to any timber harvested. These institutional constraints distort the practice of joint production but research suggests that the optimum combination would be attainable if those constraints were removed. Studies of naturally regenerated stands of cypress pine indicate that the selection of stocking levels of trees and of livestock are of critical importance, the physical rela-

tionship of production being moderately competitive. As a rule of thumb, the best stocking of trees is well below that which a forester would want if wood production was the sole consideration, but substantially more than a grazier would wish to retain if the sole consideration was grazing.

In the wetter fringes of dryland farming areas in Western Australia, clover has been sown to good effect under stands of native trees such as jarrah (*E. marginata*) and wandoo (*E. wandoo*) that have been cut selectively for sawlogs and fenceposts in the past, leaving them in a relatively open condition. Adequate data are not available to assess fully the interactions but they appear to be only moderately competitive.

Much remains to be done in investigating the use of tree species with specialty uses for timber or roundwood. In particular, species such as the casuarinas, ironbarks, boxes, wandoos, cypress pines and the like have not received sufficient attention and represent an exciting challenge for the future.

The types of products in agriculture are described elsewhere in this text. Those for forestry cover a diverse range of goods and services; from various wood products, other tree or shrub crops such as nuts, fruits and honey, trees and shrubs that provide animal fodder, services such as conservation of flora and fauna or enhancement of aesthetics, and the protection services afforded to crops or animals through shade, shelter or to water tables.

The ultimate rationale for engaging in the integration of agriculture and forestry lies in the expected increase in profitability or utility of the enterprise to the owner of the farm. Because the cash-flow effects are often less important than the satisfaction or utility derived by the owner, it is appropriate to consider the net benefits in more general terms than those of cash-flow profits alone. The difficulty then is to quantify these net benefits. Some of the benefits are exceedingly difficult to measure in any practical sense and can only be weighed subjectively. Some of the costs, such as those relating to costs borne by downstream water users of saline discharges, are also difficult to measure. Hence this review of the integration of agriculture and forestry is necessarily descriptive rather than analytical in content. As with any other example of joint production, two important characteristics of the particular system have to be evaluated, albeit subjectively.

The first concerns the nature of the physical relationship between the products or services concerned. Are they complementary or competitive? If competitive, to what degree?

Complementary production means that an increase in the production of one product is accompanied by some increase in the other. Complementary relationships offer obvious advantages because joint production is not at the mutual expense of the other product. Any farmer will seek to take advantage of a complementary relationship and to push joint production to the limit of that complementarity.

Competitive production means that an increase in the production of one product results in a decrease in the other. Competitive relationships are more common and the extent of the mutual interaction needs to be evaluated before one can establish whether joint production is beneficial. For example, if competition is severe, such that each additional unit decrease in one good results in a progressively larger increase in the other, a single use will be the optimum choice. Which of the two is chosen depends on the ratio of the prices of the products. On the other hand, if competition is moderate, such that each additional unit decrease in one good results in a progressively smaller increase in the other, then joint production will generally be the best choice with the ratio of the prices determining what is the best combination.

The ratios of the prices concerned are thus the second characteristic of importance. Clearly, any additional production has to be weighed by the benefits received per unit of product. There is no point in producing more of a good that confers no additional benefit. But many of the goods and services involved in the forestry side of the joint production have no established market and therefore no going price. Protection, aesthetic and preservation services are examples. Does this mean these are valueless? Obviously not. The fact that no market exists is an institutional defect that may simply reflect the difficulties or absence of trading the good or service. The potential economic benefits are just as real as those with monetary prices. Some of these are reflected indirectly in the prices paid for farm land when it is traded, as in the case of a farm that has obvious aesthetic benefits from scattered trees or woodlands compared with a neighbouring property that does not. Farmers exhibit the same sort of sensitivities to aesthetic benefits in the purchase of property as do urban dwellers.

Nevertheless, the absence of market prices for primary services like aesthetics does pose difficulties. In order to determine the best combination of such services and commercial farm production it may be necessary to impute a price to these services: a process that is both analytically difficult and very imprecise. More often than not, farmers must settle for incomplete information and simply explore the effects of a range of possible price ratios. Having defined a possible range in which the best combinations may lie, they proceed by trial and error with incremental changes in the amount of these services to be provided, weighing the results of each change subjectively before proceeding further.

Conclusions

Integration of trees and other revegetation measures into the system should take place after a basic layout has been defined and the needs for trees identified, ensuring that trees are likely to be established in the right place for the right reason.

The better designs will almost certainly have to be recast into several stages that span a number of years for implementation, because of the limits on the resources available. Budgets need to be prepared in order to compare and refine the designs. The budgets need to take account of both the investment to be incurred in new works and the future annual cost and revenue flows, so that an evaluation of the economics can be made.

Further reading

Beckmann, R. (1989). 'Rural dieback: restoring a balance', *Ecos*, 62, 8–15.

Bell, A. (1989). Trees, water and salt — a fine balance', *Ecos*, 58, 2–8.

Cremer, K. (ed.), (1990). '*Trees for Rural Australia*, Inkata Press, Melbourne.

Davidson, S. (1989). 'Tree clearing in the semi-arid tropics'. *Rural Research*, 144, 24–5.

Reid, R. and Wilson G. (1985). *Agroforestry in Australia and New Zealand*, Goddard and Dobson, Box Hill, Victoria.

van der Sommen, F. J. (1984). 'Trees and agricultural production — an ecosystem perspective', 24–74, *Workshop on the Benefits of Trees on Farms*, Department of Home Affairs and Environment, Canberra, ACT.

Venning, J. (1989). 'Revegetation of degraded lands', in J. C. Noble and R. A. Bradstock (eds), *Mediterranean Landscapes in Australia: Mallee Ecosystems*. CSIRO, Melbourne.

II

Operation and Management

CHAPTER 6

Crop and Crop-Pasture Sequences

P. G. Tow and J. E. Schultz

SYNOPSIS This chapter presents a comparative analysis of dryland farming rotations occurring throughout the Australian cereal belt, from the south-west of Western Australia to north Queensland. The import-ant influences of rainfall amount and annual distribution and of soil type are explained. The evolution and operation of present-day rotations are also related to available technology and plant cultivars, and the need for greater diversification, profitability, flexibility and sustainability of farm systems.

MAJOR CONCEPTS

1 Rotation sequences are adopted in response to the need to diversify products, break life cycles of disease organisms, maintain soil fertility, control weeds, prevent soil erosion and make efficient use of resources.

2 The composition of rotations is determined to a large extent by the level, seasonal distribution and variability of rainfall, by soil type and the risk of soil erosion.

3 Farmers require as much flexibility as possible in rotation sequences to deal with variability in rainfall and market prices.

4 The success of rotational systems depends strongly on appropriate research, plant cultivars, technology and management.

5 The choice between cereal-grain legume and cereal-pasture legume rotations depends on tradition and the relative prices for crop and livestock products, as well as rainfall, soil fertility and suitable cultivars.

6 Particular rotational systems are rarely sustainable indefinitely; adjustments or complete changes become necessary to overcome ecological, technical and economic limitations which develop.

There are instances throughout the world of farmers growing a single crop continuously over many years, for a reliable market. Examples in Australia are wheat and sugar cane. Over time, however, problems arise which make it desirable to discontinue growing this crop for one or more years. In this period, the land is either put into an alternative crop or pasture or is simply left fallow. Crop rotations have evolved to cope with changing ecological, economic and technological situations.

Objectives of rotations

The seven most important objectives of rotations are to:

- Diversify into a range of crops or enterprises.
- Break life cycles of disease organisms carried over in soil or in crop residues.
- Maintain or improve soil fertility (especially nitrogen) and soil structure.
- Control weeds which are difficult to manage in a particular crop.
- Control soil erosion by using pastures or crops which leave a protective cover of residues on the soil surface.
- Allow opportunity cropping: growing two different crops consecutively in one year of high rainfall; or introducing an alternative crop when markets are favourable.
- Make more efficient use of resources, e.g. of soil moisture by alternating crops with different depths of root extension or of labour by spreading the workload more evenly over the whole year.

Achievement of the above objectives through appropriate sequencing of crops and pastures are steps towards the general aims of higher productivity, profitability, stability and flexibility, as well as long-term sustainability.

Determinants of rotations

The objectives, design and management of rotations are influenced strongly in the first place by climate and soil characteristics. The most important features of climate are total rainfall, its distribution between the winter and summer seasons and its variability; and the incidence of frost and heatwave. Some of these features are shown in Figure 6.1, along with examples of commonly practised rotations. The most important soil characteristics to consider for rotations are surface and sub-surface pH, organic matter and nitrogen content, texture, structure, drainage, water holding capacity and erodibility. Differences in soil within a single farm can cause a requirement for different crops or pastures and their sequences (Chapter 15 and Case Study E). One or more of these environmental factors, as well as one or more of the seven objectives listed above may be critical determinants of the kind of rotation which is developed.

The relative importance of the determinants may also change with time and farmers may modify rotations or shift to different types if necessary.

Figure 6.1 *Some climatic factors and rotation characteristics of selected regions of the cereal zone.*

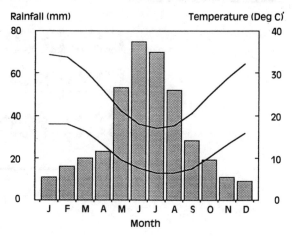

Figure 6.1a *Wongan Hills, Western Australia.*

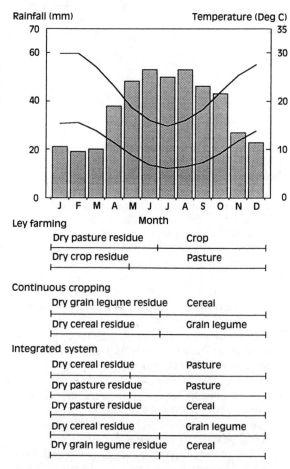

Figure 6.1b *Roseworthy, South Australia.*

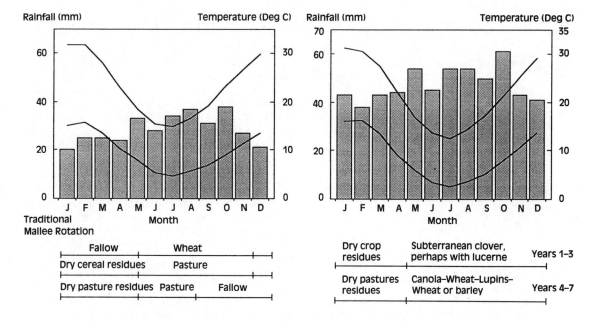

Figure 6.1c *Walpeup, Victoria.*

Figure 6.1e *Wagga Wagga, NSW.*

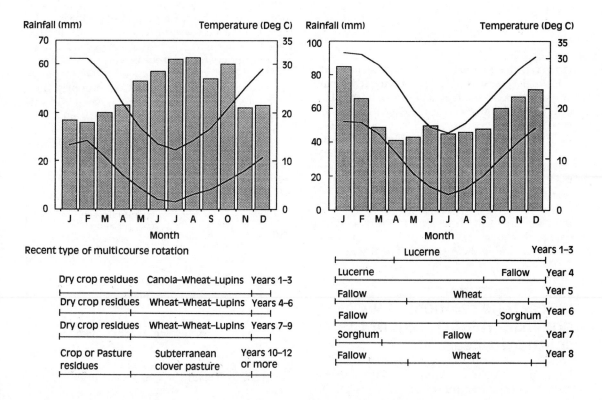

Figure 6.1d *Rutherglen, Victoria.*

Figure 6.1f *Tamworth, NSW.*

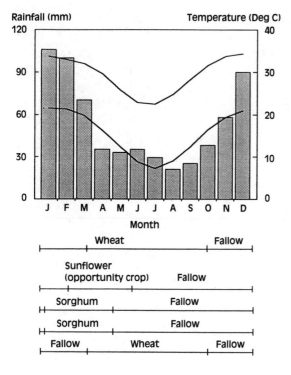

Figure 6.1g *Clermont, Queensland.*

However, the rotation or sequence alone cannot achieve the goals without appropriate accompanying resources, technology and management. These include availability of high yielding disease and pest resistant varieties, appropriate machinery as well as tillage, sowing and spraying techniques, suitable chemicals, computer techniques for financial and resource management, and techniques for conservation of soil, soil nutrients and soil moisture.

The remainder of this chapter will provide a brief analysis of the development, purpose, design, operation and management of crop and pasture sequences used in the Australian cereal zone.

Cereal zone rotations — winter-dominant rainfall areas of the southern wheatbelt

In southern Australia, with its winter-dominant rainfall distribution, the choice of crops for dryland farming is restricted to winter-growing annuals. A five-month summer drought is common,

during which the land is left as fallow or sheep and/or cattle graze dry pasture residues or crop stubbles. Despite the climatic restraints the development of new technology has allowed crop sequential practices to evolve which continue to meet the goals of higher productivity and profitability while at the same time providing stability and flexibility. The practices widely adopted from time to time have not all been sustainable in the long term and this has often been the catalyst for change.

Trends in Australian wheat yields are shown in Figure 6.2. In the early decades wheat was cropped continuously and yields declined. Clearly, this was not a sustainable system. After 1900 the commonly used wheat-fallow rotation, together with superphosphate and new varieties, led to yield increases, but soil organic matter declined and erosion became widespread, and wheat yields reached a plateau. So the wheat-fallow rotation was not sustainable in the longer term. From 1950 previously degraded soils were made productive again through the use of wheat-legume pasture rotations (ley farming). These pasture rotations largely met the goals of productivity, profitability, stability and flexibility and they were considered to be a sustainable form of land use. However, by the 1970s farmers were forced to look for alternatives because of the imposition of wheat quotas and poor prices for livestock products.

The challenge has been to devise viable farming systems which allow intensive cropping (and largely omit the pasture phase) without the risk of the soil degradation which resulted from earlier periods of intensive cropping. Fortunately, a number of technological improvements have allowed

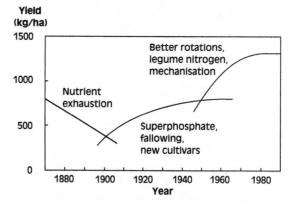

Figure 6.2 *Trends in Australian wheat yields.*
(SOURCE: Donald, C. M., 1981. 'Innovation in Australian Agriculture', in D. G. Williams (ed.), *Agriculture in the Australian Economy*, Sydney University Press, 57–86)

this to take place. These include new varieties, reduced-tillage techniques, better herbicides for weed control, stubble retention and suitable machinery.

This change to more intensive cropping rotations has been facilitated by the development of grain legume crops. Lupins (especially in Western Australia and north-eastern Victoria) and field peas (especially in South Australia and more recently in Victoria), together account for some 90 per cent of winter grain legume production. Faba beans and chickpeas are minor grain legume crops and lentils now being tried on limited areas. Oilseed rape (canola) is another crop that has found a limited niche in crop rotations in southern Australia. Cereals with particular disease resistance characteristics (e.g. Galleon barley and Molineux wheat with Cereal Cyst Nematode (CCN) resistance) have given farmers even more options on which to base their crop rotation decisions.

Use of legumes in rotations

Legumes play a central role in cereal rotations in southern Australia and the management of the legume phase has a direct effect on cereal yields.

Pasture Legumes — where annual legume pastures can be grown easily and reliably there is no doubt that they increase soil fertility and this is reflected in increased cereal yields and increased livestock production. Good results have been achieved on alkaline soils where species of annual medic form the basis of the pasture phase of the ley farming system. These species produce a high proportion of hard (impermeable) seeds which become permeable to water after one or more years in the soil. This ensures self-regeneration of the legume after one or two years of cropping. On neutral to acid soils it is necessary to sow cultivars of subterranean clover. Hard seededness in subterranean clovers generally breaks down quickly and consequently they do not regenerate well after a period of cropping and have to be resown.

The simplest guide to the input of nitrogen by pasture legumes assumes that input is related to dry matter production, and thus also to rainfall and management. Increases in soil nitrogen under pasture legumes in southern Australia have been recorded for a wide range of situations, with values falling within the range 20 to 180 kg/ha/year.

The increase in soil organic matter under pastures also has a desirable physical effect, resulting in an increase in the size and stability of soil aggregates. Increases in soil porosity, infiltration rate and available water capacity have also been recorded, contributing to the role of pastures in decreasing runoff and erosion. The trend to reduced cultivation and stubble retention is also important in achieving these benefits (Chapter 7).

The ley farming system has not been without problems. The incidence of serious diseases and insect pests of annual pasture legumes has increased, e.g. clover scorch of subterranean clover and aphids on annual medics. These and other problems of environmental adaptation have stimulated the breeding of improved cultivars, as well as the selection of suitable strains of other legumes such as balansa clover and serradella.

Difficulties have also been encountered in managing legume pastures for the benefit of both livestock and the following crop. For instance, excluding grasses from legume pastures has important benefits in terms of the control of certain soil-borne root diseases of cereals. These include Cereal Cyst Nematode (CCN) and Take-all (hay-die). However, grasses in pasture are considered a valuable source of early feed for stock. There is also evidence of more rapid improvement in soil structure under a mixed pasture than under pure legume pasture. A possible solution is to allow grasses to grow early in the season, but remove them from pastures with herbicides at the end of winter, thus allowing the legumes to dominate in spring. This could achieve both satisfactory early pasture production and the desired level of disease control (Chapter 9).

Grain Legumes — grain legumes permit diversification as a means of biological and economic risk-spreading. Thus they have an important role in the sustainability of dryland agriculture. They are a high protein, high value cash crop in their own right but their use is strongly influenced by relative commodity prices.

Grain legumes influence the yield of cereals and it has been estimated that half of the response is due to residual nitrogen. This occurs, despite large amounts of nitrogen being harvested in grain legume seed, through a 'nitrogen saving' effect. Thus, mineral nitrogen levels in the soil in the year after grain legumes are higher than after cereals or grassy pastures. However, the amount of any increase that is available to a following crop depends on many factors, such as the:

• Amount of nitrogen removed in harvested grain legume seed.
• Fate of the crop residues.
• Rate of mineralization of the organic nitrogen.
• Extent of losses by leaching and volatilization.

Selective herbicides can be used to control grasses in grain legume crops and thus remove all hosts for certain cereal root pathogens, such as CCN and Take-all. This also reduces grass weed problems for the cereal phase. From this point of view the cereal-grain legume rotation is easier to manage than cereal-pasture, because grass weeds are undesirable in both phases of the rotation.

The tap-rooted nature of some grain legumes allows them to extract nutrients from the soil which may not be available to cereal and pasture species. For example, in Western Australia lupins tend to perform well on soils where potassium deficiency limits the growth of subterranean clover. The ability of lupins to recycle potassium from the subsoil in some situations where soil potassium levels are low has been clearly demonstrated.

Current rotations in pratice

Western Australia — The agricultural areas of Western Australia are confined to the south-west of the State and can conveniently be subdivided according to annual rainfall (Figure 6.3). About two-thirds of the cleared agricultural land is composed of 'light' or sandy-textured soils, usually with gravel and/or clay in the subsoil. Throughout this area there are valleys of 'heavy' or more clayey soils. The surface pH is usually neutral to acidic. This influences the legumes that are grown in crop-pasture rotations. Subterranean clover and lupins are most suitable for these conditions and have been widely adopted in the Western Australian cereal zones.

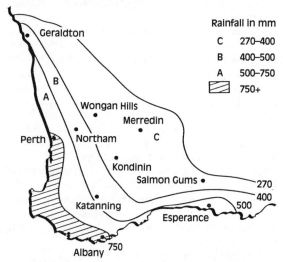

Figure 6.3 *Rainfall zones in the south-west of Western Australia.*

The ley farming system, with animals grazing annual legume-based pastures in rotation with cropping, is practised in almost all of Zones A, B and C. The exceptions are (*a*) the wetter western edge of Zone A and the high rainfall zone where the common land use is permanent annual pasture with occasional cropping (to help renovate the legume component); (*b*) the drier eastern edge of Zone C where annual legumes have difficulty in producing enough seed for survival and continuous cereal cropping is often practised; and (*c*) the relatively new system of continuous cropping using grain lupins and cereals on light soils in the north and west central areas of Zones B and C.

Fallowing is by and large not practised and crops are sown directly into pasture killed with chemicals ('direct drilling') or after a short period of cultivation. This system is flexible and the proportion of the farm cropped can be varied, depending on the physical environment or on changes in market conditions and technology (Chapter 15). The extremes can also occur where there is no equipment for cropping (permanent pasture) or no infrastructure for running animals (continuous cropping). Even on a traditional 'mixed' cereal/sheep farm there may be a place for small areas of one or both of the above extremes.

South Australia — In South Australia crops are grown within the 250–500 mm mean annual rainfall zone (Figure 6.4) and the choice of crops is largely determined by the rainfall and soil type. The relatively simple ley farming system was easily adapted to local situations, with a wide choice of pasture and cereal cultivars being continually made available by local selection and breeding programmes.

The decline in profitability of traditional farming systems from the mid-1970s encouraged farmers to search for alternatives to ensure short-term survival and/or long-term survival and prosperity. Thus, current farming systems are the net result of consideration being given to many aspects, including the most profitable ratio of crop to pasture, the range of crops suitable to the rainfall and soil type, rotations, tillage practices, stubble management, fertilizer use, pest and weed control, timing of operations, etc.

Current rotations in the South Australian cereal belt are summarized in Table 6.1 along with rainfall and soil requirements, an indication of sustainability and comments on essential inputs required for success and potential problems in the longer term.

Rotations with a cropping intensity of greater

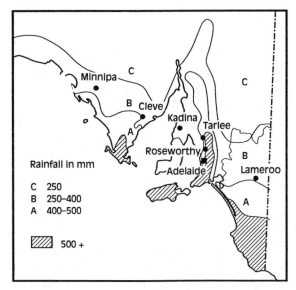

Figure 6.4 *Mean annual rainfall isohyets (mm) for southern South Australia.* (SOURCE: Australian Bureau of Meteorology)

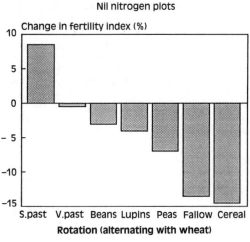

Figure 6.5 *Effect of rotation on the fertility index in the Tarlee Rotation Trial, 1977–87.* (SOURCE: Schultz, J. E., 1990. 'The impact of legumes on soil fertility and sustainability,' Proc. Pasture Symposium No. 2. Better Pasture Legumes for More Cereal-Livestock Profits, E. D. Carter (ed.), University of Adelaide, Waite Agricultural Research Institute)

than 80 per cent are considered to be of doubtful agronomic sustainability. The potential problems are attributed mainly to various combinations of high nutrient removal, loss of soil organic matter, the risk of soil structural decline and increases in diseases and weeds. In a rotation trial at Tarlee, South Australia, soil fertility declined in all continuous cropping rotations, with a loss over ten years of 14 per cent of initial fertility in wheat-fallow and continuous cereal rotations (Figure 6.5).

However, a farming system which is technically unsustainable may have a very useful short-term role, provided any negative effects on the environment are readily reversible. One hundred per cent cropping should remain the exception in South Australia in the interests of long-term sustainability.

Rotations with 50–80 per cent cropping frequency include the traditional ley farming system, such as medic-wheat or medic-wheat-barley on neutral to alkaline soils and subterranean clover-subterranean clover-wheat-barley on neutral to acid soils. Much of the success of these systems depends on the maintenance of productive, legume-dominant pastures. In the Tarlee rotation trial, soil fertility increased by 8.5 per cent over ten years in the wheat-sown legume pasture rotation (Figure 6.5). If more than two crops are grown between medic phases, it is difficult to

maintain adequate medic seed reserves in the soil and the pasture should be reseeded. Thus farmers require pasture management and sheep husbandry skills as well as those required for successful cropping. This combination has a long tradition in southern Australia.

Variants of the ley farming system with a higher frequency of cropping rely on the inclusion of grain legumes and disease resistant cereal cultivars to maintain soil fertility and prevent the build-up of soil-borne cereal root diseases.

In low rainfall areas where soils tend to be sandy and alkaline, a low intensity rotation consisting of two or three pasture years to one wheat crop is practised. Big improvements in productivity have resulted from attempts to ensure adequate medic seed reserves and the control of grass in the pasture year immediately prior to the wheat crop.

Naturally, the greatest opportunities for diversification occur in the higher rainfall areas where a range of crops can be grown. Figure 6.6 shows a 7-year rotation which has been developed from the traditional 4-year subterranean clover ley farming system (wheat-barley-pasture-pasture). Such a rotation has the following merits:
• Eelworm (CCN) control.
• Reduced haydie infestation.
• Reduced legume diseases.
• Some stubble retention on the soil surface for

Table 6.1 *Current rotations in South Australia.*

GL = grain legume; M = medic; W = wheat; B = barley; S = subterranean clover

Rotation	Crop Frequency %	Sustainability Agronomic	Grain Protein	Potential Problems	Associated Technology	Annual Rainfall (mm)	Soil Group
Continuous Cereal	100	No	No	High nutrient depletion and export. Soil organic matter decline (OMD). Soil structural decline (SSD). Disease and weed burden. Herbicide resistant in weeds.	Minimum tillage. Herbicides. High fertilizer rates (esp. N & P). Grass weed control. Disease resistant cultivars.	>250	Calcareous soil to minimise SSD and soil acidification problems.
GL–Cereal	100	?	No	High nutrient export. SDD. OMD. N deficiency in cereal. Herbicide resistant weeds.	Minimum tillage. Herbicides. Higher N & P fertilizer rates. Disease resistant cultivars.	>375	Specific for grain legumes e.g. light, acid soils for lupins.
M–W–GL–W–B	80	?	No	High nutrient export. SSD. Grassy weeds and disease in cereals. Depletion of medic seed reserves during cropping phase.	Minimum tillage. Higher fertilizer rates (esp. N & P). Grass control. Disease resistant cultivars.	>375	Neutral to alkaline for good medic growth.
M–B–W M–W–B (CCN resistant cereals)	67	No	No	High nutrient export. Takeall. SSD. N deficiency in crops. Depletion of medic seed reserves in cropping phase.	Minimum tillage. N fertilizer (2nd cereal crop). Grass control. Periodic resowing of medic.	>325	Alkaline.
M–M–W–GL–Cereal (Disease resistant cereals)	60	?	Yes	High nutrient export. SSD. N deficiency in second cereal crop (yield and protein if wheat). Depletion of medic seed reserves in cropping phase.	Minimum tillage. Grass control in second medic year. Higher P fertilizer rates for legumes. Periodic resowing of medic.	375	Neutral to alkaline.
S–S–W–B (Disease resistant cereals)	50	Yes	Yes	Weeds in second pasture year. SSD.	Minimum tillage. Grass control in second pasture year. Re-sow sub-clover after cropping.	>375	Neutral to acid.

M–W (Insect resistant medics, disease resistant cereal)	50	Yes	Yes	High N status increases weed problems.	Minimum tillage. Management to maintain medic seed reserves. Grass control in medic phase.	>325	Neutral to alkaline.
(M)–M–M–W	25–33	Yes	Mostly	Risk of cereal disease if pastures grassy. Wind erosion in poor seasons.	Weed control. Management to maintain medic seed reserves.	275–350	Sandy, neutral to alkaline.

(SOURCE: adapted from Reuter, D. J. and Dyson, C. B., 1990. 'Protein for profit. Farming practices and protein levels in ASW', *Technical Report No. 162*, Department of Agriculture, South Australia)

erosion control and improved soil water retention.
- A range of weed control chemicals and techniques reduces the likelihood of herbicide resistance.
- From 43 per cent to 70 per cent of the farm being cropped for grain production each year.
- From 30 per cent to 57 per cent of the farm being grazed each year.
- Possibility for a broad range of cereal and legume crops to be grown.

Victoria — Field crops are principally grown north of the Great Dividing Range within the 300 and 550 mm rainfall isohyets where the flat to undulating topography is well-suited to large-scale crop production. Cereals are the main dryland agricultural product, and wheat is by far the major crop. There is significant production of barley and oats, and more recently peas and lupins, with small areas of other cereals and oilseeds. The different crops are grown in rotation with each other and with annual legume pastures which sustain a livestock enterprise, mainly sheep. The cropping sequence is usually designed to maximize the yields of wheat.

In the traditional cropping areas of north-west Victoria (the Mallee and Wimmera) wheat is the dominant crop grown. Barley is the next most important crop but it occupies only about one-quarter of the area sown to wheat. Small areas of cereal rye are grown on the more infertile sandy soils of the Mallee.

Wheat has traditionally been sown on fallow in the north-west, with barley being sown into the wheat stubble. Fallowing is still an integral part of wheat production in these areas, but its role in the farming systems is being challenged (Case Study C). The traditional long (8–10 month) fallow is prepared and maintained free of weeds by regular mechanical cultivation, prior to sowing in May–June. This practice conserves soil moisture, accelerates the mineralization of soil nitrogen and

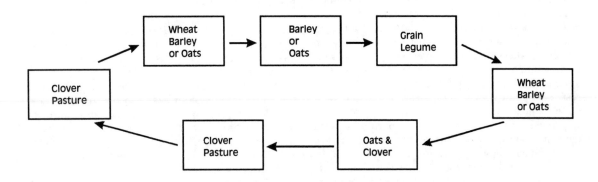

Figure 6.6 *A 7-year rotation designed to reduce plant diseases.* (SOURCE: Bartholomaeus, M. K., 1987. 'Farming Systems Revisited', *South Australian Cereal Farming Systems, Collected Papers*, South Australian Department of Agriculture, 2, 13.1)

can reduce the incidence of some cereal diseases by eliminating host weed species. Yield increases with fallow are quite consistent and are proportionally greater in the Wimmera where the soils have a higher water holding capacity than the sandy soils of the Mallee.

Adding an annual medic pasture into the previous fallow-wheat or fallow-wheat-wheat rotations was a significant advance in cropping technology in the north-west in the post-war years. Since its inclusion, the most widely practised rotation has been the fallow-wheat-pasture system. In the higher producing Wimmera the cropping intensity tends to increase towards half by including barley, oats or peas after the wheat crop. By contrast, in the exteme north-west, cropping intensity may be as low as one in four or five. At present there is a trend towards crop rotations in which grain legumes (especially improved cultivars of peas) are substituted for some (or occasionally all) of the medic pastures. However, as pea stubble is easily blown by wind, leaving paddocks prone to erosion, this system is less sustainable. Thus pasture leys remain the most effective means of replenishing reserves of soil nitrogen.

In north-east Victoria the higher rainfall (500–700 mm) provides an excellent study of the diversity and dynamics of integrated crop-livestock farming systems. The range of enterprises on a mixed farm could include crops (wheat, barley, oats, oilseed, lupins, peas), livestock (sheep for meat and/or wool, beef cattle) and other pasture products such as hay or seed. The balance of cropping and grazing responds to the relative profitability of wool, meat and grains. The most recent trends have been towards a higher proportion of cropping.

Farming systems in the north-east have evolved around the basic crop-pasture sequence but this has been extended out to become 'phase-farming', where some years of cropping alternate with some years of pasture. In the past, three or four crop years (usually wheat, but sometimes including an oat crop) alternated with up to six years of subterranean clover pasture. Crop production in this system was not really efficient, due largely to restricted plant root development, associated with waterlogging and plant nutritional problems. Improved technology involves deep ripping with gypsum application for physical structure improvement, lime application to correct acidification, direct drilling and stubble retention. This has allowed further intensification and the cropping phase might now consist of up to eight or nine crop years such as: rapeseed-wheat-lupins-wheat-wheat-lupins, and this alternates with a pasture phase of three to six years of subterranean clover. Where, for economic or personal reasons, continuous cropping without livestock production is the preferred option, a rotation of wheat-wheat-lupins-lupins has been successfully adopted by some farmers in these areas.

Cereal zone rotations — uniform summer/winter rainfall areas of NSW

In southern NSW cropping is practised where annual rainfall is 300–600 mm and a high proportion of the soils are red-brown earths. The amounts of warm-season and cool-season rainfall are almost equal. This allows perennial pasture species such as lucerne and phalaris to be grown in the pasture ley phase of the cereal-pasture rotation. Where this is done (by no means universally) it allows a more complete use of annual rainfall than by pasture composed only of annual, cool season species. However, the main component of the pasture phase is subterranean clover. As in other parts of southern Australia the traditional type of rotation involving subterranean clover is a phase of two to four years of pasture followed by a similar number of years of cereal, commonly wheat or barley.

Summer rainfall often causes 'soft' seed of subterranean clover to germinate 'out of season'. If germination is followed by high temperatures and/or dry spells as is common in the early months of the year the seedlings usually die. If the residual levels of hard seed and the rate of further softening are too low, there may be inadequate soft seed available again by the autumn for satisfactory plant populations in the pasture. Another effect of the summer rainfall is the leaching of nitrate released by the decomposition of pasture residues. Not only may this be a loss to the following pasture or crop, but it is also a major cause of soil acidification.

As in other parts of southern Australia where the ley farming system has been practised for several decades, a range of weaknesses and causes of unsustainability of the system have appeared. Research has been necessary into such matters as:
- Efficiency of utilization of superphosphate in the rotation.
- Breeding subterranean clover cultivars with higher levels of hard seededness, freedom from oestrogenic activity, resistance to clover scorch disease and winter production.

- The effect on the system of varying the proportions of crop and pasture years. Soil nitrogen and wheat yield increase as the proportion of clover years to crop years in the rotation increase (i.e. as cropping intensity decreases).
- Soil acidification (Chapter 11) and associated toxicities of aluminium and manganese (Chapter 10). Solutions have been sought in lime application, breeding of acid tolerant varieties of wheat and subterranean clover, and the introduction of new acid tolerant species to replace wheat (triticale), subterranean clover (serradella) and traditional grain legumes (lupins). A range of rotations suitable for this region is shown in Case Study B. Unfortunately the acidification process continues in all types of crop and pasture sequences because of the removal of plant products with higher proportions of cations than anions. The greatest scope for reducing acidification is through more efficient use of soil nitrogen and water by the use of summer active perennial pasture species and the early sowing of cereals.

Thus, by means of research, improved technology and adjustment, the ley farming system in this region has been sustained for half a century.

Cereal zone rotations — summer-dominant rainfall areas of the northern wheatbelt

The rainfall becomes summer-dominant around the centre of NSW, although the summer component is the more unreliable. Mean annual rainfall in cropping areas varies from 350 to 750 mm. The red-brown earths and mallee soils give way increasingly to a range of other soil types. Soils of basaltic origin are of clay to loam texture, neutral to alkaline pH, red, black or grey colour and relatively high fertility. Soils derived from sandstone or granite are relatively light in texture, acid in pH and low in fertility. They include deep sands, solodic soils and solodized solonetz. These climate and soil changes lead to changes in cropping sequences, although wheat production is still dominant.

The summer/winter rainfall pattern brings with it important features which vary somewhat with the degree of northerliness:

- There is potentially a choice of whether to grow a winter crop or a summer crop.
- Rainfall becomes more variable and thus unreliable.

- Autumn and spring are frequently dry and dry spells often occur at other times. Longer-term drought is fairly common.
- To reduce the risk of crop failure, considerable moisture must be accumulated in the soil profile by fallowing before sowing. If summer rainfall storage is inadequate sowing is delayed until the next season. Conversely, in some years soil moisture may be replenished so quickly after a crop that both a winter and a summer crop can be grown in the same year. This is called double cropping and the crop sown in response to additional rainfall is called an opportunity crop.
- Because of the need to fallow, crops can only be grown on soils with adequate storage capability, i.e. fine textured soils able to accumulate water to a depth of 1 to 1.5 metres.
- Summer rains, especially in the north, include high intensity storms which tend to be very erosive. This makes it necessary to include measures to combat soil erosion in the design of all cropping systems.
- The summer/winter pattern and variability of rainfall work against the use of a fixed rotation especially in northern NSW and southern Queensland. This includes the fairly inflexible cereal–self-regenerating annual medic rotation of the southern wheatbelt. For example, if a change is to be made from winter to summer cropping, a winter fallow would normally be needed; this would mean the loss of a legume pasture which had regenerated in autumn from seed reserves in the soil. Furthermore, there would be no replacement of medic seed to the soil reserve in that year.

On the central western slopes and plains of NSW, mixed crop/livestock enterprises are common where annual rainfall is in the range of about 400 to 750 mm. Where rainfall is lower, a simple wheat/fallow system may be used. This may still include livestock grazing of stubble and fallow areas, to utilize residues and assist weed control.

Production from subterranean clover and annual medic is variable because of the unreliability of autumn and spring rains and, in the north of the region, of winter rains. Thus where soils are deep and well drained the perennial legume, lucerne, is often included in the pasture phase. It is tolerant of dry spells and responds to light rains with the production of high quality feed. A common type of rotation is three crop years followed by five or six years of clover or lucerne for sheep or cattle grazing. Pastures may also include warm season perennial grasses such as Rhodes grass, buffel grass and *Panicum* spp. to use nitrogen provided by the legumes and help suppress weeds.

Other annual legumes such as serradella and vetch are sometimes grown.

Although summer rainfall variability is high, overall patterns of rainfall and temperature allow for a wide choice of crops and pastures. Summer crops are those grown in subtropical areas to the north and include sunflowers (the most successful), grain sorghum, cowpeas, mungbeans and maize. Winter crops are wheat, barley, oats, triticale, lupins, linseed, canola and safflower.

The marked effect that soil characteristics can have on crop and pasture sequences is illustrated by the lighter, acid soils of the Coonabarabran and adjacent districts. These occupy hundreds of thousands of hectares. With pH levels down to 4.0 and aluminium levels high, they have been unsuitable for the production of most crops and pastures, although mean annual rainfall is high (500–700 mm). It was only when research on soil acidification in more southern regions produced new cultivars of acid tolerant lupins, triticale and serradella, as well as acid tolerant cultivars of other cereals and subterranean clover, that these soils were found to have a useful production potential.

In northern NSW, summer rainfall is even more dominant and winter rainfall less reliable. Thus for growing wheat a prior accumulation of moisture by fallowing is essential. Crops are grown where mean annual rainfall is 450–750 mm. The most important soils for cropping on the north-west slopes and plains are the black earths and other heavy, grey and grey-brown clays. They are fertile and can store the 100–150 mm of moisture required in the crop rooting zone before sowing.

Wheat is grown in rotation with other winter crops such as chickpeas and with summer crops such as sorghum and sunflowers. Because of the required flexibility of rotations for moving from summer to winter crops, a relatively fixed, year-by-year rotation of cereal and annual pasture is not appropriate in this region. A rotation of two or more years of annual pasture followed by a phase of two or more years of cropping would be feasible. However, the variability of winter rainfall and, on the western slopes, the low winter temperatures, are not conducive to good winter production of available medic cultivars.

The main alternative means of including a pasture phase in a rotation on the north-west slopes is to grow lucerne for a number of years (usually three to five) followed by several years of cropping for grain. The value of lucerne in this region is for raising lambs and vealers, to increase soil nitrogen for following crops and to assist in the control of weeds of cereals, for example, wild oats. It grows well in the deep, alkaline clay soils, provided drainage is satisfactory. It has long been known that lucerne benefits subsequent crops by increasing soil nitrogen and that the effect may last for several years. In rotation studies at Tamworth a lucerne ley of three and a half years was the most efficient in improving yields and protein content of subsequent wheat crops. It provided adequate nitrogen for at least five wheat crops on a black earth and three on a red-brown earth.

However, as can be expected in dryland agriculture, the benefits from lucerne nitrogen on subsequent wheat are reduced when soil moisture is limiting. Because a lucerne ley extracts water below the −15 bar suction level to a depth of 200 cm, yields of the first wheat crop following lucerne may be depressed.

A longer fallow period would be required to replenish the profile after lucerne than after a previous wheat crop. This illustrates how species vary in the amount of soil moisture and nitrogen they leave behind and thus in the effect on the following crop. It also shows that there is sometimes a need to choose between two or more beneficial alternatives. Delaying the plough out of lucerne until January or February will maximize summer grazing, but a September plough out is normally required to replenish soil moisture for the next wheat crop.

The value of a rotation component is increased if it can be used for more than one purpose. The value of lucerne is increased by selling it for hay. Furthermore, the regular cutting and grazing over a few seasons prevents wild oats from setting seed and thus controls this weed.

Evolution of rotations in northern NSW

A good example of how rotations have evolved in northern NSW and southern Queensland and how such changes must be accompanied by new technology is summarized in Table 6.2. Since 1969 the 2400 ha property has been changed from a partly cleared grazing property to a large-scale wheat farm with summer fallow and more recently to a diversified winter and summer cropping farm with beef fattening on grain sorghum stubble and green grazing oats. Associated with these changes has been the development of conservation farming practices to prevent soil erosion from high intensity summer storms and reduce soil water loss by runoff and evaporation.

Table 6.2 *Development of rotation systems and associated technology on the University of Sydney Livingston Farm at Moree, NSW.*

Type of System, Advantages	Accompanying Technology and Management Requirements	Constraints, Problems Remaining or Developing
1 Extensive grazing. Low level of inputs to system.	Control of stocking rates.	Low level of output from system.
2 Continuous cropping. Premium quality wheat from fertile soils.	High powered tractors to clear trees and cultivate heavy soil. Storage of summer rains by clean cultivated fallow. Stubble burning or burial to control wheat diseases.	Cultivation difficult when soil wet. Bare fallow exposes soil to erosion risk and increased rainfall runoff. Weeds of winter crops increase, e.g. wild oats.
3 Continuous wheat with stubble retention. Reduced risk of erosion and rainfall runoff.	Chisel and blade ploughs to retain stubble on surface while cultivating.	Crown rot and yellow spot diseases of wheat increase with stubble retention. Soil fertility decline. Access to and cultivation of wet soil difficult.
4 Winter-Summer cereal cropping (wheat-sorghum) with stubble retention. Sorghum provides break for wheat diseases. Sorghum stubble (regrowth) can be grazed. Livestock help control weeds.	Long fallow (10–18 months) required between winter and summer crops.	Unreliability of late spring rains and need to cultivate for seed bed preparation may delay sowing of sorghum past optimum time (about Oct). Wheat seed drill gives poor establishment of sorghum.
5 'No-till' system of preparation for sorghum crop adopted to improve moisture conservation, access by machinery, sowing into moist soil. Sorghum establishment and yields improved. Opportunity double cropping of sorghum after wheat easier to achieve.	Control of weeds by herbicides. (Precision needed in application). Special 'no-till' seeder to sow through heavy residues and control sowing depth.	Soil fertility decline continues. Double cropping means late sowing of sorghum and greater risk of midge attack. Some carryover of wheat diseases in stubble still occurs.

Type of System, Advantages	Accompanying Technology and Management Requirements	Constraints, Problems Remaining or Developing
6 Grain legumes added to rotation (chickpeas the first with suitable cultivars). Nitrogen added to soil. Carryover of wheat diseases reduced.	Chickpeas require well-drained soils, early weed control. Sowing mid May–early June.	Chickpeas susceptible to *Phytophthora* root rot, requiring break of 4 years between sowings of chickpeas or *Medigago* species. Better cultivars of chickpeas and other grain legumes required. Rotation includes only stubble grazing for livestock.
7 Oats for grazing included in rotation. Follows on conveniently after grazing of sorghum stubble completed.	Sow oats March/early April, graze May/June to Oct/ Nov. Amount of oats sown should balance amount of grazing available from sorghum stubble.	In absence of vigorous pasture legume, decline in soil nitrogen levels probably still occurring.

Suitable rotation is: wheat, sorghum and livestock grazing of stubble, sorghum and stubble grazing, summer fallow, chickpeas, oats for grazing or wheat.

NOTE: Based on Crofts, F. C., Esdaile, R. J. and Burgess, L. W. (1988) *Towards No Tillage*, University of Sydney. Steps in the evolution of rotations can be followed by reading column 1 from top to bottom. The associated technology, management and problems are shown in the other columns.

Mean annual rainfall is 450–600 mm, summer dominant. The soil is fairly representative of much of the north-west plains; black, cracking, self-mulching clays, pH 7.3 at surface to 8.3 at depth, high in organic matter, available moisture holding capacity at about 150 mm per metre.

Strip cropping

Another aspect of the integration of crop rotation and conservation farming practices on the north-west plains of NSW is strip cropping (Figure 6.7). This is a system under which ordinary farm crops are planted in narrow strips across the slope of the land. It is typically used on low-slope (< 2 per cent) cropping country. The strips are arranged so that erosion-prone phases (e.g. newly planted crops) are always separated by dense, erosion-resistant crops or crop residues. Where control of wind erosion is the prime objective, the strips are aligned across the path of the critical wind. Strips well protected by crop or stubble break the flow of water and wind.

Figure 6.8 shows a rotation-strip cropping plan suitable for the north-west plains. This system, by reducing water losses from runoff and soil evaporation, is allowing summer crops to be included in rotations in drier areas than before.

Cereal zone rotations — Queensland

In southern Queensland the main area of dryland cropping is the Darling Downs with mean annual rainfall of 500–1000 mm. The deep black earths (vertisols) can store up to 200 mm of available moisture in the wheat rooting zone, before sowing.

The farming systems here have changed over the past 150 years from predominantly extensive grazing, through dairying or a simple winter cereal-summer fallow system, to more diversified sequences which include summer crops. Specialized livestock systems have also evolved. On major sections of black soil areas livestock (and fences) have been eliminated because of the greater profitability and convenience of crops. Moreover, in response to variability of rainfall and market, rotation sequences have become flexible with a strong element of opportunism.

The major type of crop diversification has been summer cropping of sorghum. A long fallow is

Figure 6.7 *Strip cropping on black soil plains, Darling Downs, Queensland.* (PHOTO: Queensland Department of Primary Industries)

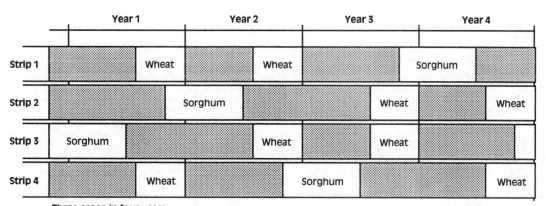

Three crops in four years.

Figure 6.8 *Strip cropping layout—north-west NSW. Because the rainfall is marginal for summer crops and they must be preceded by a long fallow, only three crops are possible in four years. Protection in early summer (end of year 1) is given by the growing wheat crop in two strips and the stubble from the previous sorghum crop in the fallow strip. In the late summer period, protection is given by the growing sorghum crop in one strip and wheat stubble in two strips.* (SOURCE: Charman, P. E. V. (ed.), 1985. *Conservation Farming*, Soil Conservation Service of NSW)

usually required from the time of wheat harvest (November) to sowing sorghum (preferably October).

Management of rotations has to relate not only to rainfall, but to many other factors. For example, sowing time for sorghum should be late enough to obtain warm soil temperatures for emergence and to avoid frosts at flowering, yet early enough to achieve flowering before the hottest time of the year and to avoid the worst sorghum midge attacks. However, in some areas, later plantings (December–January) would allow flowering to coincide with March rains and also to avoid summer temperatures. Variation in flowering time is also achieved by appropriate choice of varieties. When the whole rotation sequence is taken into account, perhaps in the context of a strip cropping design, it can be seen how readily decisions on the management of one crop can affect others.

Even more than in NSW, opportunistic double cropping of summer and winter crops is conducted when soil moisture permits. Stubble retention and 'no-till' techniques allow this to be achieved more easily (Chapter 7). Further, by drying the soil profile and making way for the entry of the next rains, opportunity cropping tends to reduce runoff and make more efficient use of rainfall.

The opportunity summer crop should ideally have a short growing season in case the summer rains cut out. As well as sorghum, suitable short-term summer crops are sunflowers and mungbeans. Chickpeas are useful as a winter opportunity crop.

Sunflowers also provide considerable flexibility in the operation of rotations. With a growing season of 14–16 weeks and tolerance of cool conditions they can be sown earlier or later than the other summer crops, in late August-September, or at the end of January. Mungbeans are of two types: green gram and black gram. They also provide additional flexibility because they can be used for grazing if soil moisture is inadequate to produce grain. Chickpeas are the most useful winter grain legume in this region but their intolerance of water logging and susceptibility to phytophthora root rot are significant limitations to existing varieties and reduce the flexibility of rotations.

Strip cropping designs for the Darling Downs take into account the possibility of opportunity cropping. The design in Figure 6.9 is a popular one for the eastern Downs.

In determining the value of an opportunity crop, one must compare the costs and returns associated with it *and* the following crop (perhaps also a summer crop) with the costs and returns associated with alternative practices (probably including a longer fallow).

On the black earths of the Darling Downs and other regions of the northern wheat zone, the fallow period has been used to accumulate mineral nitrogen as well as moisture. Although the natural capacity to provide mineral nitrogen is still substantial in these soils, responses to nitrogen fertilizers have been obtained on the longest-cultivated soils since the 1960s. A legume should be included in the rotation to replenish some of the nitrogen removed in crops over many decades. Green and

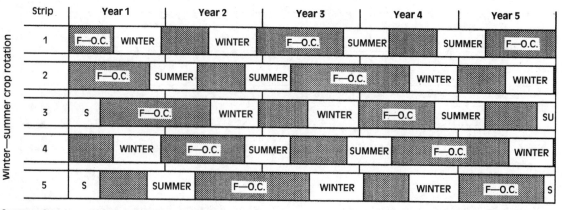

Summer Crops — Sorghum, Maize, Sunflower.
Winter Crops — Wheat, Barley, Canary, Linseed, Canola,

F — O.C. — Fallow or Opportunity Crop — Mungbean, Chickpea, Millet, Panicum

Figure 6.9 *Strip cropping layouts, Darling Downs, Queensland.* (SOURCE: Mills, W. O., 1990. *Winter Crop Management Notes: Darling Downs 1990*, Queensland Department of Primary Industries)

black gram are two legumes which have been found to benefit following cereal crops.

Occasional use is made of annual medics in rotation with crops in the northern wheat belt. Research has shown that they can give high yields over a large area of southern Queensland receiving a mean of 175–250 mm rain in the April–September period. However, the capacity of annual medics to maintain a soil reserve of hard seed and regenerate after a grain crop cannot be fully exploited in a summer/winter rainfall environment with flexible cropping rotations.

Lucerne is the only other forage legume grown in rotation with cereals on the Darling Downs. Research has confirmed its value for improving soil nitrogen and following crops.

As in other parts of the cereal zone, rotations have an important role in controlling crop diseases (Chapter 9). Examples in southern Queensland include:

- Yellow spot of wheat: rotate with other crops except triticale.
- Crown rot of wheat: rotate with oats, chickpeas or a summer crop.
- Common root rot of wheat and barley: rotate with barley, sorghum and sunflower.
- Root lesion nematode in wheat: rotate with barley, sorghum and sunflower, growing only one wheat crop in five years.
- Phytophthora root rot of chickpeas: rotate with other crops except lucerne and annual medic, growing only one chickpea crop in four years.

In the central highlands of Queensland crops are grown in areas receiving mean annual rainfalls of 500–750 mm, with a summer:winter incidence ratio of about 3:1. A wide range of soils occurs in the zone from southern to central Queensland.

Cropping is confined mainly to those which were formerly under brigalow and softwood shrubs. They are deeper, more fertile and have a reasonable waterholding capacity. The dark cracking clay soils of open downs are also used, although they are usually shallower than is desirable (60–80 cm).

The more marked summer dominance of rainfall in the north results in a predominance of summer cropping (e.g. sorghum, sunflowers and mungbeans). Wheat and chickpeas are quite well adapted and are grown as a means of diversification and to spread the use of labour and machinery.

Because rainfall is variable, farming on the central highlands of Queensland tends to be opportunistic. However, there are some important constraints which limit flexibility, and other ecological factors besides rainfall must be taken into account in system operation. For instance, high-intensity summer storms which cause high runoff, flooding and erosion are even more of a problem than in the south. Thus a vital aspect of cropping systems is provision for surface retention of stubble to protect the soil against erosion. This influences the crop sequence because the amount of stubble produced and its persistence over time varies with the crop (Table 6.3). For crop rotations to be effective in reducing soil loss, crops that produce low stubble levels should be grown following crops producing high levels of stubble, e.g. sunflower planted into wheat stubble.

Another constraint on opportunistic cropping is the decline in soil nitrogen. There is now a trend to grow cereal crops after a legume (e.g. chickpeas, mungbeans) for the residual nitrogen benefits, especially on the oldest-cropped soils. An alterna-

Table 6.3 *Variations among crops in stubble produced from average grain yields.*

Crop	Stubble produced (kg/t of grain)	Long-term average grain yield (kg/ha)	Average stubble remaining after harvest (kg/ha)
Wheat	1700	1400	2400
Sorghum	1100	1500	1650
Sunflower	1600	700	1120
Safflower	2300	600	1380

(SOURCE: Spackman, G. B. and Gilmore, M., 1982. *Conservation Cropping on the Central Highlands*, Queensland Department of Primary Industries, Emerald)

tive to this procedure is to use fertilizer nitrogen if economics permit.

Local climatic patterns involving rainfall, frost and heatwave strongly influence the operation of cropping systems in this region. An example of this is the choice of sowing times. Wheat is sown as early as late March to catch the benefit of late summer rains. However, this may result in frost damage at flowering, at least on lower-lying land. Appropriate choice of variety may also avoid flowering during the period of greatest frost risk (Figure 6.10), but winter rainfall variability still remains a problem. In contrast, the preferred time of sowing sorghum is late in its season (late December to February) in order to avoid early moisture stress, as well as heat stress during flowering.

While soil nitrogen is high, any pasture phase used in conjunction with cropping consists of perennial tropical grasses such as Rhodes grass, buffel grass, species of panic grass and purple pigeon grass, without legumes. Only when the original soil nitrogen has become depleted does the inclusion of a legume into the pasture become worthwhile. This also limits the value of experiments with grass-legume mixtures in the early years after vegetation clearing. Lucerne and annual medics are less suited to the environment than tropical legumes; but there is yet no suitable tropical legume for the rainfall regime and heavy clay soils of the cropping areas. This has limited the information available from rotation experiments such as that conducted by CSIRO over twenty years at Narayen, although the decline in soil

Figure 6.10 *Wheat sowing options, Central Highlands, Queensland.* (SOURCE: Redrawn from Keefer, G. D., 1980. 'Improving wheat cropping reliability on the Central Highlands', *Queensland Agricultural Journal, 106(6),* 525–31).

Controlling Factors

Rainfall

Frost

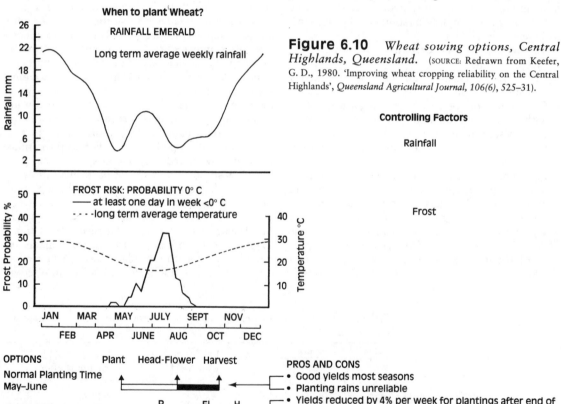

Table 6.4 *Some features of rotation systems for the far northern section of the Australian wheatbelt in Central Queensland.*

Type of System, Advantages	Accompanying Technology and Management Requirements	Constraints, Problems Remaining or Developing
1 Wheat-sunflower-sorghum-sorghum. Provides diversified crop production. Sorghum stubble provides useful cattle grazing after harvest and when regrowth occurs in spring, and continues through summer.	Large machines for timely operation over large areas. Equipment for stubble retention and reduced tillage techniques to catch benefit of late summer rains. Sunflowers sown late Jan–mid March for optimum moisture and temperatures. Sorghum sown late Dec–Feb to avoid heat and moisture stress at flowering. Grow sunflower after wheat to provide soil protection with wheat stubble. Resistance to rust (*Puccinia helianthi*) and insect control needed in sunflower.	Rainfall variability is a problem. No nitrogen input from this rotation. Price received for all commodities fluctuates. Grazing sorghum stubble in wet conditions increases soil compaction, tillage and weed control problems.
2 Inclusion of chickpeas and mungbeans in rotation. Further diversification. Nitrogen input to soil.	Control of *Heliothis* caterpillar sometimes needed. Taller varieties facilitate harvesting.	Chickpeas intolerant of waterlogging, susceptible to *Phytophthora* root rot.
3 Inclusion of annual forage crops (oats, lablab, forage sorghums) in the rotation. Provides good basis for beef production in association with sorghum stubble, lablab adds nitrogen to soil.	Sow oats Feb/Mar. Lablab sown on storm rains, Oct–Jan.	Oats risky because of unreliable winter rains and rust incidence. Risks of soil compaction and erosion higher than with perennial pastures.
4 Inclusion of perennial tropical grass pastures in rotation. Good control of soil erosion. Improvement in soil structure. Large-seeded purple pigeon grass can be sown by normal grain drill.	Grass species and variety must be chosen to suit rainfall and soil conditions.	Some small seeded grasses difficult to establish. Absence of legume excludes input of nitrogen from this source.

(SOURCES: Based on Milne, G. O., Spackman, G. B. and Barnett, R. A., 1988. 'The Central Highlands', *Queensland Agricultural Journal*, 114, 247–56; *Brigalow Farm Management Handbook*, 1976. Queensland Department of Primary Industries)

nitrogen showed the eventual need for legumes.

An alternative to perennial pasture species is annual forage crops. The rainfall regime is more suited to summer forages such as sorghum and lablab (a legume) than oats. Lablab improves the protein level of grazing from sorghum stubble or nitrogen deficient pasture.

The northern-most part of the Australian wheat belt is relatively newly developed for cropping. It has some of the characteristics of the southern Australian cereal zone some fifty to sixty years ago: declining soil fertility and structure, high erosion risk and a low level of use of legumes to improve soil nitrogen status. However, in other ways, there are great differences: larger farms and larger machines for quick and timely operations, use of stubble-retention techniques to reduce soil erosion, increasing dependence on strategic use of herbicides in fallow management, and a range of alternative crops to allow diversification and the breaking of disease cycles. It remains to be determined how far the farming systems will be able to rely on crop and pasture legumes for their nitrogen supply. Table 6.4 summarizes some of the features of possible types of rotations in the central Queensland cereal zone. The evolution of systems can be traced by reading Column 1 from top to bottom. Associated technology, management and problems are mentioned in the other columns.

The most northerly cereal-growing area in Australia is the Atherton Tableland in far North Queensland. The mean annual rainfall is 750–1250 mm in areas suited to cropping and markedly summer dominant. The frequency of seasons with a rainfall deficiency is considerably less than further south. Maize is the cereal best adapted to this moisture regime. The soils are mainly deep, red kraznozems which vary considerably in nutrient availability.

This region is of special interest because it is the only one in the northern Australian cereal zone where a technically viable ley farming system has been developed. The first impetus for change was the serious soil erosion, nitrogen depletion and low yields from the maize monoculture system within a few decades of the clearing of the rainforest. A ley farming system was researched, but was initially based on relatively low-yielding open-pollinated maize varieties and lucerne-grass pastures. Such deficiencies can often be overlooked for a considerable time in the absence of an economic analysis, which should be done at an early stage of system development. A series of economic analyses by van Haeringen showed up the deficiencies in the above system and pointed to the superior economic value of a tropical legume, glycine.

During the 1960s and 1970s a technically viable ley farming system using tropical species was developed. It comprised disease-resistant hybrid maize varieties and a ley of mixed glycine and green panic grass. The pasture species were high yielding, persistent and suitable for dairying or beef. The legume improved soil nitrogen and maintained high maize yields.

Even when such a system is available, it is not necessarily favoured by all farmers. As in other parts of Australia and overseas, the combination of factors needed for full adoption of a ley farming system includes a tradition of combining crop and livestock enterprises, reasonable prices for livestock products and the availability of highly productive and persistent pasture legumes. On the Tableland, there is a tradition of mixed maize and dairy farming originating in times of high demand for dairy products. However, in recent decades, the majority of landholders have been either dairy farmers (higher rainfall areas) or maize farmers (lower rainfall areas).

A downturn in the fortunes of the dairy industry prevented its further expansion into lower rainfall maize-growing areas. Some farmers combined maize and beef. However, small farm sizes and increasing land prices in the 1980s directed Tableland farmers even more firmly to systems of continuous cropping. The response to problems of maize monoculture has been to diversify into peanuts and to use conservation farming methods to minimize soil erosion. Yet with the risk of soil erosion still high, the question remains: are systems of annual cropping sustainable in the long term or will sustainability best be achieved either by adopting the tropical ley farming system on larger areas or by changing to perennial tree crops such as coffee, avocados and other tropical fruits which grow well there?

Conclusions

Farming systems as expressed in rotation sequences are ever changing to meet economic demands and to keep farms viable. A challenge to agriculture is to determine and adopt rotational practices that will, in the long run, result in soil improvement rather than degradation (however slow). Another challenge is to retain flexibility in the composition and operation of rotations to allow the inclusion of new species, technology and management with changing circumstances (see Chapter 21). These challenges are being met by the development of a wider range of plant species and varieties (particularly grain and pasture legumes), adoption of conservation farming practices, and use of increasingly skilled management.

Further reading

Connor, D. J. and Smith, D. F. (eds.) (1987). *Agriculture in Victoria*, Australian Institute of Agricultural Science.

'Conservation Management of Cropping Lands', special issue of the *Queensland Agricultural Journal*, Vol. 112, No. 4, 1986.

Delane, R. J., Nelson, P. and French, R. J. (1989). 'Roles of grain legumes in sustainable dryland cropping systems', *Proc. 5th Aust. Agron. Conf.*, Perth, 181–96.

Perry, M. W. (1989). 'Farming systems of southern Australia', *Proc. 5th Aust. Agron. Conf.*, Perth, 167–80.

Proc. Int. Congress on Dryland Farming, Adelaide, South Australia, 1980.

Proc. Int. Congress on Dryland Farming, Amarillo, Texas, August 1988.

CHAPTER 7

Tillage Practices in Sustainable Farming Systems

P. S. Cornish and J. E. Pratley

SYNOPSIS This chapter discusses the aims and effects of tillage in traditional farming systems and shows how these objectives can be met by other less damaging means. It considers how to develop economically and environmentally sustainable farming systems by integrating soil-conserving cultural practices into the many other aspects of farm management, including pasture and livestock management and crop rotations.

MAJOR CONCEPTS

1 Crop production in dryland farming systems has depended heavily on burning crop residues and on cultivation for fallowing and seedbed preparation. Efficient use of rainfall is a major aim.

2 Burning has led to soil erosion, particularly on light soils prone to wind erosion and on heavy soils in summer-rainfall areas where the soil is frequently bare (fallow) over summer and high intensity rains cause erosion. Stubble retention is a high priority in these areas.

3 Cultivation has damaged soil structure and led to erosion, especially in those soils with weaker structure, which are most common in the cropping zone of southern Australia. Reduced cultivation has been a priority on these soil types.

4 Techniques for fallowing and seedbed preparation have been developed that give good crop yields with less reliance on cultivation and burning. Examples are direct drilling and no-till fallowing. These are methods of crop establish-

ment, not farming systems. They rely heavily on herbicides to control weeds.

5 Soil-conserving tillage techniques also increase infiltration of rainfall and reduce evaporation of soil water. This increases potential crop yields, as well as giving long-term improvement in soil structure.

6 More intensive and more profitable cropping is sometimes possible with the new techniques because water is stored more efficiently, better surface soil conditions allow more timely planting, and no time is lost in preparing cultivated seedbeds.

7 The new techniques of tillage call for new approaches to weed management. Also, some pests and diseases demand special attention to rotations, whilst pasture and livestock management are also influenced. This integration of soil-conserving tillage practices with other aspects of farm management comprise conservation farming systems.

8 Conservation farming systems demand a high degree of technical and management skill but offer rewards of increased profitability together with sustainability.

9 Short-term profitability is vital to farms, regardless of future sustainability and profitability.

Tillage comprises: *a*) cultivation, including the soil disturbance associated with the sowing operation, and *b*) treatment of crop residues to allow the machinery for cultivation and sowing to operate.

Dryland farming systems combine crops, pastures and fallow periods with the fundamental aim of making efficient use of limited water. Cultivation and the burning of crop residues have been an integral part of these systems, enabling crops and pastures to be established and fallow options to be exercised. It has long been recognized that these tillage practices degrade soils — sometimes quite quickly. Cultivation accelerates the breakdown of soil organic matter, leading eventually to poorer structural stability and increased erodibility. Burning removes vegetative cover from the soil which otherwise absorbs the impact of falling rain. Together, these practices have led to massive soil degradation across most of Australia's farming lands (see Chapter 11). However, soil structure improves when land is returned to pasture after a period of cropping.

New practices, that rely heavily on herbicides instead of cultivation to control weeds, are less damaging and also have the potential to make better use of rainfall but they are often not economic. New tillage practices can result in profitable, sustainable farm systems when integrated with other aspects of farming (e.g. rotations, livestock management).

Fallows are periods of managed freedom from plant growth and they play an important part in dryland farming. Fallows reduce the risk of cropping in lower rainfall areas of southern Australia and allow cool-season crops (e.g. wheat) to be grown in the summer-dominant rainfall zone of northern NSW and Queensland (see Chapter 6). There are three broad types of fallows: long, short and no fallow.

Fallows have a long and often sorry history in Australia. About 1900, 'dry farming' techniques were imported from the USA. They involved deep ploughing and frequent harrowing to produce a dust mulch. The ploughing was thought to increase the waterholding capacity of the soil,

whilst the dust mulch supposedly prevented water rising to the soil surface by capillary action and evaporating. Subsequent research showed that the major loss of water from soils was through transpiration. Dust mulching, therefore, produced its beneficial effect largely through the control of weeds rather than by reducing evaporation from the soil surface.

Although the reasons for the success of dry farming were not generally understood, the technique was used to extend the limits of wheat growing into the marginal 250–400 mm rainfall zone of the South Australian, Victorian, NSW and Western Australian Mallee. Long fallows of fifteen months were employed and the frequent cultivation of these light soils resulted in soil structural breakdown, fertility decline and, ultimately, catastrophic erosion (Figure 7.1).

By the mid-1930s wheat farming in the Mallee was not a paying proposition and was replaced by sheep production. By the early 1950s wheat production occurred in rotation with pastures. Support for a 15-month fallow also waned as

Figure 7.1 *Soil erosion: the result of inappropriate tillage practices.* (PHOTO: Department of Conservation and Environment, Victoria)

results showed that fallows of 8 to 11 months produced yields equal to those of a longer fallow. The cost in terms of loss of winter grazing and soil erosion were also significant. It was also shown that much of the benefit of fallowing was from nitrogen mineralization, not water conservation as originally supposed.

Two lessons emerge from this account of the origins of fallowing techniques in southern Australia. First, widely accepted farming practices can be based on poor foundations: long fallowing very often gave no appreciable increase in water storage over short fallows whilst an unrecognized response to fallowing was nitrogen mineralization. It is therefore advisable to consider the outcome of farming practices in terms of specific effects on soils and plants and not yield alone. The second lesson is that practices can have effects well beyond those intended, emphasizing the need to understand how the specific effects interact with one another and the environment.

For winter crops, *long fallows* (> six months) usually begin in the winter or spring preceding autumn sowing, or even earlier where wheat follows sorghum in a summer rainfall area. In summer rainfall areas, the plan is to accumulate rain falling over summer and sometimes the preceding winter, before planting the winter crop. The soil is often dry at the start of the fallow and the aim is to accumulate a percentage of rain falling on the fallow (expressed as fallow efficiency). There is, however, sometimes a residue of water in the soil after previous crops. Fallow efficiency depends on maximizing infiltration and reducing evaporation of water from the soil, as well as preventing weed growth. Efficiency depends on the tillage practices used. As much as 200 mm available water can be stored in a long fallow on heavy-textured soils in the summer-rainfall zone, although fallow efficiency using traditional methods is only about 20 per cent.

Long fallows in winter rainfall areas benefit the following crop mainly by carrying-over water which accumulates in subsoil during the previous winter-spring. Effective weed control during the fallow is the key to preserving the subsoil water for the crop. Soil water at sowing can range from 40–150 mm or more, depending on soil texture, rainfall and length of fallow. Rain falling on these fallows is not stored efficiently.

The long fallow has been declining in importance since about 1965. Much of the land which would once have been fallowed is now either left in pasture or cropped instead of fallowed. This emphasizes the fact that fallowing is essentially a land use which competes for rainfall with other forms of land use, notably crops and pastures. Very long fallows (>12 months) are now largely confined to the lower rainfall fringe of northeastern Australia and the Mallee of western Victoria (in fallow-wheat-fallow rotations).

Short fallows (1–6 months) are used between successive winter crops or after a period of pasture. The main aim in winter-rainfall areas is to prepare land for cropping but water storage is often the unintended though important consequence of seedbed preparation. At Wagga Wagga, NSW, for example, which receives relatively high rainfall, an average short fallow stores about 70 mm of water, only slightly less than a long fallow. Summer falls of rain are mostly lost by evaporation and much of the water that is stored in a short fallow falls in the autumn, shortly before sowing. After a wet year, the residue of water left by a crop can be carried over to the next crop. It is often not appreciated that this residue of water, and the water that could be stored from rain between crops, is lost when even small amounts of plant growth occurs between crops. This is an important consideration when the short fallow is not cultivated and fallow weed growth is controlled by grazing: there is a trade-off between benefits to the animal and the following crop.

In summer-rainfall areas, short fallows are used deliberately to store water, resulting typically in available water storages of about 100 mm in northern NSW and southern Queensland. These fallows may start with varying amounts of water, depending upon the rain that falls near harvest of the preceding crop.

The third type of fallow before a crop is really the *no-fallow*, where seedbed preparation involves less than one month freedom from weeds. In southern Australia this is common when low summer and autumn rainfall preclude cultivation. It is also an important option in both summer and winter rainfall areas since the advent of modern herbicides. No-fallow soils can still hold available water from rain falling close to sowing.

The evolution of sustainable tillage systems

Changes in tillage practices since settlement reflect the differences in climate and soil type discussed in Chapter 6.

In more northerly latitudes, crop production depends heavily on fallowing for water conservation, particularly summer fallows for cool-season crops like wheat. Summer is a period of high rainfall, frequently with high intensity. The traditional burning at the commencement of fallowing removes protective cover and lays the soil bare to the full erosive power of rain for many months. Incorporating stubble, for example with a disc plough, does little to reduce this erosion. Continuous cropping has depleted the once high reserves of soil nitrogen on the better soils. The development of sustainable farming practices in the summer-rainfall area has focused on ways of retaining crop residues above the ground to reduce erosion. Cultivation itself does relatively little damage to the self-regenerating structure of the major soil types used for cropping, but it buries any stubble which has not been burnt and also accelerates the loss of soil organic matter.

The evolution of tillage practices in the summer-rainfall area has been from burning plus cultivation to stubble incorporation (disc implements, no burning) and then to stubble mulching, which is the main commercial practice (sweep and chisel ploughs, rod weeders, no burning). Finally no-till fallows developed in which stubble remains standing and herbicides are used to control weeds on fallows. The no-till fallow is largely confined to research. None of these tillage practices constitutes a farming system; they are techniques for fallowing and crop establishment.

During the 1980s, researchers started to examine the role of crop rotations, including pastures, in northern farming systems. Rotations are being developed to manage problems with diseases and crop nutrition that occur without burning and cultivation. New techniques of weed management are being developed to reduce costs and dependency on herbicides, and new rotations are being devised to make better use of the improved water supply found in no-till systems.

In southern areas, the main cause of soil degradation has been the cultivation of soils which generally have less stable structure than the main cropping soils to the north. Compared with northern latitudes, burning contributed relatively little to erosion because the soil was less likely to be bare over summer and the incidence of summer rain is much lower. There was also reason to burn when heavy stubbles decomposed slowly over the dry southern summer causing problems with machinery at sowing time. In these areas, the initial development of sustainable farming methods focused on reducing cultivation, which has been widely accepted by farmers, and on direct drilling which has also enjoyed some popularity where long fallowing is not common. With direct drilling as originally practised, stubbles are burnt, herbicides are used to control weeds before sowing, and crops are established with little or no prior cultivation. Direct drilling is not a farming system; it is a technique for establishing crops. Research since the mid-1980s has aimed to retain crop residues as well as reduce cultivation. During this period the importance of pastures in rotations has been recognized more widely in terms of soil *physical* regeneration. The importance has been appreciated of integrating pasture and livestock management with new tillage practices to improve both weed management and overall farm productivity.

In low rainfall areas to the south, where long fallowing is practised, the development of sustainable farming methods has drawn upon techniques of both stubble retention and reduced cultivation. Whilst burning and cultivation are still the most widely used practices, many farmers value stubble for its ability to reduce wind erosion on light soils as well as a perceived benefit for soil structure. Direct drilling is not practised frequently, particularly on the hard-setting red-brown earths of the region, because it can cause poor crop establishment and reduced crop vigour. Herbicides frequently partially substitute for cultivation.

The aims and effects of tillage

Although mankind has tilled the soil for at least 3500 years the effects which actually benefit the crop have been established with certainty only very recently. It is now known that crops can be grown satisfactorily without cultivation in many soils, provided weeds are controlled. Seedbed preparation, pest and disease control and stimulation of organic matter breakdown to release mineral nitrogen are often important aims or effects of cultivation, but the main short-term effect (within seasons) is usually through weed control: both the weeds of fallows and crop weeds which can be reduced by tillage before sowing. The residues remaining from previous crops are burnt mainly to allow the passage of cultivating and sowing machinery, although sanitation can be an important effect of burning.

Since the main benefit of tillage is weed control, the range of herbicides now available has largely released farmers from one of the most important reasons for cultivation. However, cultivation does have wide-ranging effects which need to be understood before deciding not to cultivate. The challenge is to assemble the information needed to decide what further changes must be made for farming without cultivation to remain both profitable and sustainable.

Where pastures are included in rotations to reclaim structure which has been lost through cultivation, the substitution of herbicides for cultivation could free farmers from a major reason for rotating pastures with crops. Thus land could be dedicated to cropping or pastoral uses, depending upon its suitability as a land class. Alternatively, cropping could be intensified in response to economic pressure, or to make more efficient use of rainfall, without fear of soil degradation due to excessive cultivation. Once again, of course, a farmer would need to know what effects pastures had, other than improving soil structure, so that a change in rotation away from pastures would have no unexpected effect on crop yields.

The traditional aims of tillage are: to give efficient use of rainfall; to mineralize organic matter, mainly to release nitrogen; to prepare a seedbed suitable for sowing; and for crop sanitation including pest, disease and weed control. These aims will be discussed in turn.

Efficient use of rainfall

The overriding constraint in dryland farming systems is to make efficient use of limited water. Efficient use of rainfall is central to any discussion of tillage because the profitability of dryland farming depends very much on the efficiency with which rainfall is utilized. Efficient use of rainfall is a major goal of traditional tillage practices through fallowing. The tillage practices which have been designed to reduce soil degradation can also improve the utilization of rainfall. This is the key to profitability using new tillage practices, and therefore the key to their acceptance and adoption by farmers.

Alternative fallow management techniques have now been compared in numerous long-term experiments in most States. Effects of soil and crop residue management on fallow efficiency are most clearly shown for short fallows in the summer-rainfall zone that are commenced with dry soil profiles. Compared with a clean, cultivated fallow, fallow efficiencies improve with stubble mulching (but not stubble incorporation) and improve further with deletion of cultivation in a no-till fallow.

Approximate fallow efficiencies are:
- Cultivated, burnt — 19 per cent
- Cultivated, residue mulched — 24 per cent
- Not cultivated, burnt — 25 per cent (few data)
- Not cultivated, residue mulched — 29 per cent

These improvements in fallow efficiency were obtained on heavy-textured, mostly self-mulching soils in the summer-rainfall zone. They strongly support new tillage practices in terms of greater soil water storage. Greater water storage may be used either to increase crop yields or, if sufficient water is available, to increase cropping frequency.

The results make an interesting comparison with North America where fallow efficiencies have increased from around 20 per cent with dust mulching to about 30 per cent with stubble mulching (cultivated), and to 50 per cent with the no-till fallow.

Figure 7.2 *Definitions related to water use and water-use efficiency.*

Transpiration (T)	Water lost from leaves as CO_2 enters for use in photosynthesis
Soil Evaporation (E_S)	Loss of water by evaporation from the soil surface
Crop Water Use (WU)	$T + E_S$ between two points in time, commonly planting and harvest
Fallow efficiency =	$\dfrac{\text{water stored in soil}}{\text{Rain falling on fallow}} \times 100\%$
Soil Water Balance	The change in the amount of available water held in soil between two points in time. It is given by the equation:

$$W_2 - W_1 = P - R - D - (E_S + T)$$
W_2, W_1 = water held at time 2 or time 1, respectively
P = precipitation
R = runoff
D = drainage below potential rootzone
E_S and T, as defined above.

Note that infiltration (I) = P – R

Water-use Efficiency (WUE) at three levels:

1 Plant = $\dfrac{\text{Dryweight change}}{\text{transpiration}}$

= transpiration efficiency

2 Crop = $\dfrac{\text{Grain yield}}{\text{WU}}$ or $\dfrac{\text{Crop dryweight}}{\text{WU}}$

= crop water use efficiency

3 Farm = $\dfrac{\text{Product}}{\text{Rainfall}}$

Crop yields depend on the quantity and efficiency of water use. The definition of terms related to rainfall use are given in Figure 7.2. The concepts of water-use and water-use efficiency can be introduced through an example. Long fallowing for water conservation in the Wimmera district of Western Victoria aims to save rain that falls in one season for use by a later crop of wheat. Although only one crop is grown in two years, its water use and yield are generally high. On the other hand, only about one-quarter of the rain falling on the fallow is stored (i.e. fallow efficiency is 25 per cent), the rest evaporates and is wasted. This wastage is reduced by growing a crop every year but evaporative losses will still occur during the short fallow between crops. The result of growing crops in successive years is two lower-yielding crops, but a greater fraction of the rain falling over the two years is being used by crops.

Water-use efficiency also varies. This is because water stored in fallows, deep in the soil, is less prone to evaporation later, during crop growth.

The net effect of these two rotations (fallow-wheat *versus* wheat-wheat) on water use and water-use efficiency at the plant, crop and farm level is illustrated in Table 7.1. Note that crop water-use efficiency was high in the fallow-wheat system but farm water-use efficiency was low. The profitability of these alternative rotations will vary, depending on costs and prices. A survey of farms in Victoria during the late 1980s indicated that individual crops after long fallow were more profitable than non-fallow crops. Considered over the whole rotation, however, cropping in successive years was more profitable (though more risky) than fallow-crop.

A change in rotation which intensifies cropping, as in the example in Table 7.1, can frequently improve farm water-use efficiency and profitability. In the following sections we show the potential for new tillage practices to increase farm water-use efficiency through their effect on rotations. These practices also have the potential to directly increase fallow efficiency and crop water-use efficiency. To understand these effects of tillage we need to consider the soil-water balance and discover the fate of rainfall on dryland farms.

Estimates for Wagga Wagga (Table 7.2) provide an example of where the water goes over a twelve-month period in a wheat-to-wheat rotation. A short fallow is used between crops. In this rotation the aim is to direct as much of the annual rainfall through the crop (i.e. crop T) as possible, but only about 210 of a possible 560 mm is actually transpired by the crop. New tillage practices can aim to reduce non-crop transpiration (e.g. by better

Table 7.1 *Water use and water-use efficiency in two rotations.*

	Rotation			
	Fallow	Wheat	Wheat	Wheat
Year	1	2	1	2
1 Annual rainfall (mm)	400	400	400	400
2 Fallow water storage (mm)	100[1]	–	–	–
3 Crop water-use	–	400[2]	300	300
4 Transpiration efficiency (kg/ha/mm)	–	20	20	20
5 Crop water-use efficiency (kg/ha/mm)	–	10	8	8
6 Crop yield, 3 x 5 (kg/ha)		4000	2400	2400
7 Farm water-use efficiency (kg/ha/mm)[3]		5		6
8 Crop gross margins[4]		++	+	+
9 Farm gross margins[4]		+		++

1 Assumed fallow efficiency of 25 per cent.
2 Crop water use is taken as 0.75 rain falling in crop year, plus fallow water from previous year (if any).
3 Farm WUE is calculated over two years of rotation (total yield/total rain).
4 Crosses indicate likely gross margins, the more crosses the better.

Table 7.2 *Estimates of the fate of yearly rainfall on a dryland farm at Wagga Wagga, NSW.*

Precipitation (P)	
May–November	280 mm (growing season)
December–April	280 mm (fallow period)
Total	*560 mm* (average annual)
Transpiration (T)	
Weeds in crop	10 mm
Fallow weeds	10 mm
Crop	210 mm
Evaporation (E_s)	
Fallow	170 mm
Crop	140 mm
Runoff (R)	20 mm (highly variable)
Drainage (D)	0
Total Losses	*560 mm*

NOTE: Crop ET = 360 mm comprising 260–280 mm growing season rainfall (depending when runoff occurred) and the remainder from fallow storage.

Bare Fallow

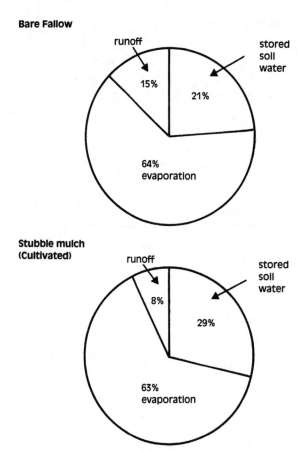

Stubble mulch (Cultivated)

Figure 7.3 *The fate of rain falling on fallows in Queensland. Tillage practices aim to reduce evaporation (E) and runoff (R) whilst maintaining good control of weeds.*

weed control) and soil evaporation, and also runoff and deep drainage where they occur.

Work in Queensland on long fallows gives a clue where to start in improving rainfall utilization in that area (Figure 7.3). On traditional cultivated fallows, about 64 per cent of rainfall is lost by runoff — only 21 per cent is conserved for the crop.

The effects of cultivation and stubble retention on infiltration and runoff, drainage, soil evaporation and transpiration by weeds are given in Table 7.3.

Effects of tillage on infiltration and runoff

Whether we examine infiltration or its complement runoff depends on our purpose. Studies of crop growth and yield require an examination of infiltration because this determines the amount of water potentially available for plants. In soil-related studies, of erosion in particular, examination of runoff is necessary because this is the driving force behind soil erosion.

Cultivation roughens the soil surface, and temporarily increases surface storage of water. Generally, maximum surface storage is small (<10 mm), even after rough cultivation, and is decreased further by weathering, rainfall and secondary tillage. The effect of cultivation, relative to an untilled surface, on runoff is likely to be due to a temporary increase in the potential rate of infiltration through the creation of more large pores which increase the effective surface area for infiltration. The breaking up of surface crusts may also be involved. Such effects are most important in soils with weak structure but they are also more temporary in these soils because slumping and crusting redevelop with further rain.

Long-term effects of *surface* cultivation on the permeability of soils are more likely to be negative and governed by the loss of structure. Maintenance of soil structure and the continuity of large pores through the topsoil are potential advantages of reduced tillage systems, including direct drilling, and this often occurs despite bulk density increases. Greater infiltration at depth can thus result.

Longer-term positive effects of *deep* cultivation ('ripping') can arise where this breaks up natural or tillage-derived subsurface barriers; thus improved infiltration and reduced waterlogging have been reported for duplex (texture-contrast) soils near Rutherglen in Victoria. Improved infiltration has also been reported on a dark self-mulching soil at Warialda and a hard-setting red soil at North Star, both in northern NSW.

Plant residues on the soil surface (i.e. mulch) can cause a major reduction in runoff, principally by protecting soil surfaces that are prone to crusting from raindrop action. Living plants in a non-fallow situation also confer this advantage. In summer-rainfall areas where the risk of erosion is high, one strategy to reduce erosion is to have a live crop cover over the soil when the risk is greatest. In grazed pastures, the ground cover must be complete and dense if surface sealing by treading is to be avoided.

Runoff reduction with straw retention on the surface has been recorded for both cultivated and herbicide-fallow (uncultivated) plots on a number of clay soils. Fallow efficiencies in Queensland have been increased from about 21 per cent to 29 per cent, almost entirely because of reduced runoff with stubble *mulching* (Figure 7.3). However, stubble incorporation by discing, although altering runoff characteristics, reduces total runoff only occasionally relative to burning and discing.

Similar increases in water storage can occur in southern Australia when crop residues are retained and infiltration is improved (Figure 7.4). Note in this example that only about 4 t/ha of residues was needed to gain the maximum improvement in infiltration. A crop of wheat yielding 2 t/ha of grain, leaves about 4 t/ha of residues. As the national average wheat yield is about 1.5 t/ha it appears that most crops produce, at harvest, sufficient residue to gain most of the potential benefit from increased infiltration. Of course, post-harvest management (grazing, cultivation) should aim to keep no less than this amount of residue on the soil surface.

Although stubble mulching undoubtedly reduces runoff and hence soil loss, it should be appreciated that stubble has less effect on infiltration and runoff as the soil becomes wetter. Towards the end of a fallow, when the soil approaches maximum waterholding capacity, residues have little effect on runoff. Under these conditions the amount of cultivation has a strong bearing on soil loss.

In summary, a reduction in cultivation should increase infiltration and reduce runoff by encouraging long-term improvements in soil structure. In addition, residue retention can improve infiltration in tilled and non-tilled soils alike. Despite these very positive effects, caution is needed. When pastures are killed with herbicides at the commencement of a long fallow the dry plant residues are quickly consumed by livestock or they break up and blow away. The flat even soil surface which remains may have low infiltration because of prior treading by animals in the wet months. In the absence of residues, runoff and soil loss during heavy rain can be quite high. Under these circumstances a primary cultivation of the fallow, followed by weed control with herbicides, appears to be the best compromise for control of runoff and erosion.

Effects of tillage on drainage

Drainage beyond the potential root zone of crops can occur whenever precipitation greatly exceeds evaporation for periods of several weeks or months. However, it is most likely to be significant in the light soils of southern Australia, especially in Western Australia where increased drainage after clearing is the primary cause of rising watertables and increasing dryland salinity. It may also occur on heavier soils where long fallow is practised in higher rainfall areas.

Cropping without cultivation can improve both infiltration and the redistribution of water in the soil profile. There is some concern, therefore, that tillage practices which improve structural stability and infiltration properties could also contribute to rising watertables and salinization in some areas. There is already evidence of this occurring in southern Queensland. Also, infiltration may cause nitrogen to be leached, possibly beyond the root zone. These potential areas of concern require further research.

Apart from these adverse outcomes of deep drainage, there is the loss to the crop of potentially useful water. Fallowing, particularly long fallows in either higher rainfall environments (e.g., Wagga Wagga, Adelaide) or on light textured soils has been shown, on occasions, to result in deep drainage. The likelihood of this occurring should increase with improved tillage practices so there needs to be a review of both fallow practices and tillage practices. Shorter fallows may be both possible and necessary with reduced tillage. This, in turn, may require a revision of fertilizer needs or crop rotations if less nitrogen is mineralized in the shorter fallows.

Improved soil surface structure may increase

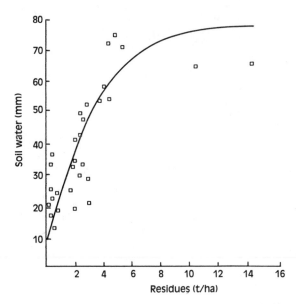

Figure 7.4 *Crop residues improve infiltration and soil water storage. Data for red-earth soil, Wagga Wagga. More water means higher potential yield and more intensive cropping may also be possible.* (SOURCE: Cornish, P. S., unpublished data)

the incidence of waterlogging on duplex soils in the high rainfall cereal zone of southern NSW, north-eastern Victoria and the Western districts of Victoria, if subsurface drainage is not also improved. Deep cultivation which improves subsoil drainage, in conjunction with tillage practices which stabilize soil structure, could lead to improved drainage in these soils.

Effects of tillage on soil evaporation

By far the greatest loss of water is from long fallows: about two-thirds of the rain falling on the fallow is lost by evaporation (Figure 7.3). Most of the water is lost from surface soil. For practical purposes, it is useful to know that over long drying periods evaporation rates fall to very low levels (< 0.2 mm/d) as the top 30–50 cm of soil approaches air dryness, provided there is no soil cracking. This happens regardless of tillage practice. Differences in evaporation due to tillage practice mostly occur in the short term.

Cultivation soon after rain, can reduce evaporation by forming a dry surface mulch. The short-term effect can be a significant increase in soil water. Over the life of a fallow the savings generally disappear as a result of further, prolonged slow drying. Although such tillage or 'dust mulches' are largely ineffective in Australia, this is not so in parts of the USA, presumably because of differences in soil type and rainfall patterns.

The effects of crop residues on soil evaporation are potentially somewhat greater than those of cultivation. Plant material on the soil surface reduces the energy available for soil evaporation because of its poor thermal conductivity. It also increases the resistance to vapour transfer from soil to air. Together, these have particularly large effects on short-term evaporation, but generally not in the longer term. Whether crop residues significantly improve fallow water storage by reducing evaporation, as distinct from increasing infiltration, depends mainly on the frequency and total amount of rain received. More water is stored, due to decreased evaporation, when rain occurs in frequent falls and the residues ensure incomplete drying of soil between rains. This permits deeper infiltration and therefore less soil evaporation.

The effect of crop residues on soil drying are illustrated in Figure 7.5. Crop residues substantially reduced surface drying to give a wetter surface the day after rain. However, after sixteen days without follow-up rain there was no net effect of residues on soil water content. This result was achieved under heavy wheat stubble (6 t/ha), so

even higher rates would be needed to improve long-term water storage by reducing evaporation. Such high rates of stubble rarely occur in farming practice.

Overall, the effects of residues on soil evaporation are small except in summer-rainfall areas with a high frequency of rain during the fallow. Of greater importance than total water storage is the potential short-term effect on water content in surface soil and thus on sowing opportunities.

Effects of tillage on transpiration

Apart from soil evaporation, transpiration through weeds is the major potential loss of water in fallows. In long fallows in winter-rainfall areas, much of the soil water stored at the commencement of the fallow (August–September) would almost always be lost to weeds by the end of the growing season (October–December) unless these weeds or remaining pasture are controlled. However, even if annual pastures are allowed to mature and set seed, the soil water remaining after

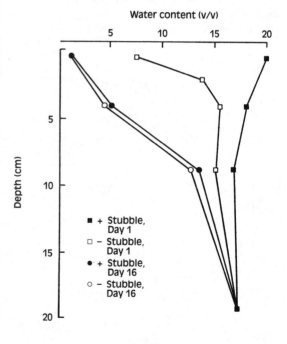

Figure 7.5 *A mulch of crop residues results in wetter soil surfaces, but only temporarily. Data for red-earth soil, Wagga Wagga. Wetter surface soil extends the period for planting, an advantage in most dryland cropping areas.* (SOURCE: Cornish, P. S., unpublished data)

a wet season could be more than that left by a cereal crop. On the other hand, a deep-rooted perennial such as phalaris or lucerne will leave the soil profile drier than will a cereal crop.

Summer and autumn rains falling on either long or short fallows can also be lost to weeds or volunteer crop unless these are controlled. This is illustrated in Figure 7.6 where stubble retention (+ *versus* –) and weed control (+ *versus* –) were combined to give four treatments, and net changes in soil water over a short fallow (January–May) were measured. Note that stubble retention resulted in the greatest storage of water (because of better infiltration) but, unless weeds were controlled, this advantage was lost.

In Figure 7.6 all of the water falling on the fallow was lost through evaporation and transpiration. This is one extreme. It resulted from rain early in the fallow which favoured the growth of weeds surviving from under the crop (wireweed and skeleton weed). It also provided a long period of transpirational losses. At the other extreme, very little water may be lost through weeds when rain falls in autumn, shortly before sowing the crop.

Cultivation and herbicides appear to be equally effective in weed control in experiments. In practice, some small differences may arise where residual herbicides can control weeds soon after

emergence, before cultivation is possible. On the other hand, knockdown herbicides require a target of sufficient size to spray. Some soil water is lost until this size is reached. Once treatment begins, however, land can be sprayed much faster than it can be cultivated.

Crop residues can interfere with the application of herbicides, although spray booms fitted to headers have been used to apply residual herbicides. Stubbles can also favour germination of some weed seeds because of wetter surface soils, but inhibit others because of toxins released from the stubble. Where germination is favoured, some farmers achieve good control of weeds which germinate in autumn by burning stubble shortly before sowing.

A farmer usually commits his land to cultivated fallows either in the spring (long fallow) or on the first substantial rain after harvest (short fallow). Once committed, a farmer must cultivate whenever a significant population or regrowth of weeds appears, regardless of rainfall or the water actually in the soil and therefore being conserved. This can be a costly exercise for only a small return in water stored. In the past, part of this cost has been debited against seedbed preparation. Herbicides, plus seeders which operate in uncultivated soil, now offer more flexibility in fallowing because land need not be committed to fallow until it is warranted by sufficient rainfall. Scientists are working on computer models which simulate the water-balance; it is planned to use these to help to make such decisions and manage fallows.

It is important to realize that the large effects on fallow efficiency of retaining crop residues and not cultivating occur only when the soil profile is dry at the commencement of fallow. Long fallows in summer rainfall areas that commence with appreciable amounts of soil water invariably have low fallow efficiencies regardless of tillage practice.

In the winter rainfall zone there have been fewer comparisons of the tillage options in long and short fallows. The comparisons which have been made suggest that herbicides are as effective as cultivation (but not superior) and that residue retention sometimes increases water storage compared with burning, but not greatly so. The smaller benefit from mulching residues, compared with summer rainfall areas, appears to result from the lower frequency of summer rains, and therefore there is little benefit from reduced soil evaporation. Also, the swelling clays of summer-rainfall cropping areas maintain continuity of cracks from soil surface to subsoil when not cultivated, enabling superior recharge of subsoil water and

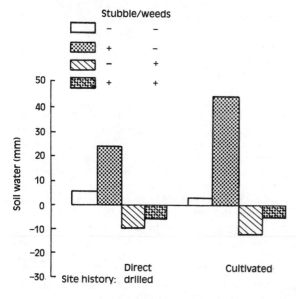

Figure 7.6 *Weed control is necessary to reap the benefits of stubble retention in a short fallow. Data for direct drill crops, Wagga Wagga.* (SOURCE: Cornish, P. S., unpublished data)

reduced runoff. Residue mulching appears to help maintain this aspect of soil structure. Because reduced tillage and stubble retention increase fallow water storage, it is likely (but not yet proven), that water is used more efficiently during the period of crop growth as well. Any effects on reduced evaporation and runoff are most likely to occur early after sowing, before full ground cover is reached.

Nitrogen mineralization and other chemical transformations

Another major reason for tillage is to stimulate the mineralization of organic matter, principally nitrogen. Cultivation alters the water content, temperature and aeration of soil, all of which can influence the rates of nitrogen transformation. These transformations can include losses of mineral N through immobilization in microbial tissue, denitrification and leaching as well as increases through mineralization. It is difficult to say whether cultivation itself greatly stimulates mineralization or whether the absence of plants in a cultivated fallow allows mineral N to accumulate. However, there is no doubt that crop yield responses to fallowing in southern Australia have been due partly to increased N supply to the crop.

Mineral N concentrations in soil at sowing time are often similar, whether the fallow has been cultivated or kept free of weeds with herbicides. Similarly, crops sown after cultivated fallows, when compared with uncultivated but weed-free fallows, give yields and N-fertilizer responses which are frequently similar. Long-term studies of fallowing do show, however, that cultivation results in lower soil organic N compared with herbicide fallows.

The trend away from long fallowing will generally reduce the mineralization of organic N and possibly increase the need for fertilizer-N applied at sowing. However, the method of fallowing seems to have no consistent effect on the amount of additional N required.

With conventional fallowing techniques, crop residues are disposed of by burning. When cereal residues, which have a high C:N ratio, are incorporated into the soil they can cause mineral N (or added fertilizer-N) to be immobilized or 'tied-up' temporarily in soil microbial tissues. Whether this increases the need for N-fertilizers is open to question, but it seems to be the case where total soil N levels in the soil are relatively low and the amounts of residue incorporated are high, i.e., from high-yielding crops. Work with stubble retention in Queensland and sometimes in NSW shows an increased response to N-fertilizer when residues are not burnt. The greater requirement does not appear to diminish after a long period of stubble retention. In southern NSW and north-eastern Victoria it seems that crop rotations will need to change if stubbles are retained unless extra fertilizer-N is applied: thus two crops of wheat can follow a crop of lupins when stubble is burnt, whereas wheat and lupins in alternate years is recommended when stubble is retained.

Legume residues are high in N compared with cereal residues. Their incorporation at the commencement of fallowing, after a pasture phase, can lead to a net release of mineral N but recovery of this N by crops is variable and often low.

Seedbed preparation

Seedbed preparation is one of the major aims of traditional cultivation. High crop yields owe much to seeding machinery which gives prompt, uniform crop establishment. Until recently, these machines had been designed to work in finely tilled soils, usually in the absence of plant residues. The fine tilth results from an initial ploughing and subsequent repeated shallow cultivation of moist soil. However, there have been remarkably few studies which show conclusively that cultivation is the only, or even the best way, to meet the needs of the plant for germination and emergence.

Moisture is usually the deciding factor in the success or otherwise of the sowing operation. Seeds can lose water as well as absorb it during germination. Therefore, cultural techniques aim to maximize the uptake and minimize the loss of water by seeds. In cultivated seedbeds, good seed-to-soil contact maximizes water uptake whilst covering the seed with soil reduces water loss. A fine tilth improves seed-to-soil contact, as can press wheels, although compaction over the seed can reduce emergence in wet, poorly structured soils. Deep sowing places seed in soil which dries fairly slowly. This maximizes germination but emergence can be reduced.

The disadvantage of a fine tilth is that the short-term improvement in structure, produced by cultivation, is offset by the longer-term loss of structural stability. This can lead to formation of a soil crust if rain follows sowing before emergence. Crusting is a serious problem for the red-brown earth soils that are the major cropping soil of Australia, extending from southern Queensland to Western Australia. The modern trend, therefore, is away from tilled seedbeds.

Sowing without prior cultivation, however, can result in poor germination or emergence. This is especially likely when soils are structurally degraded and when established weeds bind the soil into a sod. Both conditions result in cloddy seedbeds, poor seed coverage and poor seed-soil contact.

One key to successful planting in previously uncultivated seedbeds is to control weeds well before sowing. This prevents large weeds, with large root systems, from developing. Weed control with herbicides, or by grazing if sufficient stock can be mustered, will often enable conventional seeding equipment to perform satisfactorily (in the absence of crop residues). The successful operation of conventional seeders in uncultivated soil depends very much on soil moisture. Uncultivated soils are stronger than cultivated soils so they need to be wetter for machinery to operate. This can occasionally reduce the opportunity for sowing, especially in lower rainfall areas. On the other hand, uncultivated soils can carry traffic at higher water contents than cultivated soils. This can enable timely planting in wet regions or wet years.

Seeders have been developed which give good seed-to-soil contact, and good seed coverage, when operating in previously uncultivated soils over a wider range of moisture contents. Some of these seeders also operate well in crop residues (Figure 7.7). Crops established with these seeders can germinate and emerge in uncultivated soil as satisfactorily as in a finely tilled seedbed, but seedling growth rates are still often reduced. This reduction is thought to be due to physical impedance of roots in uncultivated soil, causing a 'bonsai' effect. To try and improve early root growth, some seeders cut a narrow slit beneath the seed to provide an easy pathway for seedling roots.

Crop residues on the soil surface keep the surface soil moist for longer, enabling shallower planting and an extended sowing period after rain. However, the seeding equipment must be able to plant through the residues without tynes blocking, whilst achieving uniform depth of placement and good seed-to-soil contact. Seeders capable of sowing into crop residues are mostly characterized by: increasing spacing between tynes, in both dimensions, resulting in row widths of 250 and sometimes 300 mm (for winter cereals); increased height clearance; tynes with narrow points, for minimal soil disturbance; and often press wheels. Some have elaborate mechanisms for depth control.

There are now reasonably satisfactory techniques for establishing crops in the presence of stubble without prior cultivation. Apart from soil conservation this has several important advantages. Opportunity cropping (Figure 7.12) is made easier because time is not spent cultivating between crops. For the same reason, crops in the Mediterranean zone are more likely to be sown near to the optimum time. Also, improved traffica-

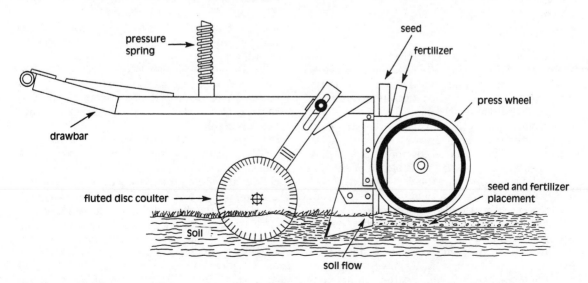

Figure 7.7 *Basic design of a seeder for use in uncultivated soil and in crop stubbles.*

bility in wet years and greater surface moisture in dry years increases the chance of sowing on time in all regions.

Crop sanitation

Both burning and cultivation can aid crop sanitation and are significant reasons for tillage. Sanitation includes the management of diseases, pests and weeds.

Burning crop residues

This is often done during autumn in southern NSW and is one option for coping with stubbles which have not broken down sufficiently for sowing machinery to operate without blockages.

Burning of crop residues can help to control diseases by destroying pathogens which survive on crop residues between susceptible crops. Burning will mostly help reduce the severity of diseases which are not dispersed widely (see Chapter 9) such as yellow spot (*Pyrenophora tritici-repentis*) in wheat and disease scald (*Rhyncosporium secalis*) in barley. Control of such diseases in intensive cropping systems where residues are not burnt, requires either crop rotations or breeding of resistant crop cultivars.

Crop residues are burnt also to aid weed control, as well as allow the free passage of machinery.

Cultivation

Cultivation aids crop sanitation in three ways:
1 By controlling weeds before sowing that would otherwise be a problem in the crop (this is the reason behind the use of cultivating tynes on Australian combine seeders).
2 By controlling weeds in a fallow that are alternate hosts for a crop disease — a long fallow which controls grass weeds for a year before sowing greatly reduces the incidence of the disease Take-all (*Gaeumannomyces graminis*).
3 By reducing the infectivity of propagules of root pathogens (Rhizoctonia, caused by *Rhizoctonia solani*, appears to be less severe after only a single cultivation).

Although burning and cultivation do play a part in crop sanitation it is not always clear just how important they are. Some pathogens are controlled by tillage, but others may be encouraged. This is illustrated for wheat in Table 7.4 . With

yellow spot, the need to remove all residues has been clearly demonstrated, and burial with tillage is not sufficient to prevent the disease. For other diseases of cereals, such as Rhizoctonia, Take-all and nematodes, the effects of tillage are less clear. What is clear, however, is that crop rotations, careful weed management, genetically-based resistance and perhaps biological control (which is a real possibility in the case of Take-all) will be determinants of success in any farming system which places less emphasis on burning and cultivation.

Weed management — a new approach

The usual approach to weed control aims at eradication, either by tillage or with herbicides. Of course, eradication is rarely feasible, either technically or economically. Successful implementation of new tillage practices depends heavily on a new approach to weeds. This is partly because we want to make less use of cultivation and partly because of the cost and environmental concerns attached to the alternative, herbicides. The simple substitution of one herbicide treatment for one cultivation to control weeds is rarely economic. Weed management is the new approach that is required.

Weed management aims to prevent rather than cure weed problems by keeping the population of potential weeds so low that they cannot cause serious losses in future crops or pastures. A number of methods are used together in a planned system to economically keep weed populations low. The principles apply to both crops and pastures.

Good weed management should lead to:
• more effective and more economical weed control with reduced chemical costs per hectare of crop;
• adequate suppression of grass weeds for which no post-emergence control is available;
• reduced risk of crop damage from herbicides through reduced use of post-emergence and possibly residual herbicides;
• reduced risk of herbicide tolerance arising in weeds (because a range of management techniques is used); and
• cleaner and more productive pastures.

An important feature of a weed management programme is the flexibility to tackle weeds whenever the need arises and by the most appropriate means. This could mean controlling weeds the year before sowing a crop, spray-fallowing or using a shallow cultivation before sowing, if necessary, to stimulate the germination of weeds which will later be killed with herbicide or in the sowing operation.

Table 7.3 *Qualitative summary of the relative magnitude of direct effects of cultivation, herbicide-fallow (not cultivated) and residue retention on components of soil water loss[a].* (SOURCE: Fischer, R. A., 1987, in *New Directions in Australian Agriculture*, P. S. Cornish and J. E. Pratley (eds), Inkata Press, Melbourne).

Component of loss	Aspect of the loss	Surface cultivation[b] short-term[c]	long-term[c]	Herbicide fallow	Residue retention
Evaporation	Short-term[e]	0[d]	0	0	+++
	Long-term	++	–	0	+
Transpiration	Weed germination	–	–	++	–
	Weed growth potential	–	0	0	–
	Weed killing	++++	0	++++	–
Infiltration	Surface storage	++	–	0	0
	Potential infiltration	++	–	0	0
	Sustained infiltration	0	– –	0	+++
Drainage	Transmission pores	0	+/–[f]	0	0[f]

[a] Expressed relative to a virgin bare non-fallow surface, with + meaning less loss or more water storage, – meaning less storage, and 0 meaning no effect.

[b] Ignores deep tillage, which may have a moderately positive effect via sustained infilitration.

[c] Within one year *versus* after several years of tillage.

[d] May be briefly negative if wet soil brought to the surface.

[e] Short-term, days; long-term (weeks or months). Terms correspond with stage 1 and stage 2 drying.

[f] There is evidence of deeper drainage where stubble is retained but the reasons are unclear. Deeper drainage may lead to water losses if water moves beyond root zone.

Crop-pasture rotations can play an important role in any weed management programme. Many weeds are encouraged by the nitrogen accumulated under pasture (e.g. ryegrass). It is important to return a paddock to crop before these weeds build up seed in the soil and their numbers become hard-to-manage in the crop. Similarly, other weeds can exploit low nitrogen conditions, (e.g. capeweed) so it is important not to excessively deplete soil N. A return to clover-based pasture can be a great help to a weed-management programme, especially if the weeds present can be readily controlled by grazing and/or pasture manipulation (Chapter 9). Crop rotations in continuous cropping sequences offer a good opportunity to use in-crop herbicides to manage difficult-to-control weeds. Alternative crops to wheat allow an extended range of chemicals to be used. In general, weeds do best in crops with similar growth requirements, such as ryegrass in wheat, or wild radish in canola. These principles have always been important in farming systems but they assume greater importance where cultivation is available only as a last resort for weed control.

On a mixed farm, economic weed management starts with the pasture. Grass weeds are among the easiest and cheapest to clean up in the pasture in the year before cropping. Of the methods available that do not depend on herbicides, hard grazing has been used often, but it is not very effective because a grazed pasture still sets seed and enough stock may not be available for hard grazing in a good spring. Burning also rarely works very well. Cutting the pasture for hay will reduce weed seed production much more effectively. But the cut should not be left too late or else live seed will be shed during curing and baling. Silage making is another possibility.

It is also possible to remove weeds from pastures using a variety of herbicides. The problem in deciding what to use and when to use it, is that costs and effectiveness vary enormously between products. The products range from the growth regulators to control broad-leaf weeds to the knockdowns for control of grass weeds. These herbicides are cheap and effective but require a clear understanding of how they affect plant growth to retain their cost advantage. Techniques include pasture manipulation, spray topping and spray/-graze. At the other end of the scale are the selective herbicides which are very effective chemicals for controlling grass weeds in clover and lucerne. They do not require a lot of skill in application but are very expensive. The techniques of weed management in pastures before cropping are discussed more fully in Chapter 9.

Soil conservation

The development of new tillage practices has been inspired by the needs of soil conservation so a primary criterion for the success of these techniques is their contribution to soil conservation. It has been recognized for many years, at least in southern Australia, that good soil management requires a period of pasture after cropping in order to restore structural stability. Even so, unacceptably high rates of erosion occur during periods of cropping, as do other problems associated with structural instability, such as crusting. The question now is whether new tillage practices can maintain the structure in good condition during a cropping sequence or even improve structure and so remove the need to return to pasture.

Tillage and structural stability

When soil is cultivated over a number of years the content of organic carbon declines and, along with it, the stability of soil aggregates. Symptoms of this are surface crusting, surface and subsoil compaction, reduced infiltration and accelerated erosion (Chapter 11).

Upon cultivation after a pasture phase, structural stability can decline very quickly even when little change in total organic carbon is detected. This decline is associated with a particular fraction of the soil organic matter that is quickly lost under cultivation but restored under pasture. Even minimal soil disturbance, such as that associated with direct drilling, can greatly reduce structural stability. This is illustrated in Figure 7.8 for a red-earth soil at Wagga Wagga which had been in pasture for many years before being cropped. (The percentage of water-stable aggregates is a measure of aggregate stability.) Note that after only four years of cropping there had been a sharp decline in aggregate stability and infiltration after both conventional cultivation and direct drilling. There was no detectable change in total organic carbon over the 0–10 cm depth but its distribution had changed. Figure 7.8 shows that soils may be structurally fragile and even direct drilling can reduce structural stability compared with long-term pas-

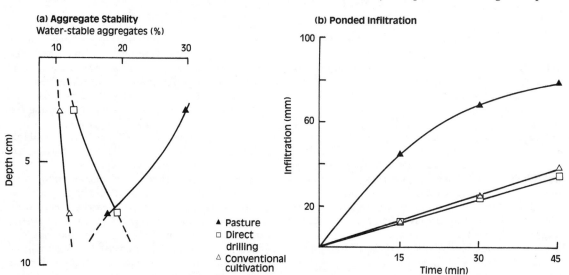

Figure 7.8 *Structural degradation of a red-earth soil, due to soil disturbance, as indicated by infiltration and aggregate stability.* (SOURCE: Cornish, P. S. and Barker, P., unpublished data)

Figure 7.9 *Changes in organic carbon and structural stability during a sequence of cultivated or direct drilled crops after pasture. Note that structural stability declines initially in the direct drilled soil. The issue is whether the longer-term stability is adequate to sustain cropping without unacceptable erosion or loss of crop yield. If not, then a return to pasture is needed.*

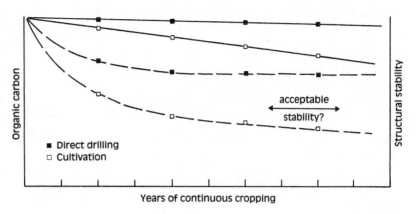

ture. In this case, a conventional combine seeder was used for direct drilling to give some weed control at sowing time. Other seeders give less soil disturbance and would presumably result in less effect on soil structure.

After the quick initial decline in structural stability following pasture there is a slower but continuing decline in both carbon and structural stability in all soil types. It appears that minimizing soil disturbance, for example by direct drilling, greatly reduces this decline in carbon and soil structure, relative to cultivation. The rapid early, then slower, changes in total carbon and structural stability are shown in Figure 7.9. The axes are dimensionless because the rates of change depend very much on soil type and climate. Compared with the red-earth soil in Figure 7.8, the red-brown earth (a common wheatbelt soil) is less stable and the black cracking clays of northern NSW and Queensland much more stable.

Many experiments have shown, on different soil types, that after a long period of cropping the soil is structurally much more stable when minimal disturbance is used. The difference usually results from a decline in the cultivated soil. There is little evidence to suggest that degraded soils can be *improved* significantly by cropping without cultivation.

From the viewpoint of sustainable agriculture it seems likely that cropping will, on most soil types, result in changes in structure regardless of tillage practice. The real issue is whether the resultant structure will lead to unacceptable future losses in soil or crop production. If so, then at what point is land returned to pasture for structure to be restored?

Returning to the soil represented in Figure 7.8, it is important to note that the infiltration rate on this soil, after four years of cropping, was still high

(40 mm/hr). Moreover, the site supported a crop of wheat producing 7 t/ha of grain the year after measurements, indicating there was no soil physical barrier to high yields. So, despite a decline in structural stability over four years, the soil was still in good condition. From other experiments we can conclude that continued cropping would be possible *on this soil type*, but only if minimal soil disturbance is used to establish the crop. Continuous cropping would not be possible if the soil was cultivated.

In summary, structural stability declines initially under direct drilling, but is subsequently maintained sufficiently for further cropping to proceed with little fear of the adverse consequences of structural decline. Soil erosion is the most serious of these consequences and this shall be discussed in more detail, for a range of soil types, shortly.

The next question to consider is the role of crop residues in maintaining soil structure. Crop residues are a rich source of carbon. However, where cultivation is used to incorporate residues there is little or no gain in organic carbon. Indeed, structural stability is often *reduced* compared with no cultivation. The message is clear that soil disturbance must be minimized in order to retain good soil structure. Where residues are retained, structure is best preserved when herbicides rather than cultivation are used to control weeds.

Hardpans are sometimes another form of degradation resulting from cultivation and wheeled traffic. Hardpans cannot be reclaimed by no-tillage cropping practices. There is some suggestion that certain crops can loosen compacted subsurface layers, but this remains to be shown conclusively. If the hardpan is serious it may be necessary to remove it with deep tillage. Unfortunately, responses to ripping are not very predictable. As a rule of thumb good responses may be obtained in a deep soil where a shallow hardpan is clearly indicated by distorted root growth or resistance to a penetrometer (a sharpened steel rod). Further evidence would be the presence of moist subsoil at harvest, other than in wet years. Removing the hardpan gives the plant access to this deep reserve of water. Of course, rainfall must be sufficient to recharge the subsoil annually to get lasting benefits from ripping. After ripping, the soil needs to be stabilized by reducing cultivation in order to delay the need for future ripping.

It is important to realize that reducing cultivation is not a cure-all for structural problems. Structurally degraded soils may need to be returned to pasture before cropping with soil-conserving methods. Gypsum or other amendments may be needed to stabilize sodic soils. Deep tillage to remove hardpans or lime to correct acidity may be needed before crop responses to reduced tillage are obtained.

Tillage and soil Loss

The traditional approach to erosion of farm lands by rainfall has been to construct earthworks and continue with the same damaging cultural practices. However, not all farmers use earthworks as recommended by soil conservation authorities. Also, extreme weather events are the most damaging and earthworks are often not designed to handle these events because of the cost involved. In southern Australia improved stock water supplies associated with earthworks allowed stocking rates to be increased. This appears to have led to degradation of pastures in dry periods with accelerated soil loss on resumption of more normal rainfall. Crop agronomists have now developed fallowing and crop establishment techniques which focus on better soil and crop residue management to reduce soil erosion. But in the areas most prone to erosion, such as areas subjected to intense summer storms on fallowed land, there is a need to complement new tillage practices with earthworks (Table 7.5) and other practices such as strip-cropping.

The estimates of soil loss in Table 7.5 are from the universal soil-loss equation. This was developed in the USA, but results from cracking clay soils in summer-rainfall areas in Australia agree reasonably well with predictions from the equation. Note in Table 7.5 the big advantage of stubble mulching over incorporation, and the further substantial gain from zero cultivation in the no-till fallow.

Table 7.4 *Pathogens of wheat that may be affected by either not cultivating or not burning crop residues*[a]

	Not burning	Not cultivating
Cereal cyst nematode (*Heterodera avenae*)		–
Root lesion nematode (*Pratylenchus thornei*)	0	+
Take-all (*Gaeumannomyces graminis*) var. *tritici*		+/–[b]
Rhizoctonia (bare patch) (*Rhizoctonia solani*)		+
Crown rot (*Fusarium graminearum*)	+	0
Common root rot (*Bipolaris sorokiniana*)	+(?)	
Eyespot (*Tapesia yallundae*)	+	–
Yellow spot (*Pyrenophora tritici-repentis*)	+	

[a] Less severe (–), more severe (+), no effect (0). Many of these responses are uncertain and remain to be proven.
[b] Effects may be related to presence of alternate hosts in year prior to crop.

These predictions of tillage effects agree generally with a number of studies. The most notable are those of Freebairn and colleagues on the Darling Downs. Working with large catchment areas on two different clay soil types they have shown progressive reductions in soil loss; from burning residues to incorporation, then mulching, then a final further small reduction with zero cultivation plus stubble mulching (no-till fallow). Crop residues had the most powerful effect on soil loss. Some 2000–3000 kg/ha of wheat stubble seems to be needed (less on light textured soils), and about three times this for sorghum stubble. In practice, only winter cereals produce sufficient stubble to reduce the risk of erosion to an acceptable point in the summer-rainfall areas, although sorghum provides useful protection. The amount of stubble left by crops has a bearing on rotation for this region (Chapter 6).

In these studies no-till fallow resulted in slightly more runoff than mulch (with tillage) but soil loss was *less* with no-till. It should be noted that under flooding conditions, which cannot always be encountered in experiments, stubble offers little or no protection to the soil. Practical observations are that soil loss is less on these occasions if the stubble remains attached by roots to soil which has not been disturbed by cultivation.

Apart from the extreme situations in which no-till is favoured because of reduced soil loss, there seems to be little to choose, from the erosion viewpoint, between cultivation and herbicides for control of fallow weeds on well-structured soils in summer-rainfall areas. What matters is the retention of a protective plant cover. The choice between methods of weed control is made basically on costs and returns (which are related to yield). Although Freebairn and colleagues reported slightly higher runoff with no-till, it has already been noted that this is an unusual result. No-till almost always results in increased fallow efficiency and soil water storage. As soil water determines potential yield, the potential advantage lies with no-till.

There have been no catchment studies in winter-rainfall areas to compare with those in the Darling Downs. The best evidence for effects of tillage on runoff and soil loss come from experiments with rainfall simulators used on long-term tillage experiments. The structure of soils covered in these experiments is inherently less stable than the self-mulching soils of the north. However, the soils (and results obtained on them) are probably similar to the structurally poorer red-brown earth soils in southern Queensland and northern NSW. In southern Australia, results of studies with rainfall simulators suggest that a distinction needs to be drawn between responses to tillage during fallows after pasture, and long-term responses in a continuous cropping sequence. Most of the evidence relates to cultivation, little relates to stubble management.

With long fallow after pasture it seems that runoff and soil loss is least with a combination of rough tillage then herbicides for subsequent weed control. Cultivation, in this instance, has a short-term benefit in reduced erosion. It should be borne in mind, however, that cultivation damages soil in the longer term.

Despite the rapid decline in structural stability after pasture caused by even small amounts of cultivation (Figure 7.8) many studies have shown that runoff and soil loss in long-term sequences is greatly reduced by direct drilling and reduced cultivation. Stubbles were burnt in these experiments so the reduced erosion with direct drilling is a direct result of reduced soil disturbance.

In summary, cropping in southern Australia which involves any soil disturbance is likely to reduce soil structural stability, but cropping can still be exercised using direct drilling with relatively low rates of soil loss resulting. Whether residue retention further reduces soil loss from water erosion in southern Australia is largely a matter of speculation because there have been too few studies. There is some suggestion from work at Cowra, NSW, that stubble retention gives some additional benefit on soils regarded as 'fragile' i.e. degraded, but improving under reduced tillage. Residues may not benefit degraded soils which are cultivated, or soils which are in good condition as a result of direct drilling.

Wind erosion is a major concern for light textured soils in lower rainfall areas. Here, the evidence is clear that crop residues, even quantities as low as a few hundred kilograms per hectare, can greatly reduce soil loss. Standing stubble is most effective.

Crop residue management — strategies for soil conservation

In terms of both erosion control and soil water storage it is desirable to retain crop residues above the soil surface. Residue retention presents the farmer with difficulties involving crop sanitation, establishment and nutrition. Where stubbles are retained and weeds are controlled with cultivation, a further problem is retaining enough residue above ground to achieve the desired benefits. Despite the advantages of the no-till fallow in which herbicides are relied on completely for weed control, most farmers who wish to retain residues still cultivate to control weeds.

Diseases which are associated with stubble retention can be managed through appropriate crop rotations. Problems with nitrogen nutrition may be approached by rotations, use of mineral fertilizers, or both as previously discussed.

Problems associated with crop establishment and reduced temperatures (which reduce growth rates) under stubble can be approached through the concept of 'target' levels. Many crops produce more stubble than is required, at least for soil conservation purposes. The 'target' level is the rate of residue which achieves the desired effect, such as control of runoff and soil evaporation, whilst allowing acceptable crop establishment and growth. As a rule-of-thumb 2000 kg/ha (the untreated residue of a 1–1.5 t/ha wheat crop) meets soil conservation objectives and can be handled by modern seeders. The management strategy is to reduce the amount of stubble over the duration of the fallow, to the 'target' level (say 2000 kg/ha) by sowing time. In this way, the benefits of stubble are preserved and the problems minimized. Residue levels can be reduced by grazing or harrowing (relatively small effects) or with an appropriate type of cultivation. For example, a one-way disc buries 50–60 per cent of stubble, a chisel plough 25–35 per cent, a sweep plough 15–20 per cent and a rodweeder 5–10 per cent. This cultivation is harmful to soil structure and should be minimized.

In winter-rainfall areas crop residues which have been incorporated may decay slowly and present a major problem at planting time. In this case stubbles are best left above ground. Towards planting time a decision is made about whether the quantity of residue remaining, and its state of decay, will allow passage of seeding equipment. If not, burning is the only recourse.

Problems with stubble are most often associated with header trailings: these need to be spread evenly across the field. Slashing, harrowing, hay removal, partial burning (burning after windrowing) and treatment with sodium hydroxide to produce stock fodder ('alkalage') are other management options for reducing stubble rates.

Grain yields and profitability of crops established with or without conventional tillage

Here we compare the techniques of crop establishment that have been most commonly used in experiments: conventional (burn, multi-pass cultivation), direct drill (stubble burnt, no cultivation before sowing) and no-till fallow (stubble retained, no cultivation). In southern Australia the equivalent of the no-till fallow is direct drilling with stubble retained, though it is not commonly practised. In most tillage experiments the conventional treatment has reproduced local district practice although this varies with district. In almost all experiments, relatively simple comparisons were made between alternative techniques. Few attempts were made to treat techniques as distinct systems or to relate them to existing systems. Thus fertilizer rates, sowing times, disease control etc. were common to all treatments.

Grain yields — winter-rainfall areas

Overall, yields of direct drilled crops in higher rainfall areas are very similar to cultivated crops, in the absence of stubble, provided equal fallowing is applied to both treatments. In the earliest experiments with direct drilling, grazing was applied up until close to sowing and this treatment was compared with one which had been cultivated for up to five months previously. Differences in water at sowing, due to differences in weed and pasture growth, account for yield reductions in these early experiments with direct drilling. That is, yields were related to water use and not directly related to tillage. The implication for farm systems is that where direct drilling (with no fallow) is going to be used it must be complemented by changes in stock management to allow the autumn weed growth to be profitably utilized. Alternatively, land can be fallowed using herbicides as this should result in comparable yields.

Where conventional cultivation is compared with reduced cultivation and direct drilling in long, unbroken crop sequences, yields from cultivated plots eventually fall below direct drilled yields. This seems to be associated with structural decline, although crop nitrogen status, weeds and diseases may be implicated.

Stubble retention, in the higher rainfall wheat belt, generally reduces yield for a wide variety of reasons which vary with site and season: reduced establishment, nitrogen deficiency, yellow spot disease, more weeds and even, in one case, greater damage from mice. Because of these difficulties most farmers still prefer to burn crop residues. This may not be serious, however, bearing in mind that simply reducing soil disturbance in this region substantially reduces soil structural decline and the risk of erosion. Moreover, burning can be delayed until late autumn by which time the risk of summer storms has passed.

In lower rainfall areas, direct drilling may reduce yield by as much as 10 per cent, even with equal fallow moisture. This seems to occur mainly on hard-setting soils. The reasons for this are unknown. On the other hand, stubble retention with cultivation in these areas generally seems to give similar yields to burning. Where yield reductions do occur with stubble retention it is most likely that the fallow was maintained free of weeds with herbicides rather than cultivation. In practice then, farmers are advised to retain crop residues and control weeds using cultivation techniques that leave most residues on the soil surface. Whilst this is not ideal for soil structure, the residues offer protection against erosion. Land can be returned to pasture if structure declines to the point that problems such as crusting occur.

Grain yields — summer-rainfall areas

The evidence strongly suggests that either tillage

practices have very little effect on the yields of wheat or, that yields are reduced, mainly because of nitrogen deficiency, yellow spot disease or root lesion nematode. The addition of nitrogen fertilizer at modest rates (<50 kg/ha) usually results in similar yields, regardless of tillage practice. However, sorghum yields have been increased with no-till.

The result with wheat is both surprising and disappointing because, as noted earlier, the fallow efficiency of stubble mulched and no-till fallows is almost always higher than that of burnt and cultivated fallows. As water is the main factor limiting yields in dryland crops, an improvement in water storage with no-till ought to increase yield. Wheat grown under soil-conserving tillage practices in summer rainfall areas is therefore not realizing its potential yield advantage, even where added N results in equivalent yields. The reasons for this are unknown. However, in both summer- and winter-rainfall areas, stubble retention reduces nitrogen supply to the crop. Therefore, nitrogen merits further study although there is little evidence so far to suggest that very high rates of N-fertilizer (>50 kg/ha) will allow no-till crops to realize their yield potential. Good responses of sorghum (another cereal) to no-till casts further doubt on N as a possible explanation of unrealized yield potential.

Comparisons of alternative tillage treatments involving stubble ought to vary N-fertilizer rates as well. Where this had not been done, failure to apply fertilizer-N penalizes the stubble-retained crop. Failure to apply sufficient N may also prevent a treatment from expressing its true potential. Because grain legumes are an important source of N in many farming systems, tillage practices involving stubble also ought to involve studies with crop rotations. This need has been addressed in studies at Wagga Wagga and Tamworth. Crop rotations also offer opportunities to manage diseases and weeds that may be associated with particular tillage practices.

The clear message is that a simple comparison of tillage techniques can penalize some treatments, or fail to allow others to express their potential. Fallow management, crop rotations and fertilizer usage are all important factors which interact with the techniques of seedbed preparation and treatment of crop residues. Working together, these techniques constitute a system (Figure 7.10). The emphasis in the system is first on the goals, and then on the techniques which achieve those goals. There is no emphasis on any of the techniques as goals in their own right; they are means to an end.

Profitability — a direct comparison of tillage practices

As in the previous section, this analysis refers to the comparison between alternative techniques, such as direct drilling and no-till fallow, and conventional tillage. The analysis takes into account changes in the mix of inputs to the crop (labour, fuels, herbicides etc.) and their costs and the output (yield) and its value. As in most tillage experiments, the analysis does not take into account changes in fertilizer usage, rotations or other aspects of farm management.

From comparisons in different regions, the general conclusion is that cash costs increase with the replacement of cultivation with herbicide application. The tractor and implement running expenses of one cultivation are lower than those of one herbicide application plus herbicide cost, although the relative costs vary with the size of the implements and the operating conditions. Therefore, a herbicide application must generally replace more than one (often more than three or four) cultivation for it to result in a reduction in cash costs.

If we assume there are no reasons for changes in seed, fertilizer, or other variable costs with reduced cultivation, the total variable cash costs generally increase with a change to reduced cultivation methods. The exceptions are where no herbicide is needed or where a single herbicide application can replace several cultivations. Such is generally the case in Western Australia and occurs periodically across areas of southern Australia. If we assume equal yields with the different techniques, then gross margins in this simple analysis will favour conventional tillage practices. Many costs are not direct cash costs, however. These include unpaid labour and machinery depreciation. Where these costs are included in the comparison, reduced cultivation is often shown to have lower total costs. The comparisons depend on the value placed on labour and the relationship between depreciation and machinery use. Nevertheless, the large savings in machinery and labour time for seedbed preparation with reduced cultivation means that it has some economic benefits for management, often outweighing the extra cash costs of the herbicide.

The labour time required for the establishment of a crop is lower with reduced cultivation than with conventional cultivation. This is one of the most advantageous aspects of reduced cultivation from the point of view of the farmer. Estimates of the savings in labour time with direct drilling vary, but generally around 60 per cent of time spent in

Figure 7.10 *The integration of tillage practices (tools) to form conservation-farming systems.*

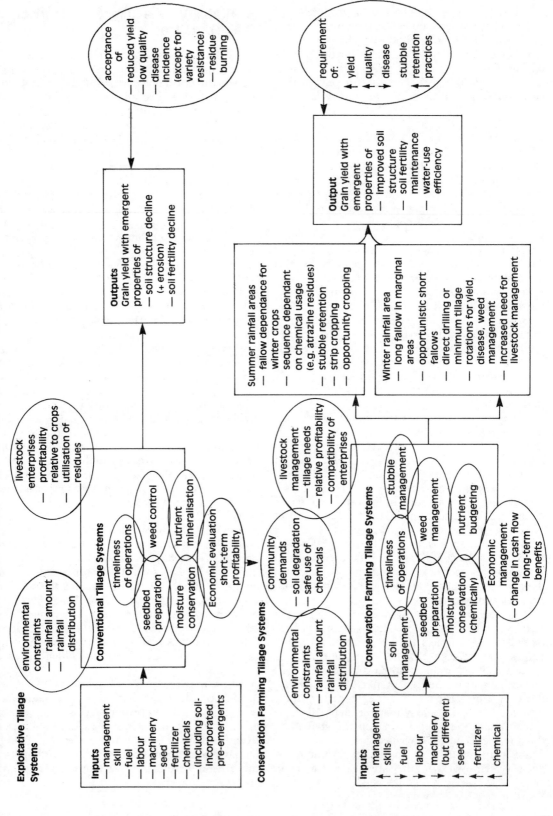

seedbed preparation can be saved by farmers. However, the need for a herbicide application shortly before sowing, and possibly a slower sowing speed, can mean that demands on labour are greater during the sowing period itself with direct drilling.

Even with benefits of reduced labour demand and machinery depreciation, the economics of simply substituting herbicides for cultivation are not particularly good. This is especially so if new machinery is required, such as a seeder for use in crop stubbles. Also, seasonal cash demand (for herbicide) may increase, causing cash flow problems for some farmers. However, when new tillage practices are set in the context of wider issues of farm management a very different economic analysis emerges.

Tillage in the context of farm systems

Soil-conserving tillage practices have important 'hidden assets' that are revealed only when they are used in conjunction with other areas of farm management.

Timeliness

We have already shown how uncultivated seedbeds and stubble retention can potentially improve the timeliness of sowing. Improved trafficability means that crop spraying (for weeds) can also be more timely. The importance of sowing time is seen in many studies showing that a delay in sowing, after the optimum for a variety in any area, incurs a yield penalty of about 1 per cent per day. With wheat valued at $110 per tonne this is a daily loss of about $2 per hectare per day when planting is needlessly delayed past the optimum. Clearly, a farmer's time is better spent sowing instead of cultivating once the optimum time has passed. This applies particularly in Western Australia where sowing traditionally has been delayed by ground preparation after the autumn break. The earlier sowing possible with direct drilling may need longer-season wheat cultivars to gain full advantage. Freedom from the need to cultivate also means that double or opportunity crops in the summer-rainfall area are more likely to be sown within the required time.

Most tillage experiments fail to demonstrate these advantages of timeliness because the tillage treatments are regarded as techniques rather than systems which need to be managed independently of one another. Therefore, treatments are sown on the same day rather than each being sown as close to the optimum for the location as the treatment allows.

Interaction with livestock

Where fallowing for water conservation is not essential for crop production, the adoption of direct drilling allows volunteer crop and autumn pasture growth to be grazed until near to planting time. This has many implications for both the crop and animal enterprise that we consider later, but for the moment let us consider what changes occur in whole-farm economics when additional forage is available. Economic models have been used for these studies. They show that total gross margins are potentially higher with direct drilling, compared with a farm depending on cultivation, since the decrease in cropping gross margin is more than offset by the increase in livestock incomes in an average year. Overhead costs are also lower with direct drilling, so that total net incomes are higher with direct drilling than with conventional cultivation. Where no value is attributed to the labour saved, the differences between the two systems are relatively small. Where labour is valued, the advantage for direct drilling is significant.

It is important to note that for direct drilling to be economically attractive it is necessary to increase livestock numbers to take advantage of the extra grazing available. Alternatively, the models show that direct drilling can allow an increase in cropping intensity, with no change in livestock numbers, and this too gives the farm system based on direct drilling an economic advantage. Considerable gains in the long run are possible as machinery investment, rotations and livestock enterprises are adjusted to the new system.

In the case of direct drilling, economic comparisons depend very much on how water is managed in a short fallow. Whilst fallowing for water conservation is not necessary for successful cropping in many areas of southern Australia, the process of seedbed preparation inadvertently stores water in many years, and this contributes to higher yields. If pasture, volunteer crop or weeds are allowed to grow and use this water, and they are not efficiently grazed, then the crop suffers a loss of water (and potential yield) with no benefit to the animal enterprise. Herbicides and direct drilling allow a great deal of flexibility in water management in these short-fallow situations, allowing farmers to essentially choose between crop and livestock enterprises. Unlike cultivated fallows where land is committed to cropping many months before sowing, with herbicides (and direct drilling) decisions to fallow can be made at almost any time depending upon seasonal conditions and the relative gross margins of crops and livestock.

When integrating crops with livestock on a farm practising direct drilling in southern Australia, two important requirements need to be considered. The first is that if grazing rather than herbicide is used to control summer and autumn weed growth, then sufficient stock must be available to keep weed growth in check. This can be difficult when crop areas have increased at the expense of livestock numbers, and in wet seasons with abundant growth. In these cases, part of the area set aside for crop needs to be sprayed for weed control and the rest grazed. Careful planning of the whole-farm operation strives to optimize the competing interests of crops and livestock. That is, to optimize benefits to the crop through fallowing, and to livestock through additional grazing.

Second, the requirements of livestock need to be considered. If heavy grazing is an integral part of the weed control programme in autumn, then a sheep reproduction cycle involving a spring joining for an autumn lambing is likely to be stressful for the ewe. Winter lambing and conservation farming are the most successful combination for areas such as southern NSW (Figure 7.11). Joining is in February/March when the bodyweight of ewes is high and often gaining with grazing on recently harvested stubble. The autumn post-joining period, when nutrition is not as critical, coincides with the need for grazing to prepare for herbicide application and sowing. The pre-lamb buildup and lamb-

ing itself can be undertaken on a range of grazing crops and pastures. Finally, the natural spring flush of pasture ensures ewes have an abundant milk supply for their lambs. Weaning can be carried out at the end of spring with weaners having sufficient wool to be shorn so as to prevent losses caused by grass seed. In any case, the grass seed risk should already have been reduced by spray-topping activities.

Crop rotations/cropping intensity

With reduced cultivation, cropping is less damaging to soil than previously. Farmers are therefore freed, in many areas, from the need to sow pastures in order to reclaim soil structure. This means that cropping and animal enterprises can be dedicated to parts of the farm which suit them most, e.g. cropping on more fertile, less steep land. This is not to say that pastures will not be rotated with crops, but the decision to rotate need not be based on reclaiming soil structure. It also means that crops can be sown on relatively steep land previously considered at an excessive risk from soil erosion. Occasional cropping of pasture land is a useful way to rejuvenate pastures, e.g. by reducing high soil nitrogen levels which favour nitrophilous weeds.

Whole-farm analyses have consistently found that if the adoption of soil-conserving tillage practices allows farmers to increase their cropping area, significant economic benefits can be realized. This is mainly because extending the area sown means that overhead costs are spread over a larger

Figure 7.11 *Livestock management may need to change in a conservation farming system. Feed demands of late pregnancy and lactation need to match feed supply. Analysis applies to the slopes of southern NSW.*

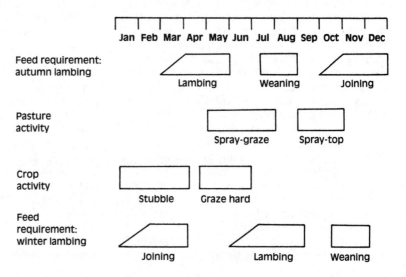

area, and the marginal costs of the extra cropped area are much lower than the marginal returns from the crop (provided it can be undertaken with the same machinery and without too great a penalty due to the loss of timeliness associated with the increased area of crop). However, in order to obtain the full benefits from this increase in cropping intensity, changes would be required in the whole farming system, not merely in the method of crop establishment. These changes could include such diverse components as type of crops in the rotation, livestock enterprise, pasture and machinery investment. Nevertheless, such whole-farm changes could lead to substantial benefits for the farmers. Where the dynamics of this change have been examined in detail, farmers have been found to move towards increased cropping intensity as quickly as possible. Of course, relative prices can change between cropping and livestock, so at other times there may be no such gains from increasing cropping at the expense of the livestock enterprises.

Freedom from the need to cultivate can also increase farm incomes where there is extra land available which, although suitable for cropping in terms of the rotation, cannot be planted under conventional cultivation methods because the limited time available for pre-sowing cultivation prevents seedbed preparation. Such areas are more likely to be common on larger farms in the Western Australian cereal belt where all pre-sowing cultivation is done immediately prior to sowing, than in the established cropping areas of southern and eastern Australia.

In summer-rainfall areas, soil-conserving tillage practices potentially allow much more intensive and profitable cropping. This results from a combination of improved infiltration and reduced soil evaporation which increases fallow efficiency, prolonged surface moisture after rain allowing more timely sowing, and freedom from the need to cultivate between crops which also allows timely planting. In the extreme, it means that crops can be sown whenever current soil water reserves, plus expected (average) seasonal rainfall, create the opportunity to grow a crop. This 'opportunity cropping' is an example of water management which takes advantage of the wide range of summer and winter cereals and legumes which can be grown in these northern latitudes. It maximizes farm water-use efficiency by minimizing runoff and soil evaporation in much the same way as increasing cropping intensity increases farm water-use efficiency in the Wimmera — except that no-till practices are crucial to its success.
success.

Opportunity cropping is most applicable to the safer cropping areas of the northern cereal belt. Figure 7.12 illustrates several options for crop rotations in these areas. They range from traditional fixed rotations of fallow-wheat, to fixed double-crop rotations using other crops, and finally to opportunity cropping. It is important to note that the least intensive cropping option, the long fallow-wheat rotation, is worst for soil erosion. Paradoxically, the very intensive opportunity cropping system has the least potential for soil erosion. This is because the frequent cropping tends

Figure 7.12 *Options for crop rotations in summer-rainfall areas.* (SOURCE: Holland, J. F., Doyle, A. D. and Marley, J. M., 1987, *Tillage—New Directions in Australian Agriculture*, P. S. Cornish and J. E. Pratley (eds), Inkata Press, Melbourne.)

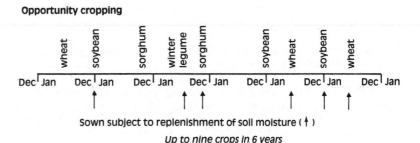

to keep the soil profile low in water. Thus, substantial falls of rain are less likely to run off causing erosion, though, of course, stubble cover offers little protection from runoff and erosion due to intense rain once the profile is full. Frequent cropping also maintains a live plant cover over the soil, reducing the risk of erosion. Opportunity cropping can also be very profitable.

We have so far considered the possibilities for improving farm profitability by using new tillage practices to alter rotations and intensify cropping, either at the expense of pastures in rotation (winter-rainfall areas) or at the expense of fallow (summer-rainfall areas). In intensive cropping systems crop rotations are always an important consideration for crop sanitation. Where cultivation is reduced and crop stubbles retained, the need for crop rotation is even greater. Rotations play a key role in disease control, they give flexibility in the weed management programme, grain legumes assist nitrogen management and they improve water management.

Improved farm management and profitability

It will be evident, then, that three approaches can be taken to the adoption of soil-conserving tillage practices. One is the rigid adoption of alternative techniques (e.g., direct drilling *versus* cultivation) in isolation of other considerations. Another approach treats tillage practices as tools of management and integrates them with all other areas of relevant farm management. A third approach is *ad hoc* replacement of cultivation with herbicides — whilst potentially reducing soil erosion this is not a viable long-term solution to the problem of soil degradation.

The first approach has been shown to be only marginally economic, and hardly likely to encourage adoption. The second, whole-farm approach, integrates soil, weed and crop residue management, as previously discussed, with soil-water management, livestock management and crop rotations. To achieve this integration is not easy and demands a high degree of managerial skill. The goal of implementing soil-conserving tillage practices on a whole-farm basis provides a focus for generally improving farm management skills. We consider that the greater level of managerial skills required, and ultimately attained, will make its own contribution to farm profitability above

that derived directly from the implementation of a conservation farming strategy.

Conclusions

Dependence on cultivation declined during the 1980s, in part because of the changing economics of cultivation relative to herbicides for weed control. Although this is welcome, this is not sustainable agriculture. Stubble burning has also declined, especially in Queensland. This trend is also welcome, but burning and stubble incorporation are still practised widely, largely because of tradition combined with the perceived risks attached to reduced cultivation and stubble mulching. Moreover, there is no convincing evidence that these new practices on their own return equal or better profits in the short term. Short-term profitability is vital to farmers, regardless of future sustainability and profitability.

In this chapter we have shown that techniques of stubble retention and reduced cultivation can give profitable crop production. Profitability requires attention to weed management (especially in pastures) to give economical weed control. Also required is attention to new crop rotations to manage pests, diseases and soil nitrogen and for optimal use of rainfall. Careful stock and pasture management and attention to new machinery are also required.

Some farmers have successfully adopted this systems approach but many have not. The challenge for the future is to give farm managers the skills required to synthesize information from diverse sources, often from simple two-factor experiments, and to integrate this into profitable farm systems. On the technical side, future research will lead to reduced dependence on herbicides. Resistance to the diseases associated with stubble retention and reduced cultivation will be developed. A wider choice of species will become available for use in rotations. A mystery surrounds the apparent failure of wheat to utilize the extra water available in no-till systems and this problem will attract research. Nitrogen cycling will also need research, leading eventually to better N management.

Water management in fallows will be improved by the use of tactical fallow-management models to decide when to control weeds. Sprayers should be developed which identify green plants and spray them selectively.

Table 7.5 *Soil erosion based on the universal soil-loss equation* (SOURCE: Holland et al. 1987)

Treatment	Soil loss (%)[a]	
	No structural works	With structural works
Residue burnt	60	25
Residue incorporated	30	10
Residue mulched	18	6
No-till fallow	3	1

[a] Soil erosion in continuous bare fallow = 100%

Further reading

Anon. (1986/87). 'New tillage systems in south-eastern Australia', *Rural Research*, 133, 8–15.

Cooke, J. W., Ford, G. W., Dumsday, R. G. and Willatt, S. T. (1985). 'Effect of fallowing practices on the growth and yield of wheat in south-east Australia', *Australian Journal of Experimental Agriculture*, 25, 614–627.

Cornish, P. S. (1985). 'Conservation farming — have we a sustainable system?' *Agricultural Ecology, the Search for a Sustainable System*, Australian Institute of Agricultural Science, occasional publication 21, 47–55.

Cornish, P. S. and Pratley, J. E. (1987). *Tillage — New Directions in Australian Agriculture*, Inkata Press, Melbourne.

Fischer, R. A., Mason, I. B. and Howe, G. N. (1988). 'Tillage practices and the growth and yield of wheat in southern New South Wales, Yanco, in a 425 mm rainfall region', *Australian Journal of Experimental Agriculture*, 28, 223–36.

Pratley, J. E. (1987). 'Soil, water and weed management — the key to farm productivity in southern Australia', *Plant Protection Quarterly*, 2, (1), 21–30.

Pratley, J. E. and Cornish, P. S. (1985). 'Conservation farming — a crop establishment alternative or a whole-farm system?' *Proc. 3rd Australian Agronomy Conf.*, Hobart, 95–111.

Steed, G. R. and Robertson, G. A. (1988). 'Alternative conservation cropping systems in other areas of Australia.' *Proc. Australian Conf. on Agricultural Engineering for Conservation Cropping*, Toowoomba.

CHAPTER • 8

Integration of Crops and Livestock

T. L. J. Mann

SYNOPSIS This chapter concentrates mainly on southern Australia where most integration of crops and livestock occurs. It considers the extent of, and factors accounting for variation in integrated dryland cropping and livestock systems. Integration brings benefits to both crops and livestock. Also there are some constraints, leading to less flexibility in both cropping and livestock operations. In some circumstances, extra feedstuffs must be supplied from off the farm system. Key management strategies that make best use of livestock for crop production and promote optimum performance of livestock are considered. The interplay of cropping and pasture practices is assessed and the robustness of the integrated system, under environmental and climatic challenges, is examined.

MAJOR CONCEPTS

1 A wide range of ways exist in which crops and livestock are integrated into the dryland farming areas of Australia.

2 The flexibility of the integrated system to change according to type, intensity and mixing of enterprises is limited primarily by climatic and soil factors, and by land capability.

3 Secondary constraints, such as personal, financial, biological and economic factors, will also affect the type, intensity and mix of enterprises in a livestock and cropping system.

4 Livestock management strategies, including the timing of operations and grazing management must cover the seasonal fluctuations in availability and quality of feed, as well as contingencies such as drought.

5 Livestock can help prepare land for cropping by removing excess amounts of stubble and by controlling weeds. Efficient utilization of such feedstuffs benefits both livestock and crop production.

6 The implications of various pasture and cropping practices need to be considered; some of these may be beneficial for livestock production while others may conflict with it. There is some loss of flexibility in timing of livestock operations and in choice of crop sequences.

7 The benefits of an integrated crop/livestock system should be evaluated after taking into account the ways in which crop, pasture and livestock interact and affect one another.

8 To maintain a viable and sustainable system the integration of livestock and crops needs to be carried out within the framework of whole-farm planning.

Scope for integration

Integration of crops and livestock has been conducted most effectively in the cereal-livestock zone of southern Australia. In most areas the pasture phase is an important component of the system, and if managed properly, promotes high levels of both crop and livestock production. The crop residues are available for livestock and careful grazing of them can help prepare the land for the following crop, while at the same time, maintain a suitable amount of cover for protection of the land. Legume-based pastures also improve soil fertility and structure. However, there is a much better choice of legumes in the southern wheatbelt than in the northern wheatbelt.

An understanding of how the present system operates and is managed will assist in the selection of effective strategies to maintain viability and sustainability. Climatic factors, particularly the amount of rainfall, its distribution, and temperature are the driving forces for the system and heavily influence the type of crop or pasture grown. Other important variables include the capability of the land for grazing or cropping, and the soil type, particularly with respect to texture, pH, depth of soil and salinity.

Table 8.1 *Types of livestock and cropping enterprises.*

Livestock	
Sheep	— wethers, breeding ewes, first-cross prime lambs, second-cross prime lambs.
Beef cattle	— breeding cows, weaners, vealers, yearlings, steers.
Goats	— angora (mohair), cashmere.
Cropping	
Winter crops	— cereals: wheat, barley, oats, rye, triticale. grain legumes: beans, peas, lupins, chickpeas, vetch. oilseeds: linseed, canola.
Summer crops	— sorghum, millet, cowpeas, mungbeans.
Pastures	— annual medics and sub-clovers, dryland lucerne, annual grasses, perennial clovers and perennial grasses.
Forage crops	— oats, vetch, forage sorghums, sudan grasses, lablab.

Due to a wide range of soil and climatic factors, topography and extent of arable land, and external factors such as infrastructure, market proximity and financial operatives, there are a number of options for integrating livestock and crops (Table 8.1). Extreme situations also occur, e.g. there may be continuous cropping, with livestock agisted on to the property to graze stubble or, at the other end of the spectrum, a livestock enterprise may be favoured with, say, a minimal amount of crop grown for fodder purposes. In some areas, e.g. in northern NSW and southern Queensland, cropping and livestock are not closely integrated. Winter and summer crops provide diversification without the inclusion of livestock except for grazing stubble and special purpose cereal grazing crops such as oats. By contrast, in southern Australia, cropping and livestock are often well integrated, with crops grown on 40–60 per cent of the farm and rotated with pasture.

The task for farmers is to select a mix of enterprises considering their peculiar situation, to make a decision concerning the intensity of cropping versus livestock, to be flexible to allow for changing circumstances and to manage their resources in a conservative manner. All this is a tall order, even when economic decision making alone is considered.

Personal factors are likely to influence selection of the type, operation and management of the system. For example, farmers may be risk averse and seek diversification of enterprises; they may have a preference for livestock production, or they may prefer a lower stocking rate than recommended for optimum economic performance. A side benefit of these less-than-optimum decisions may be that they are not operating at the edge of resource impairment. Alternatively, some farmers have chosen to reduce pasture areas and intensify cropping in response to increasing financial problems. Their actions have put pressure on their soil and plant resources. They have put sustainability at risk. Others have successfully combined these new options with risk management strategies. These include minimum tillage, improvement of existing pastures, increased fertilization of crops and adoption of rotations that are sustainable.

Benefits of crop-pasture integration

The inclusion of a legume-grass pasture phase for livestock has the added benefit of supplying nitrogen and organic matter to the soil as well as restoring the structure of cultivated soils (Chapter 7). In southern inland Queensland a decline in the return of grain crops is forcing farmers to consider raising sheep on land previously used for cropping. Here, a range of summer-growing grasses, such as purple pigeon grass, rhodes grass and green panic, together with lucerne and annual medics will help to restore fertility.

A pasture phase also allows opportunities for control of weeds and crop diseases. Judicious use of grazing by livestock and/or use of herbicides, as discussed later, assist in this type of control.

Successful integration of crops and pastures increase the reliability of income through diversification and flexibility. There may be conflicts between enterprises, however, which need to be carefully assessed, such as the harmful influence of some pasture plants on crop production, fallow management which may be more profitable than carrying stock, the adverse effect of cropping practices on pasture regeneration and the occurrence of competition for labour, capital inputs and land.

Crops provide stubble and unharvested grain for livestock and there may be opportunities for limited earlier grazing at an appropriate stage of crop growth, e.g., light grazing may reduce lodging and 'haying-off'. Also, in times of drought the crop may be used as a valuable supplement.

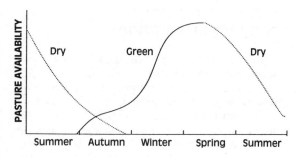

Figure 8.1 *Pasture availability (Mediterranean climate of grazed pastures, e.g. Yorke Peninsula, South Australia).*

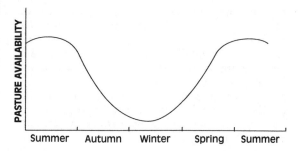

Figure 8.2 *Pasture availability (Temperate climate of grazed pastures, e.g. northern slopes and plains of NSW).*

Feed supply and livestock requirements in an integrated system

Availability and quality of feed

The primary sources of feed, considered here, are pasture and crop stubbles. Other sources, regarded as augmenting the feed supply, are discussed later.

Pasture — For most of the pasture growing areas of south-western and southern Australia the seasonal availability of feed, predominantly annual grasses and legumes, can be represented by Figures 8.1 and 8.2. For the marginal cereal areas the growing season may be 1–2 months shorter. In the higher rainfall areas, the growing season may be up to 2–3 months longer with a period of slow pasture growth during the winter; perennial grasses and legumes may be grown. In the Mediterranean environment the annual species germinate with the autumn rains, grow rapidly at first, suffer retardation of growth with declining winter temperatures and then grow rapidly with the higher temperatures in spring. Maturation occurs rapidly in late spring and the lower digestibility and protein become limiting factors to animal production. For the cooler temperate environment, rainfall is less limiting on pasture production but winter growth is often severely limited by low temperatures. Dryland lucerne is a useful option in the more northern areas where the combination of summer and winter rains allows a year-round growing season (Figure 8.2). Oats and barley are also used as fodder crops to lessen the winter feed deficiency, but there is still the problem of temperature inhibition of winter growth.

In areas with a winter rainfall regime the digestibility of the pasture annual legume can vary twofold throughout the year as shown in Figure 8.3b. A decrease in crude protein content of the herbage follows the decline in digestibility. The availability and quality of pastures reach their lowest point just prior to the autumn break of the season. Digestible energy and protein are both likely to be deficient, or in imbalance, for maintenance of body weight at this critical time of the year. The change in liveweight of young merino sheep, under these feed conditions, is shown in Figure 8.3a. The growth changes were similar for sheep on subterranean clover and medic-based pastures. The fourfold difference in the daily intake of digestible energy from winter to summer (Figure 8.3c) coincided with changes in diet digestibility. In spite of abundant feed, therefore, in early summer liveweight will decline, because of the low intake of digestible energy.

For both liveweight and wool production there is a marked response to the improvement of the quality and quantity of feed available during the growing season. The see-saw appearance of liveweight changes from one year to the next and use of body reserves do not appear to have a deleterious effect on the health or subsequent performance of dry sheep, provided the sheep are kept above a critical body weight. More caution has to

be exercised if the body reserves of breeding ewes are utilized, i.e. if bodyweights are allowed to fall.

In areas with a dominance of cool season rainfall, there is an abundance of pasture in the spring and there is often the opportunity to close up some paddocks for hay or silage. This conserved fodder can be fed back later to, say, breeding ewes prior to lambing or during lactation. The nutritive value of pasture hay is largely dependent on the proportion of legume material. Other factors such as time of cutting, degree of field spoilage after cutting and storage conditions will also affect its quality. Ideally the pasture is cut at early flowering to optimize the potential benefits of both quality and quantity. Well-prepared medic hay cut at this time will be about 60–65 per cent digestible and contain 12–16 per cent protein.

Dry legume residues in pastures are superior to grass residues; in particular the pods and seeds are a useful source of summer feed. Medic pods can

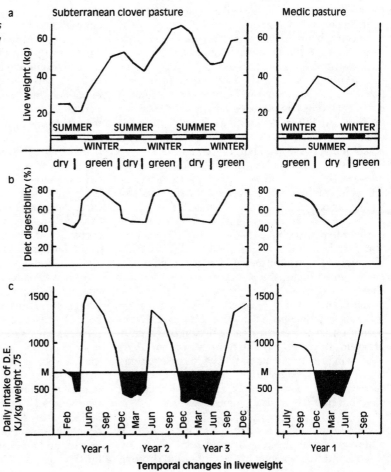

Figure 8.3 *Temporal changes in liveweight (a), diet digestibility and N content (b) and daily intake of digestible energy relative to maintenance (M) (c) for sheep grazing pastures based on T. subterranean (left) and annual Medicago spp. (right).*

constitute a drought feed after the leaves and stems have been grazed. The pods themselves are highly indigestible; they are insufficient to maintain liveweight on their own. The seeds, however, are digested and have a high protein content. By comparison a higher percentage of seeds of some of the clover species, due to their smaller size, pass through the gut undigested.

For grazed lucerne rotational grazing is essential for productivity and longevity of lucerne. Under grazing conditions, sheep select lucerne leaves in preference to the stems. If the reserves in the crown and roots are not maintained, plant vigour and survival are impaired. A spelling time of 6–8 weeks is usually sufficient to maintain satisfactory pasture density, yield and composition.

Crop Stubble — In southern Australia stubbles which are available post-harvest for livestock grazing will vary in quality dependent on type of stubbles and quantity of unharvested grain. Grain legume stubbles have been shown to be superior to cereal stubbles in grazing value, partly because of the higher protein and digestibility and partly due to the value of fallen grain.

Sheep grazing faba bean stubbles have grown faster and produced more wool than those on barley stubbles; this was associated with a higher intake of bean herbage residues and more residual bean grain. On occasion farmers have found it profitable to buy in wethers or steers for fattening on lupin stubbles. Sometimes, however, the presence of a fungus on the lupin stubble may cause poisoning of livestock. Plant breeders have selected varieties of lupins which are now fungus-resistant. These varieties will allow a longer grazing period on lupin stubbles and a decrease in the need for supplementary feed over the summer.

The decline in the value of stubble through summer may be compensated for if rain causes the appearance of edible weeds. On cereal straw stubbles alone, because of their low value in digestible energy and protein, sheep and cattle will lose body weight.

Similarly, in subtropical areas sorghum stubble progressively declines in quality. However, it has been found that in a reasonable season one hectare of grain sorghum harvested in March or early April will provide about sixty days stubble grazing for a 250 kg steer with a weight gain of 0.5 kg/day. This must be partly due to the fact that sorghum stubble produces some regrowth.

Requirements of Livestock — The nutritional requirements for sheep, goats and cattle for maintenance, growth, pregnancy, lactation and drought feeding can be calculated and are available either in tables or can be produced as required by computer, using readily available programs. Requirements for energy are normally expressed in terms of metabolizable energy, and for protein in terms of rumen degradable protein (RDP) and undegraded dietary protein (UDP), sometimes termed bypass protein. If there is an insufficiency in either the energy or protein supply, this can be overcome by the addition of supplements which can be supplied on a least-cost basis, as calculated by commonly available personal computer programs. Where the livestock are wholly ration fed (feedlotted), there is little difficulty in computing the ration. However, if the animals have access to pasture or crop residues, there is uncertainty concerning the actual amount eaten or the degree of substitution of the supplement for the paddock feed.

The energy requirements for breeding cows and ewes and their progeny are given in Figures 8.4 and 8.5; these are highest for the breeding animals in early lactation. For ewes it is 3 to 4 times the level required for maintenance. Clearly, matching the needs of early lactation with an ample feed supply of fresh plant growth is appropriate. In southern Australia this would correspond with the early spring growth to capitalize on both availability and quality. However, for other management reasons this matching may be prohibitive, as explained later, and a compromise is made. Other periods in the year where breeding animals benefit from extra requirements above maintenance are 3 to 4 weeks prior to joining and the last six weeks of pregnancy. Young livestock also have a high requirement if they are to grow rapidly, as is the case in southern Australia if they are to be finished before the feed deteriorates in the summer and early autumn. Yearling steers can be fattened, for example, on mature standing wheat and barley crops as an alternative to supplementation with grain during the summer months.

Where summer rainfall is unreliable, as on the central west slopes and plains of NSW, there may still be an opportunity to grow crops such as forage sorghums and cowpeas to supply a large bulk of green feed in a short period. This would supplement any shortfall in lucerne and any other pasture growth to overcome critical feed shortages for livestock maintenance and fattening.

To overcome a serious winter shortage grazing oats is popularly used for finishing yearlings and lambs, as well as to help prepare ewes for a spring lambing. In southern Queensland ryegrass can benefit beef producers by providing forage, high in protein and digestibility, in the cool season to help finish young cattle.

Figure 8.4 *Changes in the pattern of energy requirements for the cow, plus the calf, during the year.*

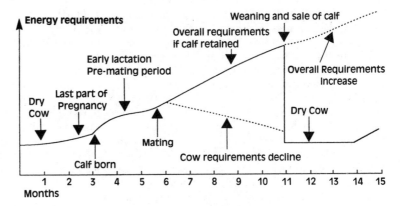

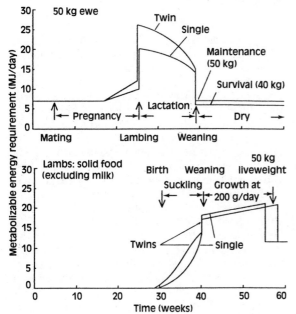

Figure 8.5 *Energy requirements (as MJME) of a 50 kg ewe and her progeny over a range of productive states (calculated from MAFF, 1975) for a ewe outdoors, on feed of 10 MJ/kg DM equivalent to good quality green pasture.*

Implications for livestock management strategies

Farmers must be able to gauge the value of their feed resources to meet the requirements of different classes of livestock during the year. Strategies for managing the livestock enterprises must be geared to optimize the use of feed, given the fluctuation in its availability and quality. These strategies include:

- Selection of time of lambing or calving; this sets the sequence for most of the other livestock operations on the farm.
- Choice of grazing management system, which considers type of stock, stocking rate and how stock are grazed.
- Plans for conserving feed (grain, hay and silage) on the farm, use of fodder crops for grazing, purchase of supplementary feed (off farm) and contingencies for drought conditions.

Selection of appropriate strategies help in ensuring livestock health and production, maintenance of breeding stock and preparation for market.

Augmenting feed supplies for livestock in integrated systems

Apart from feed supplied from pastures and crop residues, additional or supplementary feed may be required to meet livestock requirements, for survival, growth and reproduction. These sources include feed derived from the farm, such as winter or summer fodder crops, pasture or cereal hay, unharvested standing crops and grain. The cost of using these is their market value. Off-farm sources of feed may be purchased and these may include hay, cereal and legume grains, oilseed meals, minerals, urea and molasses.

The essential considerations for supplementation are to ensure efficient activity of the rumen microflora through a balanced supply of available energy, nitrogen and sulphur, and a supply of a minimum level of fibre for efficient functioning of the rumen. Also, a supply of undegraded dietary protein (UDP) may be necessary. However, there needs to be further clarification concerning the requirements of UDP for various livestock situations and the economic value of protein sources containing UDP. Vitamin and mineral supplements may also be required in certain circumstances and are considered later. When a supplement is provided there is likely to be substitution of this for part of the diet formerly grazed. The sheep prefer the supplement since it is more digestible and palatable. This non-additive effect may run counter to a farmer's intention say, to decrease the amount of stubble in a cereal paddock by grazing prior to sowing a crop. Other problems that can occur with supplementation are variability in intake, the possibility of toxicity if taken in excess (e.g. urea), the spoilage of the supplement due to weather, bacteria etc., and the likelihood of digestive problems with a changeover of diet, particularly with grain feeding.

Types of supplements

Cereal Grains — These can include wheat, barley, oats, triticale, rye and sorghum. Oat grain is commonly used; it is higher in fibre content and the livestock suffer less digestive upset. Grain feeding of sheep with oats is often used in periods of drought and for lightweight lambs during summer. The cereal grains, like other feedstuffs, can vary widely in feed quality. Feed analyses, for metabolizable energy and protein content, help define their value for feeding.

Grain Legumes — Lupin grain is valued as a supplement to sheep because of its high protein content (30 per cent) and high digestibility. It is also high in fibre, thus producing little problem with digestive upsets. Increased ovulation rate has occurred in ewes that have been fed 250 g lupins for 2 to 3 weeks prior to joining. Supplementing rams that are grazing poor quality pasture or are low in liveweight with lupin grain prior to joining will improve their fertility.

Any level of replacement of oats with lupins has been found to improve the growth rate of lambs. The cost of lupins may be prohibitive but trials in NSW have demonstrated that a diet containing 25 per cent lupins and 75 per cent cereal grains could be profitable while allowing lambs to grow at about 150 g per day. Steers supplemented with 3 to 4 kg of the above mixture, produced gains of 0.65–0.75 kg/day. In South Australia faba beans were found to be economical as a supplement for sheep on cereal stubbles.

Grain legume crops are used in many areas to extend the cropping phase of rotations. The unharvested grain (200 to 400 kg/ha) represents a valuable source of stubble feed for livestock, par-

ticularly for weaner growth and increasing reproductive performance (e.g. trials in central NSW have shown an increase of 10 per cent more lambs for ewes joined on lupin stubble compared to ewes on cereal stubble).

The extra value of grain legumes in a cereal rotation system includes complementary advantages for both cereal and livestock production. However, these crop-livestock systems have to be carefully evaluated with respect to long-term effects on soil fertility and structure, particularly as the percentage of cropping increases (Chapter 6).

Supplements based on urea

Urea can be used as a source of non-protein nitrogen for microbial use if the intake of nitrogen is likely to be insufficient to maintain optimum microbial activity. Conflicting results for urea-based supplements can usually be explained by the presence or absence of sufficient energy and sulphur for protein production. Thus, field responses to urea-mineral blocks for sheep grazing stubbles or dry pasture residues are usually poor; urea-molasses mixtures give better responses for sheep on low quality roughages. Urea-based blocks with additional energy and protein have been popular for fattening lambs; response to them will depend on the basic diet. Urea plus straw treated with alkali has resulted in increased intake and liveweight gain. Improvements in feed value of the straw, though, have not been sufficient to justify the cost.

Stock blocks and mixes containing ingredients such as urea, molasses, cereal grain and grain legumes need to be evaluated in terms of liveweight change, stubble reduction and cost per head. Stubble reduction aids preparation for cropping; excess removal, however, may lead to wind or water erosion (Figure 8.6).

Fodder crops

This term refers to crops which are grown to supplement pastures and crop stubbles and can be grazed (forage) or harvested and fed to stock later (e.g. hay, silage and grain).

In areas of higher rainfall and colder winters there is frequently a period of slow pasture growth and low availability. This shortage of feed can be overcome by grazing livestock on cereals, particularly oats. This practice of early grazing of winter cereals is commonly used on the tablelands, western slopes and plains of NSW and southern

Figure 8.6 *The aim of summer management is to protect the soil with crop residue but there is a need to reduce the trash before sowing. Sheep or cattle can use the crop residue. In this photo the level of ground cover after grazing is about 50 per cent.*

Queensland, and the higher rainfall areas of Victoria; it is rarely used in areas where the growing season is shorter, such as occurs in most of South Australia. Normally, the crop is allowed to regrow and is harvested later for grain. Other winter crops include barley, triticale and cereal rye. They can be used to help finish off store lambs or steers or to improve liveweight and condition of ewes and heifers before joining.

Lucerne has been regarded as a prime forage source in some areas, such as the central and north-west slopes and plains of NSW. Lucerne rotationally grazed is an alternative to wheat where declining soil fertility and infestations with wild oats temporarily reduce profitability from the wheat enterprise. Other summer crops include the forage sorghums, sudan grasses, millets and summer legumes (e.g. cowpeas, mungbeans, lab lab). These are important in northern NSW and southern Queensland to supplement crop stubbles. For example, in the eastern Darling Downs grazing of beef cattle is carried out on oats and summer crop residues (e.g. sorghum) in winter and grazing crops (e.g. forage sorghum) and native pastures in summer.

Cereal or pasture hays are regarded as valuable supplements, particularly during a drought. The higher protein content of legume dominant hays (e.g. 12 per cent crude protein compared to 7 per cent for grass or cereal hay) make this roughage particularly important for weaned lambs and breeding ewes in late pregnancy or in lactation. Hay is easy to feed out with little risk of digestive upsets. A mixed ration of hay and cereal grain (e.g. 30 per cent hay/ 70 per cent oats to weaners) is preferable to feeding grain or hay alone; livestock requirements are more easily satisfied and the feed more efficiently utilized.

The value of fodder crops is assessed in terms of the productivity of the whole crop-livestock system, performance of livestock, dual-purpose value of cereals for forage and grain production and opportunity cost for the period in which land is used.

Minerals and vitamins

Where cereal grain is provided as a major part of the diet extra calcium is required, often given in the form of finely ground limestone. The calcium/phosphorus ratio (1:1 to 2:1) is also important in these diets.

Trace element deficiency may occur in some circumstances with cobalt, copper and selenium being the most common. There are various ways in which to supply the supplement, e.g. as a drench or in the form of a 'bullet' (Figure 8.7). Vitamin A may need to be given to sheep, especially weaners, breeding ewes and rams if they have been on dry feed for more than about four months.

Figure 8.7 *Cobalt 'bullet' and grinder. The grinder ensures a slow release of the microelement.*

Protein meals

Sheep and cattle on dry paddock feed benefit from a protein meal, such as cottonseed meal. These meals are often a useful source of UDP and help to balance the ration (other nitrogen and energy containing sources may be needed as well).

Guidelines for supplementary feeding

In most situations green pasture has the capacity to supply the needs of livestock. Energy and/or protein are likely to be in shortest supply when the lambing or calving time do not coincide with the flush of green growth. Young stock have difficulty in growing out when there is a late break/early close to the growing season. High stocking rates exacerbate shortages of energy and/or protein and cause undesirable pressure on soil, plant and livestock resources.

In principle, therefore, the farmer should assess the quality and availability of both dry and fresh plant matter for grazing, and the condition of stock. This should be matched to requirements of stock throughout the year. A time of lambing or calving must be determined carefully, giving an appropriate weighting for the feed situation. Farmers should also determine the availability, quality and cost of supplements. An economic evaluation is particularly important for drought feeding.

In a drought situation, the effects of overgrazing in the livestock/crop system should be carefully assessed in relation to soil structural decline and botanical changes. Where little paddock feed is available in a drought farmers will have to make decisions concerning the sale of some of their stock, type and quantity of rations to be fed, and method of feeding. Wind erosion may be a serious problem once surface vegetation has been removed by grazing. A farmer anticipating this problem should remove the livestock from the paddocks before this situation arises and feed them a balanced ration for survival in small enclosures. In drought conditions lot feeding becomes an increasingly important strategy to retain plant cover and structural stability in the surface soil in order to sustain resources in the long term.

Livestock management in integrated farming systems

Making best use of livestock for cropping

Livestock can be stocked heavily on pasture at the right times to stop weeds from growing and setting seed. This may be difficult to accomplish if sufficient numbers of livestock are not available. The spraytopping and spray-graze techniques (Table 8.2) combine low dose herbicide spraying of weeds with grazing to offer better prospects of weed control for the following crops. In this way improved crop yields are expected to come about from less weed competition and from the breaking of crop disease cycles which rely on grasses as a host. Sprayed pastures have been found to be very palatable and land prone to drift needs to be grazed with caution, to leave the soil protected.

Grazing of stubbles prior to cultivation or sowing helps to reduce the amount of residue to a manageable level, in preparation for seeding of the following crop. In the case of cereal stubbles, grazing may be combined with spreading and chopping the straw and, particularly when the quality of the straw falls, by providing feed supplements. It is preferable to use wethers for heavy grazing of stubbles since bodyweight losses are less deleterious compared to other classes of livestock. However, farmers may not have sufficient wethers for this purpose and it may not be profitable to purchase them. The reduction in stubble allows for better emergence of crops and pasture, promotes efficiency of pre-emergent herbicides, reduces

Table 8.2

Spray or pasture topping: a technique for reducing grass seed set in a pasture year before cropping. For good results paddocks need to be grazed heavily in winter and left free in early spring to ensure that all grasses come to head at the same time. Low rates of knockdown herbicides (such as Roundup) are then applied to burn off the seed heads before visible seed is set.

Spray-grazing: involves application of low rates of hormone-type herbicides (such as 2, 4-D ester/amine of MCPA) early in the pasture's growing season. This causes an increase in plant sugars, making the weeds highly palatable to stock. Heavy grazing a few days later reduces the number of weeds, and should stimulate vigorous growth of pasture legumes.

problems with toxins in the straw material and reduces the tie-up of nitrogen during the early growth stages of the crop or pasture. A minimum amount of stubble should remain to protect the soil from wind and water erosion. This will vary from place to place depending on soil type, slope, etc. An amount of 1.5 to 2 t/ha of cereal stubble is enough to offer protection for most soils (Figure 8.6).

In general, more information is required concerning the relationship between the amount and type of stubble and carrying capacity, and also the effect of treading by cattle and sheep on plant residues and surface soil characteristics.

Type of livestock enterprise

The type of enterprise and mix of enterprises (Table 8.1) will largely depend on expected sale price, variable costs, capital investment required and interest rates, and how well the enterprises integrate with each other and with the cropping system. In the higher rainfall areas there is more scope to meet market requirements, such as fattened steers and prime lambs. Near Ballarat in Victoria for example, Border Leicester-Merino cross ewes are joined to Dorset rams for a July–August lambing, the lambs being sold in prime condition at weaning in December–January. Beef producers could turn off weaners, vealers, yearlings or steers as appropriate. By contrast, in the drier areas self-replacing breeding ewe flocks are commonly run; surplus ewes may be sold as prime lamb dams. Frequently a wether flock is run as well. While this is primarily for wool production it also gives flexibility in management to cope with variations in seasonal conditions. Wethers running on poorer feed can be allowed to lose body weight. If they do not produce tender wool (due to a severe stress), they act as a good buffer against poor seasonal conditions. They can be sold as an initial strategy in a drought situation or bought in, if cheap enough, to deal with excess feed or stubble.

It has been shown that combining livestock enterprises may have advantages with the possibility of running more total livestock on the farm. The benefits to livestock production from running sheep and cattle, or sheep and goats together, arise from overall better utilization of feed, especially control of rank growth and weeds, and fewer problems from internal parasites. A combination of livestock enterprises also helps to spread the risk. In Victoria many sheep properties also run cattle. Capital investment, or preference for one type of livestock may, however, preclude such changes.

Sheep are preferable to cattle for reducing crop stubble; they are able to graze closer to the ground and pick up fallen grain; they also cause less treading damage through hoof compaction of the soil. Larger numbers of sheep can be used to graze more evenly over a shorter period of time.

Timing of operations

The timing of the livestock operations is influenced, to a large degree, by the choice of the lambing (approximately 2 months) or calving period (approximately 3 months). Once these periods have been fixed then other operations such as marking, weaning, shearing, sale of surplus livestock and joining period fall into place. The choice of lambing or calving time is more flexible than sowing time which sets the scene for crops (Figure 8.8). Conflict may arise for the use of labour and/or feed resources if land preparation or seeding operations coincide with lambing. This may not be serious where lambing requires little supervision but where difficult birth is a significant problem in lambing ewes, or where there are other factors contributing to high neo-natal mortality, close supervision could be required. Also, where lambing occurs in poor feed conditions, such as prior to the break of the season in southern Australia, hand feeding can be time consuming.

This conflict is only one of several factors influencing the choice of lambing or calving time (Table 8.3). For example, the availability of green feed for ewes during lactation and weather conditions at the time of lambing are major considerations. Other factors such as growing out the weaners before summer (in southern Australia), seasonal changes in ovulation rate of ewes, feed conditions prior to joining, and market prices and requirements, may also influence this choice.

With the adoption of practices such as stubble retention and minimum tillage, feed may be available for joining ewes from January to March. Lambing would occur on high quality pastures in winter with both lambs and ewes thriving on the spring flush. Lambing percentages are likely to be higher in winter than for an autumn lambing due to seasonal differences in ovulation rate (Figure 8.9). However, against this, lamb losses could be severe due to cold, wet and windy conditions in the winter. Lambs would be younger when pastures die off in late spring and may not be ready for market. Shearing at the end of spring may also conflict with hay making or harvest preparations.

Figure 8.8 *Pasture and stubble availability and calendar of operations, South Australia, 6 months growing season.*

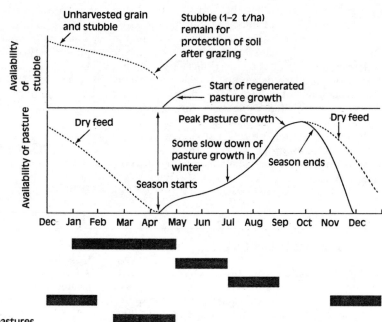

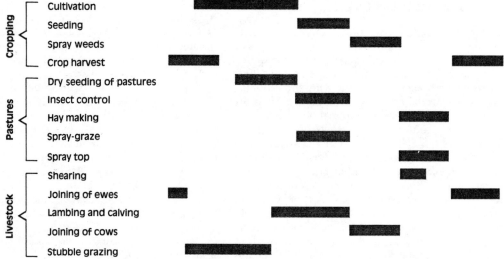

Blowfly-strike activity may increase because sheep are in full wool in early spring. The important final choice, then, is a compromise and depends on the weighting given to these factors.

Shearing is a labour-intensive operation and should not conflict with any cropping operations; e.g. shearing in early spring in South Australia, would not clash with tillage, sowing and harvesting. The choice of date for shearing may depend on the availability of shearers; if this is not a problem then likely weather conditions should also be considered — a cold, wet and windy snap post-shearing can cause severe mortalities. Other factors include shearing to lessen grass-seed injury and to assist in control of blowflies and external parasites. In some situations shearing could coincide with the potential break or 'tenderness' of

wool. Following shearing the appetite of sheep is stimulated and livestock performance on pastures, grazed crops and residues may be improved, through increased liveweight gain and wool production.

Grazing management

A grazing management plan comprises three variables: the type of stock, the grazing method and stocking rate. Of these, stocking rate can have a strong influence on the system, especially through the composition and stability of pastures and conservation of the soil resources. It is a key factor influencing production per unit area and profitability. Welfare aspects also need to be con-

Table 8.3 *Time of lambing.*

Factors favouring autumn or early lambing:

Better weather for lambing
Weaners older when feed dries off
Improved feed conditions for joining
Fewer problems with grass seeds and blowflies
(combined with spring shearing)

Mar	Apr	May	Jun	Jul	Aug	Sep	Oct	Nov
				------- 11				

Factors favouring spring or later lambing:

Higher fertility and fecundity of ewes
Improved pasture feed, less handfeeding
Less stress on ewes
Faster growth of lambs.

Other factors

Farm operations e.g. seeding may clash with lambing
where extra labour is needed for mothering-up or
handfeeding.
Finishing of lambs for market e.g. lambs may need to be
finished before feed deteriorates.
Shearing time — this should be chosen to allow
sufficient length of lambs' wool and maximize returns
for wool quality.

Breeding season of ewes in Australia (ovulation rate).

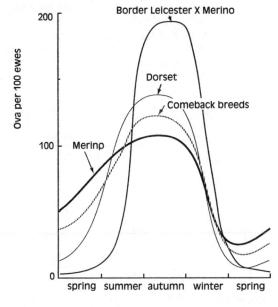

Figure 8.9 *Graphs of relative reproduction rates (as indicated by ovulation rate) for various breeds and crosses through the season.*

sidered in setting a stocking rate to avoid poor condition and stress to the animals.

A continuing assessment of the feed supply should be made so that decisions can be made to modify the grazing plan. These decisions then recognize an overall objective for an efficient utilization of a productive pasture for improved crop and livestock production, while maintaining stability in the system.

In practice a paddock may be grazed for livestock production and for other purposes, such as limiting the production of weeds and keeping plants in their vegetative stage to avoid seed injury to animals. Alternatively, reasons for not grazing a pasture include, allowing pastures to establish or recover from grazing or drought, avoiding problems of internal parasites, building up legume seed reserves and conserving feed.

There are other problems in pastures which can lead to unsatisfactory performance in livestock. For example, the following may occur:
- Animal diseases due to pasture plants otherwise considered useful, e.g. 'clover disease', annual rye grass toxicity and phalaris staggers.
- Animal diseases due to plants normally considered undesirable, e.g. soursob poisoning.
- Nutritional deficiencies, e.g. cobalt, selenium and copper.
- Nutritional disorders, e.g. lush grass dominant pastures may precipitate hypocalcaemia or hypomagnesaemia.
- Physical injury by seeds, e.g. barley and brome grasses allow seeds to penetrate eyes and skin of sheep; very small seed (shive) may give rise to processing difficulties of the wool.
- Contamination of wool by clinging spiny seed pods.

Thus, strategies such as selection of suitable pasture cultivars, integrated weed management, fertilization of pastures, control of problem grasses and adoption of a suitable grazing management plan are needed for both livestock and crop production and their integration in the system.

A move towards increased cropping can put more pressure on the efficiency, flexibility and stability of a grazing management system, e.g. larger paddocks for cropping may conflict with optimum paddock size for livestock and thus pasture management. Legume seed reserves may decline under increased stocking pressure. If carrying capacity is maintained there may be limitations of insufficient stock, or of time and labour involved for moving stock, to allow adequate grazing of stubbles. Heavy stocking rates in wet conditions may cause soil compaction.

Part of the grazing management plan includes the method of grazing livestock; it here refers to the type of movement of stock from one paddock to another. It includes:

- Continuous grazing, e.g. on non-arable land.
- Set stocking, i.e. grazing for an extended period, e.g. lambing to weaning (2 to 5 months).
- Rotational grazing, usually short periods of grazing on a rotation basis, e.g. 1 week on and 6 weeks off for grazing lucerne.
- Deferred grazing, to allow 'spelling' of pastures.
- Strategic grazing, a combination of the above to meet the needs of livestock and pastures, e.g. spray graze technique.

The advantages and disadvantages of each type of grazing method should be weighed up for the benefit of the total system. For example, with set stocking the preferred pasture species may decline, the number of internal parasites may increase, nutrients may be concentrated on camp sites but there is less stress on the stock. Deferred grazing may be beneficial for the pasture if it occurs at the break of the season, or at flowering to encourage seed setting or during part of the summer to avoid excessive grazing of the seed pods.

The type of livestock enterprises control the type of stock and flock structure, i.e. sex and age composition, ratio of breeding sheep to dry ewes and wethers. The differing requirements for these sheep in maintenance, production and reproduction must be matched as well as possible to the qualities of feed available. For example, weaner sheep, if taken through the summer, should be given the best feed to allow continued growth for as long as possible. This can be done by offering the weaners first pick of the stubble or legume dominant residues.

Stocking rate

Selection of a stocking rate is a key determinant of the profitability of a livestock enterprise. It is a complex management variable interacting with other variables such as joining time, reproductive performance, application of fertilizers and herbicides, and fodder conservation (Chapters 4 and 14).

Many experiments have been carried out in Australia to investigate the effects of stocking rate on meat and wool production. Most of these were in the high rainfall areas and are not very relevant to the dryland farming zone (250 to 500 mm); also few studies have encompassed pasture/crop/animal systems. Accordingly, it is only possible to offer guidelines for stocking rates as part of a grazing plan. In general, farmers stock conservatively, compared with optimum levels defined in stocking rate trials. The reasons for this may be associated with reduction of risk, appearance and health of livestock, labour availability, costs of supplementary feed, conservation of fodder and stability of the system. Certainly, there is a combined ecological and economic risk attached to increasing stocking rates, which results from adverse effects on soil, pasture and livestock attributes (e.g. surface structure, pasture composition, pasture seed production, seed reserves and wool quality).

Primarily, the optimum stocking rate (Figure 8.10) depends on variable costs of production (depreciation of stock, interest, deaths, running costs) and the prices received. Higher stocking rates usually require a provision of conserved feed to maintain dry stock above critical liveweights (e.g. 45 kg for strong-wool Merinos) and ensure proper condition of ewes and cows in late pregnancy and lactation. With better knowledge concerning pasture production and livestock performance in the system, models of pasture can assist in determining suitable stocking rates and other options for manipulating livestock strategies. However, models for crop livestock systems are still being developed (Chapter 15).

Despite generally low stocking rates, the legume content of pastures in southern Australia has declined in most years. The results of surveys show low levels of seed reserves are a prominent factor associated with grass-dominant, low legume-content pastures. Here, grazing management needs to be re-examined to determine how much the intensity of grazing can be controlled in early winter, at flowering and during summer/autumn.

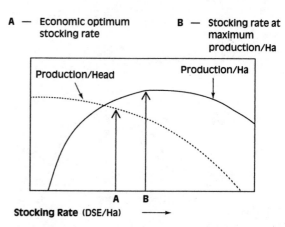

Figure 8.10 *Relationship between stocking rate and production.*

Specifically, the important issues are the effect of stocking on seed reserves and pasture regeneration after cropping.

Livestock health

With any change in the system there are likely to be implications for the health of livestock. For example, a move towards grass-free farming may result in an early feed deficit where the condition of livestock is likely to suffer. Legume-dominant pastures may cause bloat in cattle and a 'redgut' condition (due to excess ammonia production) in sheep. Problems can also arise with grass-dominant pastures or with specific pasture species. Balanced grass-legume pasture swards often counteract or cushion these potential adverse effects.

Where growing conditions are suitable (length of rainfall season, soil moisture, etc.) the level of animal nutrition may be improved by using later maturing varieties of medics and clovers, perennial species such as lucerne or by growing shrubs such as tagasaste. These strategies are particularly important for maintaining health and fertility of the breeding animals. Nitrogen fixation and input to the soil are enhanced and this benefits crops.

To assist in comparing feed requirements of different classes of livestock the dry sheep equivalent (DSE) is used throughout Australia as a standard unit. One DSE represents a wether (45 kg) maintaining its present weight; the DSE equivalents for some of the classes of livestock are given in Table 8.4. DSEs are a useful guide for estimating potential income, comparing different areas of land and classes of livestock, and for shedding light on optimum stocking rates.

Table 8.4 *Dry sheep equivalents for different classes of livestock.*

Class of Livestock	Dry Sheep Equivalent (DSE)
Dry sheep (45 kg) maintaining bodyweight	1.0
Weaner sheep (25 kg) (gaining 200 g/day)	1.9
Lactating ewes (50 kg)	3.1
Rams	2.0
Dry cows, steers (500 kg)	7.0
Cows with 4–6 month calves	17.0
Bull (800 kg)	14.0

Crop and pasture management in integrated farming systems

Cropping intensity and rotations

An indication of the amount of nutrients removed by crops and sheep is given in Chapter 17. Of the nutrients returned by livestock some are lost to the atmosphere (e.g. urea), and others are distributed patchily and may cause harmful effects (e.g. camp areas may contribute to rank crop growth and lodging). The decision relating to types of crop sown, cropping frequency and rotational sequence are an important part of the whole farm planning.

Increased intensity of cropping without reduction in stock numbers may be accommodated by methods involving reduced tillage or direct drilling. This allows livestock longer access to paddocks prior to seeding. Sheep on stubbles that are low in digestible energy and/or protein may require supplements.

Weed control

Livestock benefit from the control of crop and pasture weeds through less damage to carcasses, fewer facial infections and less vegetable matter in the wool. Controlling weeds in crops can be assisted by using high stocking rates in the pasture phase. The spray-graze technique and spraytopping (Table 8.2) are two ways in which this can be accomplished. Chemically topped pasture plants maintain their digestibility and protein, making them attractive to livestock. Alternatively, grasses can be killed using selective herbicides. The removal of disease-host grasses from pasture in late winter can reduce the level of Cereal Cyst Nematode and Take-all present during the following year's growing season. However, the remaining pasture relies heavily on the presence and persistence of legumes for livestock feed. Grasses usually provide a more rapidly growing feed resource at the break of the season than do legumes in a mixed grass/legume pasture, and they also help to add bulk and balance to the diet, with corresponding less nutritional upsets.

If legume plant densities are low, grasses and broad-leaved weeds play a dominant role as a feed source. The elimination of some grasses though may be beneficial for sheep health, e.g. barley grass where the seeds can work their way into the skin. Other seeds or pods may become attached to the wool and cause problems with the processing of the wool. Removal of the offending plants could improve livestock performance. A clearer picture needs to emerge from research concerning the monetary benefits of sheep production on different pasture swards, ranging from legume-dominant to grass-dominant pastures.

If farmers move towards grass-free farming additional grazing could be provided early in the growing season by sowing species such as oats and vetch. In the drier areas, such as in the Mallee of north-west Victoria, fodder crops such as rye, oats/peas, canola and barley, sown dry into stubble or fallow are being evaluated. Three conservation methods (hay, haylage and hayfreeze) are also being evaluated to determine the most economic method of fodder conservation. The conserved fodder is fed back to livestock at the time of traditional feed shortage (February/March).

In general, research is needed to show the impact of grass control on grazed pasture yield, botanical composition, feed quality and seed production of legume-dominant pastures and show the potential for direct drilling additional legume seed and also crops, such as oats and barley, for grazing.

Stubble handling and utilization

Methods of handling straw at harvest include flail mowers, choppers and spreaders; these may reduce the ability of the sheep to select out better components compared with the conventional straw-walker trails, but information is lacking.

Broadcasting or feeding grain legumes, such as lupins or beans, can improve straw intake while being a valuable supplement. However, the practice of providing a nitrogen supplement in the form of urea to sheep grazing stubbles is not supported by field trials.

The carryover of plant diseases from one crop to the next, in some instances can be reduced by grazing of stubbles, e.g. near Moree in northern NSW the grazing of wheat stubbles has reduced the level of crown rot fungus due, most likely, to a combination of feeding and trampling.

A huge research effort overseas has been directed to increasing the value of straw to sheep. Invariably the straw is harvested then treated by an alkaline solution. The application has proven worthwhile to a limited extent in Europe's protected economy, and in some instances for treatment of rice straw in south-east Asia. In Australia, improvements in feed value have not been suffi-

cient to justify the cost. There is also evidence to suggest that cereal crops can be bred for improved straw quality, to benefit sheep, without compromising yield. However, selection on the basis of improved resistance to many diseases as well as grain quality and yield mean that such straw will continue to be treated as an incidental by-product of grain production in Australia.

Overgrazing of grain legume stubbles may result in increased dust penetration in the fleeces as well as make the paddock more prone to erosion. Pea stubbles present a particular risk. Stubbles treated with mulchers (flail mowers) may also leave the ground more prone to erosion when these areas are heavily grazed. For grazed lupin stubbles in Western Australia it has been shown that paddocks should be grazed uniformly and at least 1.5 t/ha of lupin stubble should be retained to protect the paddock and subsequent cereal crop.

Pasture improvement

All the mainland States reported a deterioration in the legume content of annual and perennial pastures in the 1980s. Pastures are now being dominated by weeds and inferior annual grasses. This situation is regarded as a major constraint for both livestock and crop production. There are a number of factors which are believed to be associated with low legume persistence and these are related to climate and soil, pasture, crop and grazing management, and plant factors. Insect pests and diseases are also important.

Plant breeders have selected improved varieties of legumes and grasses for each of a range of environmental conditions and niches. In Western Australia cultivars of burr medics are resistant to blue-green aphid; balansa clover in South Australia has good persistence under continuous grazing; seradella and *Medicago murex* are adapted to acid soils. Some species are drought hardy, e.g. lucerne grown in the Coonabarabran district of NSW grows well throughout the year except in June and July when it is too cold. The benefits of these various practices and adoption of improved varieties have been recognized in general terms for legume pastures. The monetary benefits, arising from the responses in animal and crop production and stability of the system, are not readily apparent to farmers. Scientific evidence is needed to clarify these specific benefits.

Sustaining a crop-livestock system

Many options have been tried and tested for the integration of crops and livestock. Some of these have opened up new avenues for increased profitability; generally these have involved a trend towards increased cropping. Technologies have been improved to cope with problems such as weeds, pests and diseases, and more importantly the increased pressure on soil and plant resources. Notwithstanding this, many farmers, knowingly or unknowingly, may be operating close to the margin of sustainability of the system. In particular, farmers may not be willing to suffer loss in short-term profitability to ensure maintenance of stability of the system for the long term. Monitoring the effects of enterprise changes is needed to predict trends in the livestock-crop system for structure, organic reserves, nutrient status, salinity, residues and acidity of the soil (Chapter 17).

Over time, scientific investigations improve the ability to quantify the behaviour of many basic components of the crop-livestock system. Using computers and appropriate software it is possible to obtain an understanding of many important interactions and feedbacks that occur between these components and so evaluate them in a system context. The level or scope of application at which this integration of knowledge can be made is changing. As techniques for estimating key information at a more local level become available, at, say, a farm or paddock level rather than a district level, so the known basic relationships can be applied at that more detailed level (Chapter 18). As this process continues there will be less reliance on recommendations that are often subjective, relatively global, embody considerable extrapolation of the original observations, are infrequently revised and are not integrated into a whole system. Greater use will be made of more local and more timely, quantitative, information that will be examined and presented within a framework of scope and nature (e.g. biological, economic) relevant to the question/s concerned (Table 8.5).

Producers, or their advisers, will increasingly use local data and computer models as additional tools, together with existing recommendations and their own experience, to assist in assessing the most appropriate mix and management for their integrated pasture-livestock enterprises. With the development of more extensive models this assist-

Table 8.5 *Robustness of livestock-cropping system.*

Factors affecting Robustness*	Activities to improve Robustness
Economic	Short- and long-term economic analysis; costs of soil erosion; comparison of land uses and land management systems.
Social	Welfare husbandry of livestock; farmer group meetings, societies, hobbies, education, health and sport.
Informational	Improvement of knowledge relating to livestock and cropping practices, land capability; improvement of extension programmes.
Personal	Upgrading of skills, farm courses in livestock and cropping.
Technological	Research — specific regional problems. Technical packages; livestock and crop improvement; efficiency and careful use of resources.
Governmental	Land resource surveys, land-use policies, land legislation; environmental impact statements, monitoring and evaluation of land practices, financial assistance.
Community	Awareness of improvement, landcare, soil and water conservation projects; total catchment management, community-involved projects for farms (e.g. tree planting).

*Robustness refers here to the soundness of the system, incorporating such characteristics as flexibility and resilience for sustainability.

ance will apply to the whole crop, pasture and livestock system (Chapter 15).

More important are the attitudes, skills and qualities of the farmers, and how they interact in the social community. With positive and supportive assistance from the community, the farmer, as the main operator in the livestock-crop system, can participate in activities to improve the robustness of the system. Some of the ways in which this can be achieved are given in Table 8.5.

Support from the Government and research funding bodies is also a vital requirement for sustainable agricultural systems. Improving the level of education in the farming community and livestock-crop systems research are seen as priorities. Graduates, with sound training in agricultural science and associated disciplines, are required to effectively carry out new projects dealing with research and extension on sustainable livestock and crop production.

Conclusions

The integration of crops and livestock in a changing economic environment requires an understanding of the interrelationships in the system and how they can best be utilized. To maximize returns and to operate the system within sustainable limits, farmers must:

- Devise livestock management strategies to achieve this objective; these include matching of feed supply to livestock requirements, timing of operations, grazing management plans and provision for seasonal differences and drought.
- Efficiently utilize on-farm feed resources and supplements if required.
- Adopt appropriate cropping practices.

The main problem in the livestock-crop system arises where a cropping programme prevents improvement of soil structure and fertility by pastures from taking place. The decline in the quality of legume-based pastures and poor grazing management may exacerbate the situation.

Further reading

Alexander, G. and Williams, O. B. (eds), (1986). *The Pastoral Industries of Australia*, Sydney University Press, 2nd edn.

Allden, W. G. (1980). 'Integration of animals into dryland farming systems', *Proc. of International Dryland Farming Congress*, Department of Agriculture, South Australia, 342–78.

Cottle, D. J. (ed.) (1991). *Australian Sheep and Wool Handbook*, Inkata Press, Melbourne.

Egan, J. P. (1979). 'Role of nutrition in small sheep flocks', *Breeding Coloured Sheep and Using Coloured Wool*, Peacock Publications, Adelaide, 61–82.

Jefferies, B. C. (1989). *Sheep Husbandry in South Australia*, Department of Agriculture, South Australia, 2nd edn.

Wheeler, J. L., Pearson, C. J. and Robards, G. E. (eds) (1987). *Temperate Pastures*, Australian Wool Corporation/CSIRO, Melbourne.

Acknowledgements

I appreciate very much the valuable comments of Dr M. Round and Mr I. Cutten, Senior Research Officers of the Department of Agriculture, South Australia.

CHAPTER 9

Pest, Disease and Weed Control

P. Pittaway

SYNOPSIS This chapter outlines the factors regulating abundance of organisms such as plant pathogens, insect pests and weeds. It describes the characteristics of modern farming systems which render them more prone to outbreaks and contrasts traditional methods of pest and disease control with more recent trends in Australian farming practices. Changes in the management of pest and disease outbreaks in the age of pesticides are outlined and strategies to cope with outbreaks and the role of strategic intervention are discussed. Current developments in commercial pest control including integrated pest management are reviewed.

MAJOR CONCEPTS

1 Design of any successful pest management strategy requires an understanding of the interrelationships between climatic factors, host or community proneness to attack and the dynamics of natural enemies.

2 Successful pest and disease control requires an ability to predict when an economically significant outbreak is imminent and to select appropriate intervention strategies.

3 Successful intervention must be framed in the context of a particular farming system.

4 Integrated pest management is a multidisciplinary approach which relies on an analysis of the interaction between host, pest species and the environment.

5 Farming systems can be designed to maximize pest and disease control options. Multiple land use in space and time can cause disruption of pest life cycles and enhance the role of natural enemies.

6 Monitoring has an important role to play in assessing the timeliness and extent of strategic intervention. Prediction of outbreaks also requires good monitoring.

7 The phenomenon of resistance to chemicals must be countered by selection and breeding (including genetic engineering).

8 Modern technology can be a great ally in combating pests and disease.

In the quest to produce food for human population there is competition with a huge array of other species also intent on utilizing the resource. Those competing are referred to as pests and diseases.

The challenge today is for agricultural technologists to evaluate pest, weed and disease control in the context of the whole farming system. Total reliance on chemical pesticides is biologically and economically untenable. Prevention of economically significant pest outbreaks by enhancing (or introducing) those factors which suppress or prevent pest activity, forms the basis of integrated pest management (IPM) in its broadest sense. Absolute eradication of pest species is no longer the ultimate goal, except in the case of very recent exotic pest outbreaks where local quarantine still remains the most effective control measure.

Technologies in use today must avoid the presence of chemical residues in produce, and counter the evolution of pesticide resistance in target pest populations. Thus, the emphasis on this chapter will be to evaluate why agricultural production is prone to damage by pest species. Analysis of proneness to invasion and those factors which suppress populations are examined in the context of dryland farming systems and the potential to maximize management strategies to avoid economic losses.

Changes in the management of pest and disease outbreaks in the age of pesticides

The 'Age of Pesticides' dates from the commercial introduction of the first synthesized organic insecticide (DDT) in 1946. The efficacy of lower doses of broad-spectrum biocides in providing relief from human parasitic insects was observed during the First World War. Their very nature as broad-spectrum biocides fuelled the optimism of the people for the concept of total eradication of insect pests. This optimism promoted the change to higher yielding, insect-sensitive crop varieties, and the consumer demand for unblemished, pest-free produce. To achieve this it was common practice to spray crops repeatedly throughout their growth. The phenomenon of insecticide resistance to the older inorganic insecticides had been recorded as early as 1914, and to the newer generation organics by 1946.

It was not until the phenomena of pest resurgence, secondary pest outbreaks and pesticide resistance became widespread, that both agriculturalists and ecologists re-evaluated the costs and benefits of indiscriminate pesticide usage. Under these constraints some crop yields were actually reduced as pesticide applications increased.

Initially non-target species other than insects were protected by the lower doses applied. However, these early chemicals were extremely persistent (residual) in the environment. Repeated exposure and concentration up the food chain resulted not only in the depletion of natural enemies, but also in pesticide residues in animal carcasses and even in human fat tissue and milk. With the occurrence of multiple resistance to insecticides in key pest species, it became obvious that reliance on syn-thetic organic biocides to eliminate pest problems was untenable. The pattern of resistance is equally applicable to weed control and microbial pathogens.

The emphasis now is on a systems approach, analysing management on a whole-farm basis to optimize as diverse a range of preventative and control measures as possible.

Integrated pest management (IPM) restricts the use of synthetic organic chemicals to rapid-acting, short-term intervention strategies for those occasions when cultural and biological control methods have not been sufficient. The cornerstone of this new approach is to define objectively tolerable levels of pest activity, and to predict when an unacceptable population outbreak is imminent.

Decision making in the era of integrated pest management

Although in the final analysis it is the farmer who implements a pest management procedure, many interacting groups influence and condition that decision (Figure 9.1). The most binding influences are legislative for the control of high priority key pests, e.g. rabbits, skeleton weed.

The interrelationships between the farmers, researchers and government are a key feature of the dryland farming system in Australia. Effective implementation of new policy must take into account these interactions if advances in IPM are to be made at the farm level.

Economic thresholds and injury levels

Provided the pest species is not notifiable or under other legislative control the decision when to intervene is related to the economic threshold. Ideally the aim of IPM is to prevent pest population outbreaks, but intervention becomes necessary when the attack immediately reduces quality or quantity, or when current seasonal trends encourage population increases above the economic injury level. Intervention demands a rapidly acting strategy, and this can be provided by chemical pesticides. The economic injury level is based on past experi-

Figure 9.1 *Participants in decision-making processes defining the problems, solutions and farmer adoption of integrated pest management in the Australian context.*

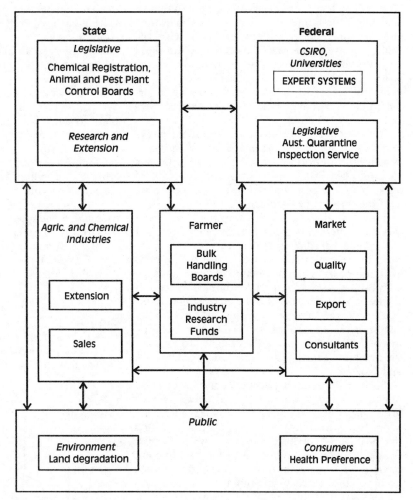

ence of the host reaction to the pest, defined as the pest population level above which the cost of damage caused would be greater than the actual cost of intervention. Intervention involves the time of the farmer, the cost of the chemical and of the fuel and equipment required to disperse the pesticide. To be economic this total cost must be less than the returns received for the portion of the crop or produce saved by intervention. The key concept in pest management has changed from one of pest eradication (except in the case of quarantine issues) to that of tolerating non-economically damaging pest levels.

Farmers cannot wait until the economic injury level is reached before they intervene, instead they require an operational lower level, guaranteeing sufficient time for the application and action of the pesticide before produce losses become too great. This lower level is called the economic threshold.

As it is predictive, detailed information of the relationship between the pest, its host and their interactions with management and the environment are required to confidently recommend a pest density at which intervention is necessary. Collecting or gaining access to this type of information has two benefits: it provides a reference point (economic threshold) for the farmer to intervene in the short term, and it provides a guide to the critical management factors that could be changed to avoid the pest problem in the future.

Factors determining pest-induced injury include the
• Timing of injury.
• Host part injured.
• Injury types.
• Injury intensity.
• Influence of the environment.
Both the host and the pest will have phases in their

life cycle which are most vulnerable or most resilient. If the farmer is able to avoid synchronizing the vulnerable phase of the host with the resilient phase of the pest, the economic threshold may not be reached. More importantly, if the farmer can also adjust management to exploit the vulnerable phase of the pest by weakening it further, then a pest outbreak may be avoided altogether.

At its simplest, the economic threshold is the pest density set by the research units of the State agricultural authorities at which intervention is recommended. These rely on farmer sampling of the host for the pest, and would typically consider only the stage of development of the host and the current pest density. If the threshold has been exceeded, chemicals registered for both host and pest would be administered at the recommended rates to the prescribed growth phase of the host. Such information can be obtained from State agricultural authorities or specialist books on crop or livestock management. If the demographic information on the pest and the host is sufficient to enable research groups to produce simulation models, expert systems can be designed to improve the flexibility of the decision making (Chapters 15 and 16).

Such models not only account for factors determining pest-induced injury but also can include host vulnerability and the density independent and dependent factors operative on both the pest and the host. Input of local variables by the farmer enables the expert system to produce different scenarios as the season progresses, and can even evaluate the efficacy of different intervention options and the timing of intervention.

The advantage of such packages is the opportunity for farmers to reassess control options and the consequences of different intervention strategies, prior to physically implementing control. Examples of decision assistance packages in Australia are the RUSTMAN and RUSTRON programs for rust epidemics in wheat, and the SIRATAC pest management package for cotton.

Factors regulating the abundance of organisms

Climate

If climatic conditions were optimal, and access to primary compounds (water and nutrients) unlimited, population growth of a species should be

exponential (Figure 9.2). However, unfavourable factors such as extremes of temperature can disrupt life cycles and population growth. As climatic factors affect small and large population groups in the same way, they are referred to as density independent factors.

In temperate zones, climatic extremes may dictate the pest status of a species entirely. For example, although leaf and stem rust of wheat can readily occur in Queensland and northern NSW, the cooler winters and hot, dry finishes to the growing season of South Australia and north-western Victoria seldom favour epidemics of these diseases. To accommodate these adverse seasonal conditions, many arthropod species and nematodes have adapted their life cycles to include climatically triggered dormancy. In southern Australia dormancy is most common over the hot, dry summer. If false breaks to the season occur such dormancy will be broken, with the newly

Figure 9.2 *A comparison of the density-dependent and independent factors on the reproductive success and survival of a population.*

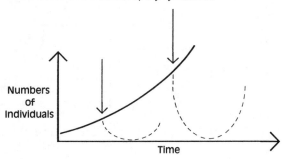

Numbers of Individuals

Time

a) Climate (density independent)
Solid line shows exponential growth under optimal conditions. Dotted lines show population crashes after adverse conditions (arrow).

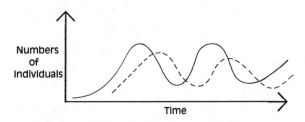

Numbers of Individuals

Time

b) Natural enemies (density dependent)
Solid line shows host population trends, dotted line shows the interdependent dynamics of a predator or pathogen population.

emerged young extremely vulnerable to dessication once the normal conditions return.

Migration

Most other factors regulating the abundance of organisms vary with the population size of the species, and are referred to as density dependent. In the real world, just as the climate is not uniformly optimal neither is the availability of primary compounds, nor is the space to grow unlimited. A major constraint on population size is competition between individuals to utilize that resource, or in its absence to migrate (or disperse) to a more accessible and available alternative. Most animals are capable of independent mobility, often equipped with highly sensitive receptors to locate appropriate hosts. However, with the exception of the mammals, most species reproduce from eggs, produced in multiple batches. The effects of competition within a species is thus most extreme at the immature phase, when the individuals are least mobile. As a consequence, population reduction is greatest at this phase of development. This is particularly true for insects where egg clutches are commonly over one hundred, the immature are flightless, and mortality due to dispersal exceeds 80 per cent.

Most foliar pathogens and weeds disperse passively by wind, or in soil, or by adhering to animals or machinery. These species typically produce hundreds (microbes thousands) of reproductive units to compensate for dispersal to unfavourable habitats. Such 'migratory' losses are overcome in some specialized pathogens by utilizing more mobile, host-selective arthropods as intermediate hosts. For these vector-transmitted diseases, the factors regulating the abundance of the pathogen and therefore the disease risks depend on the population dynamics of the arthropod vector.

In agricultural ecosystems, the growing of uniform genetic types in even-aged stands (or herds) over large areas minimizes losses by dispersal in adapted pest populations. With higher yielding more nutritious genotypes, the fecundity and survival of adapted parasitic populations is much greater. A similar trend has occurred in weeds, selection enhancing those species adapted to modern cultural methods. In grain crops, the most prevalent weeds are annuals, possessing rapid germination, synchronous ripening and shattering. Provided the species can compete with the crop, population losses due to dispersal are minimal. The key factor reducing such populations would therefore be that of predation or parasitism of the seeds.

Natural enemies

Natural enemies are of three main types: competitors, predators and pathogens. The reproductive success of each type is strongly related to the population size of the host or rival, and thus natural enemies are referred to as density dependent factors. Predator and pathogen populations are most intimately associated with the abundance of their hosts. When the host is at low density, their successful dispersal is also more difficult. Thus under natural conditions, a predator or pathogen population will always be less abundant than its host and will never totally eradicate the host (Figure 9.2). This is a particularly important concept in classical biological control. A predator or pathogen is introduced into the pest population; it should be self-perpetuating to minimize the need for reintroduction. Thus, the pest will not be totally eradicated, but will be reduced to non-damaging and stable population levels.

Natural enemies are the primary factor stabilizing host populations, provided they can respond reproductively as the host population increases and can exploit the same geographic range and climatic conditions. Predators and pathogens may be highly specific, as in the case of classical biological control agents where the host range is restricted to one species, or they may be generalists. In the arthropod world, hoverflies, lacewings, predatory mites, spiders, ladybirds and wasps are beneficial generalist predators. Insectivorous birds are also highly beneficial in pest control, as indicated by the title of 'Farmers' Friend' bestowed on the ibis. Apart from sufficient prey, natural enemies often require permanent shelter afforded by perennial vegetation.

In agriculture, growing annual crops and pastures involves repeated disturbance unfavourable to generalist natural enemies. Furthermore, as predators favour exposed positions for prey location, application of non-selective insecticides often more effectively reduces predator populations than the pest species. However, in Australia there is a more fundamental reason why natural enemies are not always effective at controlling pest species: most of the agricultural plants and animals have been imported from Europe, contaminated with their co-evolved pests and pathogens but in the absence of predators of these pests. The likelihood of indigenous natural enemies switching to these exotic pests is very low because of the extreme dis-similarity between the European and Australian flora and fauna.

Occasionally shifts in host preference in natural enemies have occurred, as in the case of control of the Portuguese millipede. A nematode parasitizing native millipedes has broadened its host range by exploiting the more abundant Portuguese pest species. The corollary can also occur, when changes in habitat management elevate the status of a native species from harmless to key pest, e.g. the pasture cockchafer evolved to feed on low nutrient, sparsely populated native grasses. A shift in host-preference to clover (both more abundant and higher in nitrogen) has enabled this native species to become a key pest. Dispersal losses are minimal, with the new ecosystem not favourable to natural enemies but enhancing the reproductive success and survival of the pest species. Similarly, weed species contaminating grain or animal hides during importation to Australia have arrived with their co-evolved crop rivals, in the absence of their natural enemies. The original environment of these weeds is similar enough to the agricultural en-vironment in Australia to promote rapid exploita-tion of the habitats; they are pre-adapted. However, despite these complications the potential influence of extant natural enemies on agricultural pests should not be overlooked. This is particu-larly pertinent to insect pests, e.g. records from a potato crop grown in south-eastern Queensland in the absence of insecticides revealed a high diversity of natural enemies.

Strategies to avoid invasion by plant pests

Role of quarantine

Quarantine still remains the most effective way to avoid the introduction of exotic diseases into this country, and for the containment and eradication of recent introductions. For many non-notifiable and unproclaimed pest species, quarantine at the farm level can be equally effective. Not all pest species have efficient, independent migration or dispersal mechanisms. This is particularly true for specialized animal pests lacking arthropod vectors, large-seeded or exclusively vegetatively reproduc-ing weed species, and some soil-borne nematode, bacterial and fungal diseases. It is also true for plant viruses where unsanitary vegetative propaga-tion techniques provide the major mode of disper-sal. Depending on the host range and survival mechanisms of the species, some of these pests, even if widely distributed or endemic, may be effectively eradicated at a local level. Thus, if the pest is absent either due to local eradication or due to its slow rate of dispersal, the simplest way to avoid outbreaks is by stringent local quarantine. At the whole-farm level, an inventory of the local movement of machinery and produce likely to spread pests and stringent testing and observation of agricultural commodities entering the farm should be considered.

Importation of hay contaminated with weed seeds is a major source of local introduction of weeds, and also (in South Australia and Western Australia) the nematode-bacterial-ryegrass associ-ation causing annual ryegrass toxicity in stock. Likewise, inadequate inspection of agisted or recently purchased livestock prior to release into existing stock and pastures brings a very high risk of introducing new diseases and weeds.

With sound pasture management and strategic-ally timed chemical drenches, it is possible to mini-mize the sheep stomach worm burden on lambing paddocks. However, introduction of infected ewes into that paddock will rapidly undo a season's good management. Similarly, confining new ani-mals in a holding yard and hand feeding for sever-al days prior to release will promote the passing of viable seed in the faeces, thus containing possible sources of infestation to one easily inspected and controlled paddock. Noxious weeds, such as Noogora and Bathurst burrs, are also commonly introduced on the hides and tails of livestock. A short period of local quarantine and curative chemical treatment of new animals is an extremely effective way of avoiding weed and disease out-breaks, and is highly recommended.

The use of certified, pathogen-tested seed and propagating material, and purchase of tube stock from registered nurseries utilizing partially steril-ized potting mixes, is an efficient way of avoiding the introduction of new diseases. It may be feasible to curatively pickle seed to minimize the impact of diseases, but once seed is infected the health and vigour of the seedlings may already be impaired, and prevention is always better than cure!

Vulnerability of organisms to attack

A fundamental consideration of sustainable agri-culture and integrated pest management is to ensure that the crop or animal is adapted to the local climatic, topographic and soil conditions. Unthrifty organisms are always more vulnerable to attack. The worst situation is when environmental conditions are unfavourable for the host but promote optimal activity of a parasite or competitor (the 'opportunist') (Figure 9.3). If such conditions occur routinely it is highly unlikely that the enterprise will be economically viable.

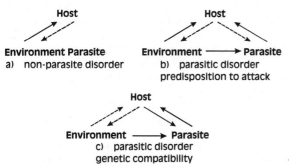

Figure 9.3 *The role of the environment in parasite:host interactions.*

Concept of barriers to invasion

Despite the appropriate selection of a crop for a specific location, the nature of current management practices still renders a crop prone to weed and vermin invasion. The dynamics in a native plant community result in a highly diverse species interaction over space and time. The cycles of disturbances within the community promoting recruitment of new individuals are also diverse, and characteristic of that community. The species comprising that community have evolved to service and exploit such disturbances, and as a result, invasion of exotic species (weeds) into the community is minimal. However, as the species' temporal and spatial diversity is reduced, the previously existing 'barriers to invasion' are removed and the ingress of exotic weeds at the expense of preferred species increases. It is not possible to reintroduce all of the barriers to invasion operative in an 'undisturbed' native community into the typical disturbance patterns of agriculture, but by analysing the characteristics of successful pest species and the disturbance they are exploiting, it may be possible to increase the barriers thereby minimizing outbreaks.

When environmental conditions are more marginal, opportunistic parasites may attack and create damage otherwise resisted by the host under more optimal conditions (Figure 9.3b). Such disorders are best remedied by improving the health of the host.

Some diseases are characterized by a high level of co-evolution between the parasite and the host. Typically the parasite is totally dependent on the living host, existing outside the host only as dormant survival units. The environment is critical in both triggering parasitic activity, and remaining conducive for attack of the host (Figure 9.3c). Success is determined by how well the parasite avoids, masks or overcomes the host's resistance.

The outcome of such disorders depends on the extent of genetic compatability or 'gene matching' between the host and parasite, e.g. wheat varieties partially resistant to rust effectively slow the rate of disease spread by reducing the reproductive potential of the fungus.

Strategies to avoid weed outbreaks

The advantage of using the ecophysiological approach to weeds is that the attributes of the weed are considered in combination with that feature of the plant community providing the opportunity for invasion. Identification of the source and size of the weed reservoir is essential. In most cases the reservoir will be either a seed bank or seed rain. If the weed species is well adapted to long distance dispersal (therefore seed rain), prevention of the reservoir is impractical, but modification of management to reduce the gap or improve the competitive success of the preferred species may effectively suppress the weed. Optimizing crop and pasture densities, enhancing early crop vigour by fertilizer application and early sowing will reduce the opportunity for invasion. In pastures, over grazing will create gaps which allow entry of weeds.

For many of the worst weeds of crops and pastures, the existing seed bank in the soil is the major factor that determines the severity of infestation. Such seeds may number thousands per square metre. The best strategy for controlling such species is to prevent the seed bank from increasing by preventing seed-set. For species closely compatible with the crop itself, control within the crop will be very difficult. However, a feature of the dryland farming system is the use of cereal and legume crop rotations, and commonly a pasture phase (Chapter 6). Weeds difficult to control in the cereal phase can be targeted in the pasture or legume phase. Commonly monocot- or dicot-selective herbicides are used for control in the alternate cropping cycle (refer to later section 'Strategic Use of Pesticides').

If such weeds are allowed to replenish their seed bank in the pasture phase, repeated applications of selective herbicide in successive cropping phases will rapidly induce herbicide resistance. It is therefore critical to consider weed control within the total farming system, aiming at long-term exhaustion of weed seed reserves and minimization of the opportunities for weed invasion. Integration of crops and livestock management offers a unique opportunity for the cultural control of crop weeds. It also offers the only management alternative (via grazing pressure augmented with non-selective herbicides) for farmers plagued by weed species known to be resistant to several selective herbicides.

It is equally important for farmers to realize that major changes in management techniques (i.e. the opportunities for invasion) also change the spectrum of weeds affecting their enterprise. Victorian researchers have identified the key phases in weed flora evolution as:
- The cultivation and late sowing of the 1940s.
- The dependence on phenoxy herbicides of the 1940s–60s.
- The development of the early post-emergence herbicides of the 1960s–70s.
- Herbicide armoury 1970s–80s.
- Crop diversification and reduced tillage 1980s–90.

Each phase is characterized by selection-driven changes in the weed opportunists as the 'weaknesses' in the management system change. As one would expect with the availability of a complete herbicide armoury, the opportunities for invasion have been drastically reduced, placing very strong selection pressure either on weeds very similar to the crop (e.g. wild oats), or on populations of weeds resistant to herbicides (in south-east Australia, barley grass and ryegrass).

To avoid selection of herbicide-resistant weeds, more diverse whole-farm system methods of weed control must be considered. The emphasis remains on exhausting weed seed reserves in the soils, preventing weed seed-set, and on maximizing crop/pasture vigour and densities to minimize the opportunities for invasion.

Many weeds also have a vegetatively reproducing phase, regenerating from root fragments or bulbs. Poorly timed cultivation of infested fields will only serve to exacerbate the problem. Similarly, herbicide application must be matched with the growth phase of the plant most likely to result in effective translocation of the poison to the roots and bulbs. For example, if a dessicant herbicide is applied to soursob (*Oxalis pes-caprae*) the foliage will temporarily be removed, but the contractile root and any bulbs are unaffected. However, if a slow acting, translocated herbicide is applied just before flowering, when downward transport to the contractile root and developing bulbs is greatest, control will be much more effective.

It is important to stress that weed control does not necessarily imply eradication of all non-agronomic species. Tracts of perennial vegetation can serve as important refuges and reservoirs for natural enemies. Likewise, some weed species not economically damaging to the crop may enhance the activity of natural enemies within that crop, or may serve as an in-crop alternative host reducing the pressure on the crop itself.

Invasion by vermin

Since European occupation Australia has had a history of deliberate exotic animal introduction. Notorious examples are the rabbit, house mouse, feral pig and feral cat. Part of the early success of imported species was the local eradication of native predators, lack of pathogens and parasites, and their pre-adaption to the altered agricultural landscape. Some species, such as the sparrow and house mouse, continue to restrict their range to human urban or agricultural environs (commensals), whereas others like the pig and rabbit have extended much further.

The economic impact of pest animal species falls into three categories:
- Reduction of agricultural produce via herbivory or predation (direct).
- Reduction of agricultural produce via the transmission of infectious diseases harboured by the vermin, to agricultural stock (indirect).
- Displacement of native or preferred species (both animal and plant).

Situations where control measures are required fall into three categories:
- Those posing potential threats if quarantine breaks down.
- Naturalized species reaching plague proportions occasionally.
- Those requiring routine control strategies.

All three depend on knowledge of their present and potential geographic range, and conditions known to favour population increase, to implement control strategies. Designing control strategies and implementing legislation is more difficult than with weeds, as vested interests have conflicting opinions on the value and impact of the species. Some species such as house mice, are generally considered as pests, but for wild pigs, deer and even rabbits, hunters and agricultural land owners have a major conflict of interest. The economics, feasibility and implementation of management strategies must accommodate these viewpoints.

Attack by arthropods

The main differences between arthropod pests and microbial pathogens are their size, mobility and level of sophistication of chemical receptors. As a consequence of their size arthropods tend to remove plant tissues, either by chewing or piercing the plant cuticle and sucking up contents. In agricultural crops, the problem of attractiveness to insects is compounded not only by the desirability of high yielding varieties but by the fact that the very characteristics rendering plants repellant to arthropod herbivores are also toxic or repellant to humans. Thus, if plant breeders select varieties which are repellant to insects, care must be taken to ensure they are not toxic or repellant to humans also.

It is possible to avoid pest outbreaks by exploiting the host preference strategy. It can be exploited in two ways: manipulation of the host preference of the pest species, and manipulation of the natural enemies preying on the pest species. Multiple cropping is a very efficient mechanism for disrupting pest species. Strategies include the use of trap crops, sown prior to the crop to stimulate hatching of the pest species.

Strip cropping with species differing in both their attractiveness and tolerance to the pest is also very effective. If the second crop is more attractive and more tolerant, the higher value first crop can thus be protected.

Figure 9.4 *Two neighbouring farmers used stripcropping to minimize erosion. The strips reduce the concentration of water and its velocity.*

In south-eastern Australia paddocks are traditionally in squares to accommodate a grazing phase. However, with the trend in contour and soil-type farming (Chapter 10), it may be possible to include strip cropping systems for pest control purposes (Figure 9.4).

The use of non-host rotations over time is also classed as an example of multiple cropping. For arthropod (and nematode) species surviving in the soil as eggs, inclusion of non-host rotations depletes the population due to the lack of mobility of the immatures. However, care must be taken to control weed species known to host the pests if rotations are to be effective. This is particularly true when using a legume crop to reduce Cereal Cyst Nematode populations. Grassy weeds in the legume crop act as effective hosts.

Attack by pathogens

Viruses, bacteria, mycoplasms and fungi are sufficiently small to invade at the cellular level. The plant has co-evolved with these specialized invaders by triggering specific defense reactions. For the vast majority of microbial pathogens, the generalized physical and chemical defenses are sufficient to render the plant immune (complete resistance). Those able to overcome or avoid these defenses fall into two categories, opportunistic pathogens exploiting weakened hosts, and specialized pathogens genetically matched to the host (Figure 9.3). The opportunists possess generalized enzymes and toxins and kill the host cells as they advance. As they are not dependent on living host cell processes they are often found on the older tissues first, and can persist and build up in plant residues. They are thus referred to as necrotrophs.

Necrotrophs have a wider host range and can survive and build up on alternate weedy hosts and residues in the absence of their agronomic hosts. Depending on their degree of host specialization, crop rotation employing non-hosts is an effective way of reducing the pathogen population in the soil, but care must be taken to eliminate closely related weed hosts within the alternate cropping phase. The efficacy of alternate non-host rotations relates to the limited dispersal of most necrotrophic pathogens. Most are either soil or splash-dispersed. Although classed as endemic they will only reach epidemic proportions if numbers are very high at the beginning of the host crop phase. Although capable of growth within crop residues, in the presence of more vigorous saprophytic competitors typically necrotrophs are suppressed.

Thus, two strategies can be used to avoid losses due to necrotrophic pathogens: maximize plant health and vigour to minimize the opportunity for attack, and manage the crop residues to enhance the activity of saprophytic competitors (the natural enemies of the microbial world). The practice of adding appropriate organic amendments to increase the level of saprophytic competitors is a major feature of organic gardening. Care must be taken in the selection of the organic amendments in case the residues only serve as a food base for the pathogen.

Examples of more specialized necrotrophs are barley scald (*Rhynchosporium secalis*) and speckled leaf blotch (*Septoria nodorum*), characterized by a narrower host range. More generalized necrotrophs include the root rots Take-all (*Gaeumannomyces graminis*) and *Rhizoctonia solani*. The disease phomopsis of lupins is also a necrotrophic pathogen but differs from the other diseases in that its economic impact is not on the host plant but on livestock grazing on infected stubbles. If summer rainfall occurs, the fungus infecting the residue recommences growth, producing a highly toxic chemical. Keeping sheep off lupin residues after summer rains avoids the problem.

The second category of pathogens is characterized by a very precise co-evolution with their host, resulting not only in extreme host specificity, but also the ability to exist only within a living host cell. These pathogens are referred to as biotrophs. Gene matching can be so precise that the pathogen has been able to mask or evade host detection, parasitizing the living host cell and actually stimulating cell respiration (Figure 9.3c). By enhancing the production of plant energy compounds and hormonally prolonging the active life of parasitized cells, the pathogen can maximize its own reproduction. Fungal biotrophic pathogens reproduce by wind-dispersed spores, typically produced at or very close to the plant surface to minimize damage to the lower cell layers.

Due to the extreme host specificity, the easiest strategy to avoid biotrophic diseases is to select for resistant varieties within the crop species. In comparison with the necrotrophs, the host range is very limited thus non-host rotations is also a very useful strategy. Because biotrophs cannot build up to cause epidemics in residues, elimination of local volunteer crop or weedy hosts is essential to counter their prolific spore production and ease of dispersal. The precise nature of the interaction between the host cell and the pathogen also imposes physiological limits on when the pathogen can invade. A young, vigorously growing plant will rapidly support a biotrophic epidemic within a few weeks, whereas the pathogen may not successfully complete its life cycle on a senescing plant. Because most biotrophs invade the foliage, temperature and humidity within the crop, synchronized with a favourable growth phase of the host, is critical for pathogenic success. For example, in South Australia leaf and stem rusts seldom cause epidemics because by the time temperatures are high enough, humidities are too low. In contrast the lower temperature requirements of powdery mildew (*Erisyphe graminis*) and stripe rust (*Puccinia striiformis*) more frequently synchronize these pathogens with the earlier growth phases of wheat and the higher humidities of autumn and winter.

In temperate climates timing of sowing can be used to desynchronize vulnerable crop growth phases and environmental conditions conducive to biotrophs but because of their very rapid reproductive potential, avoidance by the use of resistant varieties is a much safer strategy. Any factor reducing the humidity within the foliage will reduce the time available for pathogen germination and invasions. Multiple cropping can also have a major influence on the rapidity of an epidemic simply by temporally and spatially reducing the proportion of vulnerable hosts available, thereby increasing losses due to dispersal.

Researchers have considered sowing multi-lines of a single crop to increase the diversity of resistance within a crop, but the practical problems of uniform quality and other agronomic factors still need to be overcome before it is a viable option for farmers.

Strategies to avoid invasion by animal pests

Strategies to avoid pathogens

In contrast to the more generalized reaction by plants to pathogens, the immune system of animals constitutes an extremely specific pathogen recognition and reaction system. As a result, provided the animal does not die at the onset of an invasion, typically pathogens within the body have a limited time span before the immune system deactivates or kills them. Exceptions to this are the highly specialized pathogens evading or deactivating the immune system (e.g. the AIDS virus), or essentially 'external' parasites of the gut or skin surface. If an animal is able to successfully deactivate a pathogen on its first encounter, then all repeated attacks by that pathogen will be ineffective. This is because of the remarkable memory component of the immune system.

Particularly in the case of pathogens residing in the gut — where it is difficult for the immune system to effectively destroy them — it is possible for immune, older animals to be carriers of infectious diseases. In the case of sheep the most effective strategy to protect vulnerable young is to separate lambing ewes from wethers and rams etc., and to use curative drenches (or treatments) before the lamb is born. If the pathogen is a gut parasite transmitted in the faeces, it is important to guarantee that the parasitic load in the faeces is minimal at lambing. For other tissue or body fluid parasites vaccinating the mother can also transfer across the placenta and protect the foetus. A vaccine is essentially injecting the fingerprint identity of the pathogen into the animal to activate the memory component in advance of the pathogen itself. It is a very effective strategy to protect high value livestock, but too costly for widespread use.

In addition to curative drenches for gut pathogens, it is also possible to rotate paddocks for use by lambs, to minimize the egg survival in faeces. For example, helminth pathogens require either the eggs or hatched (but relatively immobile) larvae to be ingested by the new host from pasture infected with faeces. If farmers can avoid placing sheep on a prospective lambing paddock over the summer (in the case of sheep stomach worm), hot summer conditions in the absence of new hosts will dramatically reduce the worm load available to infect the lambs in autumn. The use of adverse seasonal conditions to destroy free-living phases of pathogens is an important management option in avoiding disease in vulnerable and young animals.

Strategies to avoid arthropods

Because of the mobility of animals and the efficiency of the immune system, most arthropods do not attack animals. However, the exceptions are the blood-sucking and biting parasites, and sheep blowfly. Strategies to avoid pest attack will depend on the relative mobility of the pest. Many lice, ticks and fleas are relatively immobile (i.e. flightless), and can thus be locally eradicated.

Inspection of new stock, brief quarantine before release on to the property and possible curative treatments are all useful ways of avoiding the local introduction of pests on to a property. For many of these parasites, unless the animal is severely attacked or its immune system is badly weakened, these pests may not be economically damaging. If they act as vectors for microbial pathogens they are far more damaging as modes of transmission of diseases. It is often much easier to avoid outbreaks of these diseases by locally eradicating the arthropod vector rather than to attempt to eradicate the disease itself.

In the case of biting or sucking flying insect pests, local eradication at the farm level is more difficult. This is true also of the sheep blowfly. However, as with insect attack of plants, proneness of animals to attack depends on their attractiveness to the insect. In the case of generalist disease vectors such as mosquitoes, it is difficult to find 'unattractive' breeds of animals, but with blowfly strike of sheep the principle of attraction is very important, and a critical tool to use in avoiding attack. Fleece rot, urine stains, daggy wool and unprotected wounds are all highly attractive to the sheep blowfly (*Lucilia cuprina*). Thus, the management practices of mulesing, crutching and tail docking are very effective methods in reducing the attractiveness of the sheep. However, if mistimed they can also lead to strike. The sheep blowfly has a peak adult activity regulated by the climate. If management practices resulting in wounding are timed outside this season, avoidance will be even more likely. Animal breeders are currently making progress in selecting sheep less prone to fleece rot, thus aiding in the availability of 'resistant' breeds.

Role of monitoring and options for strategic intervention

Monitoring and risk assessment

All the strategies mentioned in the preceding section on avoidance of pest outbreaks are based on a detailed knowledge of the life history of the pest species and its interaction with its host and the environment. Such knowledge can only be gained by methodically monitoring the life cycle and population fluctuations of the pest, and causally relating these to host dynamics, natural enemies and the climate. Managers need to monitor the host to minimize the opportunities for invasion, but for key or occasional pest species they should have a sufficient knowledge of population dynamics to predict when outbreaks will become economically significant, and to identify the most accessible life phase of the pest to effectively target for control. The time scale and flexibility for controlling an imminent pest outbreak will depend on the rate of increase of the pest.

Endemic diseases and pests typically require favourable conditions over at least a season for population increases. Thus, observations on the existing weed populations in the pasture or alternate crop phase, or of volunteer or alternate disease hosts prior to sowing the actual crop, will serve as a very useful index of the risk factors of particular pest species affecting the crop. For example, a high proportion of grassy weeds in a pasture phase preceding a cereal crop substantially increases the risk of losses due to cereal cyst nematode and root diseases.

For some diseases, commercial monitoring or bioassay systems are available to assess the risk of disease prior to the crop being sown. Diagnosing the risk before the season begins is very important for many of the root diseases and gap-grabbing weeds as the only time available for chemically protecting the plant is a seed treatment (or pre-emergent herbicide) at sowing. Routine use of fungicidal seed dressings is not recommended because a) over time pathogens will become resistant (not to mention the deleterious effects on beneficial micro-organisms; and b) some of the fungicides are phytotoxic to seedlings and can retard emergence.

Monitoring populations of weeds or alternate hosts prior to the crop can therefore serve as a forecast, enabling the farmer to decide if a bioassay or seed treatment is required. Bioassays are currently available for cereal cyst nematode, seed testing legumes for seedborne diseases, and some root diseases. There is also a test to screen annual ryegrass at flowering, to check if the bacterium responsible for stock poisoning (known as annual ryegrass toxicity) is infecting the seed head. There are also commercial insect monitoring services advising farmers of imminent pest outbreaks and recommending strategies for intervention. Most of these services depend on accurate observations and crop records supplied by the farmer to improve the recommendations on both chemical intervention and longer-term avoidance strategies.

Some pests and diseases are either too mobile and/or reproduce too prolifically within a season for bioassays or previous seasons' forecasts to apply. Fortunately, most organisms falling into this category require critical temperature and humidity conditions before population explosions can occur. They are also often restricted to critical developmental phases of the crop. Monitoring crop development and keeping accurate temperature and rainfall data allows a farmer to pinpoint when the critical periods coincide, hence when to check the crops most rigorously.

Categories of biological control

Biological control focuses on the pest species, using natural enemies to suppress them. In contrast, cultural control focuses on the host, using agronomic and management practices (which may include biological control) to minimize the impact of pests and diseases.

- Augmentation — Management of Existing Natural Enemies to Enhance or 'Augment' their Efficacy.

This may include the application of attractants to the crop to increase the density of arthropod predators, or the addition of organic matter to the soil to enhance the activity of saprophytes and other microbial antagonists of pathogens. Augmentation of natural enemies by increasing the available shelter or by providing alternative host species is also included in this category.

- Inundation — Swamping or 'Inundating' a Localized Pest Population with a Specific Natural Enemy.

This strategy applies both to pre-existing natural enemies and specifically introduced enemies. The approach is very similar to chemical pest control, with limited dispersal from the point of application (or release) and the potential for self-perpetuation minimal — as the local population is eradicated. Arthropod (and nematode) predators and parasites are particularly useful as their mobility, accurate host location and host specificity improve the potential for selective pest eradication. Commercially available examples are the parasitic *Aphytis* wasps for red scale control in citrus, and nematodes parasitizing the grub phase of weevil pests of strawberries.

- Microbial Pesticides — a special Case of Inundation.

They consist of a pathogen and/or its toxic products marketed and applied to an animal or crop using conventional pesticide equipment. The characteristic resilient spore or cyst phase of fungi and some genera of bacteria and protozoa provide an improved 'shelf life' analogous to chemical pesticides, but with improved target specificity. The products Mozkill (targeting mosquito larvae) and Dipel (caterpillars) are selected strains of the bacterium *Bacillus thuringiensis*, and Nogall is an antibiotic-producing bacterial antagonist controlling the crown gall pathogen *Agrobacterium tumefaciens*. Although yet to be registered in Australia, the fungus *Colletotrichum gloeo-*

sporoides is marketed in USA as a mycoherbicide controlling joint vetch and some mallow weeds. Strains of the fungus *Metarhizium anisopoliae* are being commercially developed in Australia to control soil-dwelling beetle grubs, including the cane beetle and the pasture cockchafer.

- Classical Biological Control — Selection of a Specific Natural Enemy to Co-exist with and Suppress Populations of the Target Pest Species. Unlike the inundation strategy, classical biological agents (CBA) do not locally eradicate their target. Instead the strategy aims at the long-term self-perpetuation of the agent, dispersing over the entire range of the pest species, reducing the pests' impact below the economic threshold. Neither is the strategy commercially lucrative as control extends beyond the point of initial release, and hopefully reintroduction is unnecessary. Because of the exhaustive research effort required to produce an appropriate CBA, pest species targeted are typically scheduled (or proclaimed) species affecting large regions, often in more than one state. Examples of successful control programmes in Australia include the cactoblastis moth targeting prickly pear (*Opuntia stricta*), the rust fungus controlling skeleton weed (*Chondrilla juncea*), and the nematode and wasp species controlling the sirex wood wasp (*Sirex noctilio*).

Strategic use of pesticides

Theoretically it is possible to control pests by maximizing cultural and biological control, but in the real world of market-driven cropping sequences, the demands for high quality produce, the constraints of broad-scale production, mechanization and labour costs, it is seldom a reality. Thus, the emphasis is on the strategic use of chemicals to complement existing cultural and biological control, to minimize the risk of pesticide resistance, and to return optimal profits to the farmer. However, most chemical pesticides are not selective, and if used unwisely will not only destroy otherwise effective biological control systems but may also adversely affect the crop itself, and even the operator. These properties of broad-spectrum efficacy and activity in very small quantities also have advantages.

Provided they are handled safely, chemical pesticide concentrates take up little storage space, can be used against a wider range of species, rapidly affect target species, and require less stringent application methods than the safer, more selective microbial pesticides or biological control agents. Although in theory the microbial pesticides are more compatible with the IPM methodology, their limited availability and narrower activity spectrum act as a barrier to adoption by farmers. As more products become available and formulations to improve their efficacy are developed, microbial pesticides will become a more prominent component of the pest control armoury. However, current practice is to use improved formulations of less persistent chemicals, timing the application to coincide with the most vulnerable phase of the pest and the most resilient phase of the host — Building in Selectivity. This is referred to as *Rate-Dependent Selectivity*. In weed control, differences in the development of the crop and weed species, such as presence of an impermeable waxy cuticle, sheathing of vulnerable plant meristems, and differences in metabolic rate (and therefore rate of active uptake of a pesticide) are exploited in 'selective' weed control.

The most commonly exploited difference is targeting broad-leafed weeds in a narrow (or preferably vertically oriented) leafed crop. Droplets of pesticide will accumulate on a broad, horizontal surface at a much greater rate than on the latter (hence the difference between dicotyledonous and monocotyledonous plants). Choice of a larger droplet size in applying the pesticide spray will ensure that even more pesticide will roll off the vertical foliage, minimizing the dosage absorbed by the monocot crop. It is for these reasons that the recommendations on product labels specifying application rates and timing coincident with crop and weed development must be adhered to for 'selectivity' to be achieved.

The potential for biochemical selectivity in targeting arthropod or microbial pests in plants is a little greater, because of the greater dissimilarity between the host and pest at the biochemical level. The most selective are grouped as 'systemics', compounds active at sufficiently low concentrations and targeting specific pest biochemical pathways, to enable uptake and circulation within the plant. However, if used excessively even these pesticides can adversely affect the plant, and can reduce the emergence rate of seedlings even when used at recommended rates. Their advantages are that being systemic, the plant does the redistributing and only pest species feeding on the plant accumulate a toxic dose. They are therefore much more compatible with the continued presence of natural enemies using the crop for shelter. The disadvantages are that by targeting more specific bio-chemical sites it is easier for pest species to evolve resistant types, if the chemical is applied repeatedly. The second problem is that chemicals differ in their affinity for transport in the xylem and phloem systems of the plant. Most are mobile in the xylem, but if applied as a foliar spray and the pest is attacking the roots, unless it is also mobile in the phloem, the pesticide will not reach the target site.

Similarly with herbicides, if translocation to roots and bulbs is required to kill the plant the herbicide selected must not only be phloem mobile, but must be sufficiently slow-acting to avoid killing plant cells before reaching the target site. Increasing the dosage of such compounds above the recommended rate may result in a more immediate 'contact' kill of tissues, guaranteeing that the compound will not be translocated. If repeated applications of pesticide are unavoidable, rotation with chemicals of different, unrelated modes of action can minimize the development of pesticide resistance. However, in South Australia and Victoria populations of barley grass and rye-grass resistant to one herbicide also show multiple resistance to a very broad range of other chemicals with very different modes of action. In such cases, farmers have to resort to diversifying rotations and the introduction of a pasture phase to broaden the selection pressures on weed populations. Therefore, to maintain the very advantages offered by chemical pesticides integration of chemical control with cultural and biological control methods is essential.

Chemical control of soil-dwelling pests and diseases is even more problematical than targeting the foliage. Unlike foliar surfaces, the rooting zone is extremely active, depending on the presence of many beneficial organisms for the recycling of plant nutrients. Pesticides typically are not selective (the exception being microbial pesticides), and provided they are not immediately bound to soil particles or deactivated by soil microbes, they may adversely affect both pest and beneficial species. Persistence is greatest in soils low in organic matter and clay, with low rainfall, e.g. in the drier regions of South Australia on alkaline sandy loams, the herbicide Glean has persisted for up to three years! However, for the control of soil-dwelling pest species, especially if targeting germinating species, higher doses must be applied to guarantee that a sufficiently high concentration will persist over the period of pest germination (typically weeks). As most pesticides depend on reduced rates to build in selectivity, these initially high dosages pose a major threat to non-target

species. For some microbial root-attacking pathogens it is possible to improve selectivity by using phloem mobile foliar-applied pesticides. If the behaviour of the pest species permits, it is also possible to broadcast the pesticide in an attractive 'bait' on the soil surface and improve selectivity by having the pest (typically an arthropod or vertebrate pest) come to the bait. However, there are still many species requiring application to the soil itself for efficacy.

Chemical companies are resolving the problem through the development of slow-release granule formulations. This technology allows the use of less persistent pesticides released at a consistent rate over the time required to match the germination characteristics of the pest species (Figure 9.5). Lower rates can then be applied, improving the margin for rate-dependent selectivity. By selecting the type and size of granule to carry the pesticide, the rate of release can be matched to local weather conditions and the duration of activity can be matched to the hatching or germination characteristics of the pest species. Formulations are currently being developed for herbicides for weeds, and 'insecticides' targeting soil-dwelling grubs and cyst-forming nematodes. Researchers are also exploiting the phenomenon of microbial deactiva-

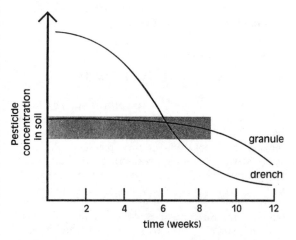

Figure 9.5 *The concentration characteristics of a pesticide applied to the soil as a drench, and as a slow-release granule. The shaded area represents the 'window' of time and concentration of pesticide activity required to match the hatching or germination characteristics of the target pest. Note the much higher concentrations required if applied as a conventional drench. Choice of size and type of granule used to carry the pesticide can match the pest's life cycle. The rate of release is dependent on soil water contents, therefore on rainfall and soil type.*

tion of persistent pesticides, developing the concept of pesticide 'safeners'. Soils differ markedly in the persistence of some herbicides, depending on the type of bacterial populations present. By selecting bacteria capable of rapid deactivation of a chemical pesticide it may be possible in the future to market two products, the pesticide itself, and its bacterial antidote or 'safener'. Once the pest problem has been reduced to a satisfactory level, the farmer can apply the safener and deactivate the chemical.

As a final note, the role of chemical intervention and resistant varieties will be discussed. Typically, resistant varieties are perceived as the ideal replacement of chemical control, however, it is important to note that in some microbial host-pathogen systems resistance may not extend to total immunity. For example, with many nematode pathogens resistant plant species may be invaded by the nematode, but the nematode will not be sufficiently compatible with the plant variety to enable completion of the life cycle (Figure 9.5).

In contrast, a completely 'immune' resistance reaction implies that the pathogen is incapable of invading the resistant variety. When resistance is incomplete, if very high numbers of the pathogen invade it is possible that the damage inflicted on the host is sufficiently great to break down the resistance. The outcome for the farmer is not only substantial crop loss, but also even higher numbers of the pest — and on a 'resistant' variety. Therefore, if a farmer suspects that a particular pest species is at very high levels and truly immune non-host rotations are unavailable or unfeasible to guarantee the resistance of a host variety, chemical treatment to reduce the pest pressure is necessary. Similar situations may also occur in weed control where initially weed numbers may be too high for cultural control to be effective. The appropriate timing of chemical intervention, at rates to selectively target the pest species, integrated with

follow-up cultural control to minimize future seed-set, is a very important strategy in IPM.

When wisely used chemicals do play a critical role in guaranteeing that pest populations are maintained at sufficiently low levels for cultural and biological control methods to take over, and in protecting a farmers' income during those times when pest outbreaks are unavoidable.

Further reading

Anon (1990). 'SIRATAC: death and rebirth', *Rural Research*, No. 147, 8–12.

Amor, R. L. and Kloot, P. M. (1987). 'Changes in weed flora in S.E. Australia due to cropping and control practices', *Plant Protection Quarterly*, 2, 3–7.

Combellack, J. H. (1989). 'The importance of weeds and the advantages and disadvantages of herbicide usage', *Plant Protection Quarterly*, 4, 14–32.

Fenemore, P. (1982). *Plant Pests and their Control*. Butterworths, Sydney.

Francis, C. A. (1989). 'Biological efficiencies in multiple cropping systems', *Advances in Agronomy*, 42, 1–42.

Groves, R. H. and Burdon, J. J. (1986). *The Ecology of Biological Invasions*, Australian Academy of Science, Canberra.

Hoy, M. A. and Herzog, D. C. (1985). *Biological Control in Agriculture IPM Systems*. Academic Press, London.

Metcalf, R. L. (1980). 'Changing role of insecticides in crop protection', *Annual Review of Entomology*, 25, 219–56.

Norton, G. A. and Pech, R. P. (1988). *Vertebrate Pest Management in Australia*, CSIRO, Melbourne.

Palti, J. (1981). *Cultural Practices and Infectious Crop Diseases*, Springer-Verlag, Berlin, Heidelberg.

Rodriguez-Kabana, R. and Curl, E. A. (1980). 'Non-target effects of pesticides on soilborne pathogens and disease', *Annual Review of Phytopathology*, 18, 311–32.

CHAPTER 10

Soil Management and Fertilizer Strategies

I. Grierson, B. Bull and R. Graham

SYNOPSIS This chapter explains the complex interaction within the soil between the physical, chemical and biological activities. Measures to improve the soil fertility and structure are outlined. The variation in plant response to soil nutrients is considered.

MAJOR CONCEPTS

1 Australian soils are among the oldest in the world and relatively infertile being particularly deficient in the major nutrients phosphorus, nitrogen and sulphur. The advent of agriculture into the Australian terrestrial ecosystem caused both physical and chemical soil degradation.

2 Soil structural decline can be arrested by the introduction of organic matter which can be furnished by pasture leys, green manuring and stubble mulching.

3 To facilitate the increase of organic matter, green manure crops need to be grown and this requires inputs of phosphate, nitrogenous and sulphuric fertilizers. Micro-nutrients are also required in many situations.

4 The amount of fertilizer required can be determined by soil testing in conjunction with environmental factors.

5 Better crop nutrition increases disease tolerance.

6 Utilizing better-adapted plant varieties can effectively complement the fertilizer strategies and increase plant growth.

Australian land forms are among the oldest in the world, being relatively untouched by recent tectonic activity, vulcanism or even quaternary glaciation. These events serve elsewhere to renew the surface rocks from time to time. The consequences of continental old age for Australian soils, terrestrial ecology and agricultural potential are quite profound. Australian soils are mostly (there are obvious exceptions) formed not from the weathering of fresh rocks, but from the particular contents of older soils that have been reworked by the forces of wind or water (erosion and deposition) with a net loss of nutrients to the oceans each time. At least nine cycles of erosion/deposition are recognized as occurring in the last one million years. The nutrient elements in the original parent rock have slowly but inexorably been lost to the salty basins or the sea. More rarely are there soils known to have survived destructive erosion by virtue of their particular location in a landscape. These mature soil profiles are none the less infertile through the less drastic, but still relentless, attrition of nutrients by percolating water entering the ground-water systems.

The native Australian vegetation has adapted to generally low and variable rainfall, fire and these infertile soils. Fire tolerance and low harvesting pressure from native animals and humans allows native vegetation to adapt to the low nutrient level of the soils. Perenniality, low growth rates and exceptionally efficient nutrient recycling are the principal mechanisms. The summer-deciduous habit allows escape from the excessive water demands of much of Australia's hot, dry summer, but before a leaf is dropped, 90 per cent or more of the limiting nutrient elements it contains is remobilized and withdrawn into the stem, the remainder being cycled through the soil microbial biomass and eventually returned to the shoot in subsequent growing seasons.

In this way, an exceptionally high percentage of all nutrients in the ecosystem is preserved in the biomass. Leaching and erosion losses are small. The same small pool of nutrients serves each succeeding flush of leafy growth and the subsequent depositions of carbohydrates in woody stores, making the final biomass of a mature ecosystem impressive considering the nutritional bankruptcy of its soil. Minor leaching and erosion losses from such a system may be compensated by the very low rates of weathering of the remaining (resistant) minerals in the soil. More significantly, small accretions of nutrients from wind and rain make for a generally stable ecosystem. The potential for catastrophic losses of nutrients by fire however, is a major threat to stability. One must surmise that the Australian terrestrial ecosystem was slowly but relentlessly running downhill, although the time scale for this may be almost 'geological'.

If a serious bushfire was a threat to the impressive nutrient economy of the natural Australian terrestrial ecosystem, the advent of agriculture was a catastrophe. Not only were the felled mallee trees or shrubs fired, producing immediate losses in smoke, but the pool (however small) of nutrients released by their demise was then subject to erosional and leaching losses, as well as removal in produce (Chapter 17).

Annual cropping exposes the land for a significant period (Chapter 7). The paltry root systems of crop seedlings are no match to the leaching power of heavy rains which move nutrients down through the profile. In particularly wet years, nutrients are leached right through to ground water and out of the ecosystem.

Not only the chemical fertility but also physical fertility is severely affected by continued and intensive cultivation which destroys structure, creates hard surface and subsurface crusts. Wind and water erosion usually follow. To redress this situation requires a closer examination of soil fertility and its component parts.

Physical factors affecting soil fertility

The physical factors can be considered as the soil structure and are concerned with the organization and arrangement of soil particles into aggregates and the stability of such an arrangement to stresses. The stresses may be external, caused by the mechanical action of raindrops, cultivation machinery or even plant roots (Figure 10.1). The stresses may also be caused internally by the differential swelling and pressure caused by entrapped air when a dry soil is wetted.

Internal forces may also occur with the osmotic stress occurring between the clay particles: if the wetting solution has a low electrolyte concentration then swelling will occur, where there is a

Figure 10.1 *Soil structure relationships.*

Internal Soil Factors (which determine arrangement of aggregates and pores)	Exchangeable sodium Clay type Cation exchange Texture Metallic oxides Organic matter Roots
External Factors (which may change arrangement of aggregates and pores)	Raindrop impact Burrowing animals (worms, ants) Plant growth Tillage Soil compaction management

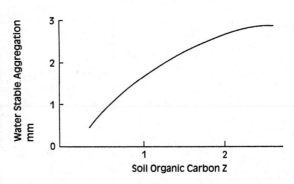

Figure 10.2 *Relationship between soil organic carbon and water stable aggregation.*

high exchangeable sodium presence dispersion will occur.

In terms of management of the factors affecting soil structure economically, little can be done in the broad-scale economic sense about texture, metallic oxides, clay type or cation exchange. However the problem of exchangeable sodium can be alleviated by the addition of an external calcium source to replace the sodium on the clay exchange sites. This calcium source is generally gypsum ($CaSO_4$ $2H_2O$) a common mineral to be found in dryland situations.

The most important variable controlling the stability of the soil aggregates is the overall soil biota (organic matter, roots, burrowing animals) and as can be seen from Figure 10.1 this factor is principally affected by the external factors summarized as soil management. Figure 10.2 illustrates how increasing soil organic matter improves the stability of a red-brown earth soil to water movement, measured as water stable aggregation.

Organic matter is thus vital to the structural stability of dryland soils, but it is also an important reservoir of nutrients and an important absorptive surface for the retention of nutrients against leaching. Figure 10.3 indicates the carbon cycle in the soil.

Organic matter and its replenishment in the soil

The losses of organic matter through the external management factor of cultivation promoting its breakdown and oxidation is particularly serious in dryland situations. Its replenishment via biomass production is limited by the climate. The essential requirements of moisture and temperature are out of phase in dryland agriculture of the seasonally humid zone. For example, in the winter rainfall areas of Australia's southern wheatbelt there are two short periods favourable to growth per year: late autumn and early-mid spring. At other times losses of organic matter exceed gains. Net losses accrue which are significant for soil stability and fertility may accrue in the space of a single generation of farmers. It is in this double context of low soil organic matter and low soil nutrient status that fertilizer agronomy is so important.

Organic fertilizers supply both nutrients and organic matter together, being an ideal slow-release form which minimizes losses to leaching, denitrification and various forms of fixation. In an

Figure 10.3 *The carbon cycle.* (SOURCE: White, I. D., Mottershead, D. N. and Harrison, S. J., 1984. *Environmental Systems: An Introductory Text*, George, Allen & Unwin, London.)

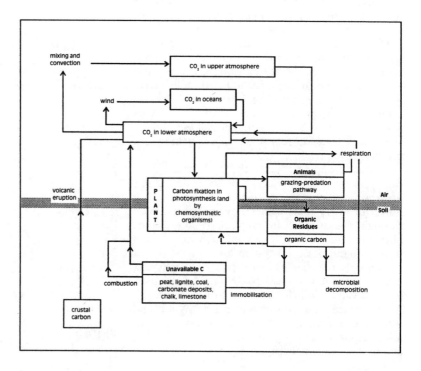

export-oriented agriculture, there is a major net loss of nutrient overseas (Table 10.1) which must be replaced from external sources. Currently little effort is made to recycle off-farm organic nutrient wastes. The benefits of organic manures have been overlooked in modern times, foolishly so in this Mediterranean context, and much greater efforts to recycle sewage and slurries from intensive animal producers must be made. They can however, benefit only a small percentage of farmers; for the rest, the double jeopardy of low soil organic matter and low nutrient status must be addressed with a combination of mineral fertilizers, green manuring and greater efficiency in the return of stubbles to the soil organic phase.

Table 10.1 *Amounts of the principal nutrient elements in wheat grain, based on crops in southern NSW and South Australian grain.*

Nutrient element	Conc. in grain NSW	SA	Amt. in 8×10^6 t tonnes
N (%)	1.95	2.44	156 000
P (%)	0.31	0.25	24 800
S (%)	0.15	0.18	12 000
K (%)	0.43	0.44	34 400
Mg (%)	0.11	0.13	8 800
Ca (%)	0.03	0.04	2 400
Fe (mg/kg)	42.3	—	338
Mn (mg/kg)	54.8	37	438
Zn (mg/kg)	22.8	18	182
Cu (mg/kg)	4.9	6	39
Mo (mg/kg)	0.3	—	2.4
B	trace	—	—

Green manure crops, like animal manures, have long been part of agriculture. They have lost favour in current Australian production systems because of the short-term economic losses when a paddock is left out of production. Yet, a single large green manure crop (including roots) turned in could constitute up to 1 per cent of the top soil dryweight, which after decomposition and incorporation into the *(quasi-) stable organic fraction* could increase the size of the latter several times. This may represent a reversal of accumulated losses under cultivation of 5 to 25 years depending on circumstances.

Moreover, a green manure crop, especially if incorporating legumes, will also contribute meaningfully to the accretion of soil nitrogen. Further, there may be minor improvements to the availability of other nutrients though virtually no increment in the total pool. A smaller increment in soil organic matter may accrue under pasture, the rate depending on approach. The modern southern Australian philosophy of pure medic or clover pastures (i.e. grass-free, in order to harvest seed and/or reduce the carry-over of root disease via the grasses from the previous cereal crop to the next) is the least effective. The root systems of these legumes are small and constitute little organic matter, while the standing herbage may be completely removed by summer grazing. The only benefit of grazed legume pasture for organic matter accumulation is that self-regeneration of the legumes saves the annual cycle of tillage with its consequential accelerated loss of organic matter. Nor is there any gain where pastures are deliberately sown for seed production. Thus volunteer legume/grass mixed pasture is the grazing system which most favours an increase in soil organic matter (Chapter 7).

Stubble return is an obvious source of organic matter for the soil, but is poorly regarded in winter rainfall regions because of slow breakdown over summer when temperatures are favourable but moisture is not. This low breakdown leads to a further complication due to carryover inoculum of various soil-borne diseases. In northern NSW and Queensland where there are heavy summer rains, stubble is often left on the surface to prevent erosion (see Figure 10.4).

Figure 10.4 *Large quantities of stubble retained as a mulch and reduced tillage help to minimize erosion even when severe storms occur.*

Little advantage is currently taken of occasional heavy summer rains for incorporation, perhaps because breakdown is still slow. But breakdown could be accelerated with a light application of nutrients to the stubble before incorporation to stimulate microbial activity, the latter being intrinsically poor because of the double barrels of low nutrient status of the soil and especially of the stubble itself. Both nitrogen and trace elements are needed. Nitrogen levels are low in stubble because of remobilization to grain during maturation. Copper and manganese are essential to microbial enzymes involved in breakdown of the relatively inert lignins and polysaccharides which dominate the composition of stubbles. These nutrients may be cheaply and effectively sprayed on stubbles immediately prior to incorporation. These soils therefore, contrast with the rare, deep, fertile alluvial soils elsewhere in this country, or perhaps more pertinently, in Europe or North America on which the principles of organic farming were developed.

The role of fertilizer use in sustaining farming systems

There are four stages of fertilizer use in a functioning cropping system, from virgin land to full replacement of nutrients. These are shown in Figure 10.5.

External sources of nutrient inputs

The major elements that are essential in relatively large amounts to the growth of plants are carbon, hydrogen, oxygen, nitrogen (N), phosphorus (P), potassium (K), sulphur (S), calcium (Ca) and magnesium (Mg). Tables 10.2 and 10.3 indicate the amounts of the major and minor nutrients removed by crop and animal products.

Of these only nitrogen, phosphorus and potassium are generally added in the form of fertilizers. The other elements essential to plant growth and required by plants in large quantities are sulphur, calcium and magnesium. In dryland areas the latter three are usually present in the soil in sufficient amounts for crop production or where deficient can be added in combination with commercial fertilizers supplying nitrogen, phosphorus or potassium.

Most dryland soils have large reserves of potassium and so fertilizer inputs have focused on the

Table 10.2 *Nutrients removed by 1 tonne of grain.*

	N	P	K	S	Ca	Cu	Zn	Mn
	Kilograms					Grams		
Wheat	23	3.0	5	2	0.4	7	16	40
Barley	20	2.7	5	2	0.4	7	19	15
Peas	41	4.5	10	4	1.0	8	26	12
Lupins	57	4.5	10	4	2.5	8	32	16
Canola	41	7.0	9	10	4.0	8	43	27
Safflower	25	4.3	9	4	2.0	14	26	13
Medic Hay	30	3.0	25	2	9.0	8	20	15
Medic Seed	64	8.4	12	5	2.0	7	23	13

Abbreviations for nutrients:

N	= Nitrogen	P	=	Phosphorus
K	= Potassium	S	=	Sulphur
Ca	= Calcium	Cu	=	Copper
Zn	= Zinc	Mn	=	Manganese

NOTE: The above figures were calculated by Mr Reg French of the South Australian Department of Agriculture.

Table 10.3 *Nutrients removed in the course of producing selected animal products.*

	N	P	K	S	Ca
1,000 litres milk	6.0	1.0	1.4	0.6	1.2
One fat lamb	2.3	0.2	0.1	0.2	0.4
One wool fleece	0.7	trace	0.1	trace	trace
Nutrient to grow feed to support:					
One sheep per year	25	3.0	20	2.0	7.0
One cow per year	110	30.0	130	12.0	45.0

addition of phosphorus and nitrogen, particularly on phosphorus because of the large use of leguminous pastures in the dryland rotations. Where cropping systems have intensified during the 1980s the use of nitrogen has increased to maintain crop yields.

Thus, the loss of nutrients from the farm as produce, with the possible exception of nitrogen, must be replaced as mineral fertilizers since there is virtually no prospect of replacement from weathering of soil minerals. Whether these fertilizer minerals are in a soluble or insoluble form is a matter of debate between exponents of conventional and organic farming practices. On the infertile soils which dominate the dryland region of southern Australia, the level of productivity necessary for economic survival can at present only be obtained by further, balanced additions of nutrients. These increase the immediately available pool

Figure 10.5 *Four stages of fertilization as a function of cropping system.*

Role and Utilization of nutrients in soil	Cropping System	Soil Fertility	Additional Fertilization	Yields
1a. exploitation	exhaustion cropping (exploitation cropping) with regeneration.	degeneration	none	decrease
1b. utilization	utilization cropping	maintenance by natural capacity	none	constancy in 'steady state' equilibrium
2. replacement	permanent stable sequences with soil improving plants	maintenance through replacement	replacement fertilization to compensate for losses	
3. enrichment	permanent stable demanding crop sequences	increase through improvements	differentiated fertilization according to supply state with aim of optimal supplies of all nutrients	increases to high yield levels
4. substitution	arbitrary	optimization or substitution	artificial combination of optimal nutrient substrate	high yield levels

to levels which can sustain sufficient growth and not merely replace the amounts of nutrients lost at the farm gate.

Phosphorus and its availability in the soil

The mean content of total phosphorus in Australian soils is only about 240 ppm compared to 400–900 ppm for soils in Europe and North America. The phosphorus contents vary considerably with soil type, e.g. lateritic podzolic soils 70 ppm, solonized brown soils 140 ppm, red-brown earths 265 ppm, and while not all soils have low phosphorus contents (e.g. black earth 815 ppm) the general situation is one of phosphorus deficiency. However, total phosphorus content has only limited value as an index of the phosphorus status of a soil.

Satisfactory growth of plants depends on the ability of the soil to provide an adequate concentration of phosphate in the soil solution at the surface of the root, and to maintain this concentration during the growth of the plant. This ability depends on many factors other than the total amount of phosphorus in the soil. The chemical and physical nature of the soil and the reactions in which the phosphorus-containing compounds participate are very important. They help explain why soils differ in the ability to supply phosphorus to plants and why they have different requirements for fertilizer, even though they may contain the same amount of total phosphorus.

Phosphorus occurs in soils in both organic and inorganic form. The phosphorus in organic compounds is not directly available to plants but has to be converted to inorganic form. This conversion (mineralization) is brought about by the activities of micro-organisms and is highly dependent on climatic conditions. Hence, release of phosphorus from soil organic matter is difficult to predict. Figure 10.6 shows a schematic diagram of the phosphorus cycle in the soil.

Many different inorganic phosphorus compounds are present in soils. They include phosphates of iron, aluminium and calcium, and in many soils much of the phosphate is adsorbed on the surfaces of other minerals and compounds. The phosphorus can be conveniently put into dif-

Figure 10.6 *The phosphorus cycle.* (SOURCE: White, I. D., Mottershead, D. N. and Harrison, S. J., 1984. *Environmental Systems: An Introductory Text*, George, Allen & Unwin, London.)

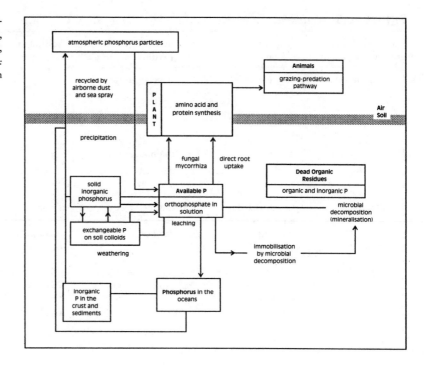

ferent categories according to the ease with which it can enter the soil solution:

Relatively insoluble phosphorus 'unavailable'	Exchangeable phosphorus	Phosphorus in the soil solution 'available' to plants

The distribution among the different categories varies considerably from soil to soil, but (except in very unreactive soils like the siliceous sands) the amount present in the soil solution is a very small fraction of the total. In any event it is far less than is required to supply a crop. Exchangeable phosphorus (i.e. that which readily releases phosphorus to the solution) is generally regarded as being 'available' to plants and is what most phosphorus soil tests aim to measure. However, exchangeable phosphorus is not all equally available. Soils with the same content of exchangeable phosphorus are not necessarily able to maintain the same concentrations of phosphate in solution. They may vary in the rate at which the phosphorus is released when the phosphorus content of the solution is depleted by plant uptake. Hence, soil tests must be calibrated for each type of soil.

The properties of the soil are important, determining not just the extent to which native soil phosphorus is available, but also the effectiveness with which phosphorus fertilizers are used. When

fertilizers are added to the soil they react to a greater or lesser extent with the soil. It is the reaction products rather than the fertilizer itself which determine the availability of phosphorus and the response of plants to the fertilizer.

When superphosphate granules dissolve in soil an acid solution is formed. This solution reacts with and is neutralized by the soil constituents. The reaction products, which may include iron, aluminium and calcium phosphates, are all less soluble than the original superphosphate. Some of the phosphorus remains exchangeable so that the immediate effect of applying fertilizer is to increase the available phosphorus, but much of the added phosphorus is converted to relatively insoluble forms, i.e. it becomes 'fixed' and is unavailable to plants. Fixation of phosphorus can be particularly severe in acid soils containing high amounts of iron and aluminium. Such soils may contain so much reactive material that very large amounts of phosphorus fertilizer have to be added to produce small increases in available phosphorus. Figure 10.7 indicates how the availability of phosphorus varies with changing soil pH.

Because of the reactions of phosphorus fertilizers with the soil, usually only 10 to 25 per cent of the phosphorus added as fertilizer is taken up by the crop in the year of application. However, the phosphorus remains in the soil and it does have a

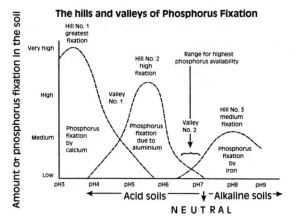

Figure 10.7 *Availability of phosphorus varies with soil pH.*

Table 10.4 *Adequate soil available P levels — cereal/sheep areas (300–500 mm rainfall zone)*

Soil Type	Desirable Soil Test Value (ppm)*	Annual Increase in Soil available P (ppm) at 100 kg/ha superphosphate
Sandy mallee	20	3
Shallow sand over clay soils	25	2
Loamy mallee	30	4
Red-brown earth	30	5
Dark brown cracking clays	35	4
Highly calcareous sands and loams	40	nil (fertilize for current crop only).

residual value, i.e. it contributes to the phosphorus supply and yield of subsequent crops. Monitoring levels of residual phosphorus is recommended to avoid over-application.

The desirable soil test value should be interpreted as follows:

- P Value *low* (more than 10 ppm below table value)
  Action: Apply as much as affordable or can be handled.
  Drill all phosphorus with the seed. *Use* the annual increased figures above to determine when it will be useful to again check the F level.
- P Value *medium* (within 10 ppm of table value)
  Action: Increase the application rate and check P level as above.
  Drill all fertilizers.
- P Value *high* (at or above the table value)
  Action: Reduce to 'maintenance rates' based on:
  1 tonne cereal grain removes the equivalent of 4 kg phosphorus.
  Fertilizer may be *broadcast* or drilled with crops.
  1.5 kg of phosphorus are required for each DSE carried sub-clover or medic pasture.

Fertilizer strategy

Phosphorus

It is most important that the responsiveness of the soils to phosphate is known. Knowing when to cut back to maintenance rates is the key to efficient phosphate fertilizer use. Where records show that cumulative phosphorus application amounts to 200 kg/ha or more on heavier soils it is likely that the soil is at 'maintenance level'. Hence responses to currently applied phosphate are very small and the only requirement is to replace that removed in produce.

The available phosphate test (Table 10.4) is most useful in determining this point. In extremely dry and unpredictable areas or highly fixing (calcareous) soils the strategy to adopt is to fertilize for the current crop only!

Surveys have shown that at least 35 to 40 per cent of Australia's major cropping soils are at or above maintenance level, i.e. with available P values at or above those shown in the Table 10.4. The farmer has the choice of applying all fertilizer with the crop (current practice), or splitting the application, which has been shown experimentally to give the optimal results in both the crop and pasture year.

Hence a maintenance programme for a two year rotation would be:

	Phosphorus Maintenance Rate (kg/ha)
Average cereal yield (2 tonne/ha)	8
Stocking rate in pasture year (DSE/ha)	4.5
Total phosphorus maintenance rate (2 years)	12.5 kg/ha

*SOURCE: Colwell's method.

Nitrogen

Despite the low levels of nitrogen present in dry-land soils the use of nitrogenous fertilizers to increase production has been much less than the use of phosphorus fertilizers. The reason for this is the use of leguminous pastures within the South Australian rotations and hence the corresponding input of nitrogen. Some soils, such as the black earths, were high enough in nitrogen for decades, especially when (as still practised) the soil was fallowed before each crop (Chapter 6). The risk of dry finishes to a season may give a negative yield response to applied N. As cropping rotations intensified in the 1980s so did the use of fertilizer nitrogen.

Plants usually absorb nitrogen from the soil in the nitrate (NO_3) or ammonium (NH_4) ion form. Ammonium ions are present on the cation exchange complex and nitrate in the soil solution. Figure 10.8 shows the nitrogen cycle and its reaction in the soil. Particular emphasis is placed on the soil biological reactions important in the mineralization of the living and dead plant remains into the ionic forms.

The use of nitrogen fertilizers always involves some risk because of the large influence that climatic conditions have on mineralization of soil nitrogen reserves, leaching and potential crop yield (i.e. the demand for nitrogen). The aim should be to satisfy the needs of the current crop only (Table 10.5). Some of the reasons for using fertilizer nitrogen could be:

• Very poor nitrogen input in pasture years (absence of legume).
• Second or third crop following pasture.
• Turning in heavy stubbles with a low nitrogen content (see above).
• Drought in the year prior to cropping.
• Very late break to the season and/or wet conditions at sowing.

A total nitrogen soil test (Table 10.6) may be helpful where paddock records are poor or unknown.

Table 10.6 *Range of values for total nitrogen in soils of various textures.*

| Soil Texture | Total Soil N (%) | |
	Low	Adequate
Sand	0.03	0.06
Sandy loam	0.07	0.12
Loam	0.09	0.14
Clay loam	0.12	0.18

Low values are an indication of likely response to fertilizer N.

Figure 10.8 *The nitrogen cycle.* (SOURCE: White, I. D., Mottershead, D. N. and Harrison, S. J., 1984. *Environmental Systems: An Introductory Text*, George, Allen & Unwin, London.)

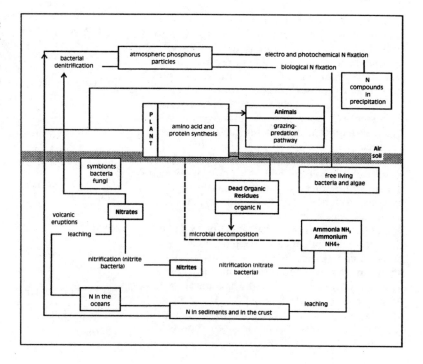

Table 10.5 *Suggested rates of N application (kg/ha) of nitrogen for use on cereal crops*

These rates will give highest profit over a period of years:

| | Previous Year | | | | | |
Rainfall	Good Legume Pasture	Poor Legume Pastures or Sandy Soil	Grain Legume Good Growth Poor Yield	Grain Legume Fair Growth Good Yield	Cereal	Continuous Cropping*
Less than 350 mm	0	10	0	0	10–15	20
350–400 mm	0	10–15	0	0	15–20	20–30
400–450 mm	0	15–20	0	15	20–30	30–40
Above 450 mm	15–20	20–30	15–20	20–30	25–40	40–60

These rates may be split between sowing and tillering applications.

Increase rates by 5–10 kg/ha if poor conditions for mineralization exist before sowing, e.g. soil not wet until within 2 weeks of seeding.

In winters where heavy rains occur in July or August and soils become cold and wet then topdress with another 15–30 kg of Nitrogen. Generally this is best applied as solid urea.

*In continuous crop situations where rotations that have not included a legume pasture or legume hay crop for over 3 years the rates of Nitrogen usage must increase. This will include the use of 5–15 kg of Nitrogen when sowing legumes. Even if sowing a cereal following a grain legume use the rates of Nitrogen in the table.

Sulphur and its availability in the soil

Sulphur lowers the soil pH and increases the availability of other plant nutrients such as phosphorus, manganese and zinc. Sulphate leaches easily in sandy soils with higher rainfall (above 500 mm). However, on heavier soils little leaching occurs. Some sulphur can be lost from the soil as gases, e.g. the smell given off by onions. Most cereal-growing soils in southern and eastern Australia have sufficient sulphur released by cultivation to meet the plants' needs.

Sulphur is present in both inorganic and organic forms. Only inorganic sulphur, occurring as sulphate dissolved in the soil moisture, is readily available to plants. Inorganic sulphur is derived from the weathering of rocks, from the breakdown of organic sulphur or from added fertilizers. Organic sulphur is sulphur combined with organic matter derived from the debris of plants, animals, insects and micro-organisms. It is broken down to sulphate slowly, depending on the temperature, moisture content and pH of the soil. Soil sulphate can be used by plants or soil micro-organisms, or can be absorbed ('fixed') by the clay in the soil.

Interaction between nutrition and disease

Adequate, balanced nutrition of the crop is fundamental to the full expression of its resistance and tolerance to disease. Better nutrition generally favours the crop over its pathogens with the notable exception that high nitrogen levels favour some leaf diseases such as powdery mildews and rusts. In this latter case, the yield of the crop may still be enhanced by overcoming a deficiency with fertilizer even though the growth of the pathogen is likewise enhanced; this response is known as tolerance, whereas resistance denotes suppression of the growth of the pathogen. Thus, nitrogen fertilizers often enhance the tolerance (yield) of crops to mildews and rusts, while at the same time decreasing their resistance.

These effects of enhanced tolerance and reduced resistance to disease resulting from fertilizer use generally occur only over the deficiency range, i.e. addition of more fertilizer beyond sufficiency is without further effect except in the case of nitrogen already mentioned. Phosphorus fertilizers may enhance tolerance especially and may also improve resistance to a lesser extent. With a number of fungal pathogens, especially wilts such as *Fusarium, Verticillium* and *Gaeumannomyces*, potassium fertilizers markedly suppress infection through an increase in host resistance, and may lead to large increases in yield.

Especially important to the southern Australian soils is the major role micro-nutrients play in disease resistance. Because trace elements have quite specific roles in higher plant and microbial metabolism and because the host pathogen relationship is often a delicate one, there is some specificity in these interactions. Thus Take-all (*Gaeumannomyces graminis*) is more common in cereals on high pH, manganese-deficient soils and is suppressed by manganese fertilizer additions to these soils, whereas Crown rot (*Fusarium graminearum*) is more common on zinc-deficient soils and is suppressed by zinc fertilizers on these soils. *Rhizoctonia* root rot is also strongly favoured by zinc deficiency in soils and markedly suppressed by adequate zinc additions. Copper and zinc are also known to suppress Take-all, and copper and manganese strongly inhibit powdery mildews on a number of crops whereas zinc and nickel have been reported to suppress rusts. These indicate the importance of achieving an adequate and balanced micro-nutrient nutrition if diseases are to be controlled. In the southern Australian environment, which for both micro-nutrient deficiencies and root diseases is regarded as one of the worst in the world, proper attention to the micro-nutrient of crops is the first priority.

Because of their low cost and fundamental importance in disease resistance such an action is quite sensible. Only then should fungicides be used to 'top up' control where this remains an economic proposition. Such an approach is conservative from several points of view and will help to achieve minimal levels of toxic residues in both produce and soil.

Fertilizer strategies and the interaction with better adapted varieties

Fertilizer strategies in the grossly nutrient deficient southern Australian environment may be effectively complemented by the use of better adapted varieties. Varieties better adapted to soils of low nutrient status are termed nutrient-efficient and may be combined with fertilizers in a 'belt and braces' approach to overcoming severe deficiencies.

Copper deficiency caused severe losses in grain yield of current cultivars of wheat due to deficiency-induced male sterility. Copper-efficient varieties perform much better. Generally speaking, efficiency factors for copper, zinc and manganese are not linked in wheat or barley. However, rye is markedly more efficient at extracting all three elements from soil than either wheat or barley. Rye has a place, along with selected triticales (some of which inherit a useful degree of efficiency — but not all of it — from their rye parent), in production systems on marginal soils. Rye and triticale, being more efficient and tolerant, have stronger tissues and therefore greater resistance to sand blasting which is a feature of drier and sandier soils.

The advantages of the 'belt-and-braces' ap-

Table 10.7 *Nutrient balances in a wheat-growing soil of South Australia.*

Nutrient	Amount removed in grain		Total amount in deficient soil		Equivalent number of crops
	(ppm)	(g/ha^{-1})	(ppm)	(g ha^{-1})	
N	20 000	30 000	1 200*	2×10^6	67
P	2 000	3 000	250	3.8×20^6	1 250
Cu	2	3	3	45 000	15 000
Zn	20	30	5	75 000	2 500
Mn	33	50	10	150 000	3 000
Mo	0.1	0.15	1	15 000	100 000

The data of this table are based on a uniform profile 1 m deep (except for N which declines rapidly with depth from a 0–5 cm value as shown [*]). Grain yield was taken as 1.5 ton ha^{-1} (about the Australian average). Bulk density was taken as 1.5 g cm^{-3}. The equivalent number of crops is simply the ratio of amounts of nutrient in the soil profile to the amounts removed in the grain. (SOURCE: Graham, R., 1978. *Nutrient Efficiency Objectives in Cereal Breeding*, Proc. Eighth Int. Coll. Plant Analysis and Fertilizer Problems, Auckland, NZ, 165–170)

proach of combining fertilizers and nutrient efficient varieties is clearly demonstrated in field results. The reason stems probably from the fact that trace elements can be added only to the top soils (5–7 cm) and roots may not grow into lower soil layers acutely deficient in certain elements (especially phosphorus, manganese, zinc and boron), regardless of the adequacy of supply to other parts of the root system. Resistance of roots to diseases may be similarly affected. The Take-all fungus-penetrated wheat roots more effectively when the environment was low in manganese regardless of the manganese status of the root itself (though the manganese status can have additional effects). Deep penetration of roots to access subsoil water is vital to achieving the production in this environment, and by utilizing this water may also minimize the contribution of cropping to the creep of dryland salinity.

The question is often raised whether efficient genotypes will be a sustainable approach as they may hasten the exhaustion of nutrients in the soil. Data from South Australia suggests that depletion of micro-nutrients is so slow that it is not a reasonable concern. Indeed replacement may come from wear and tear on implement tynes and tyres and 'contaminants' in macro-nutrient fertilizers, all of which contain some boron, copper, zinc, manganese, cobalt, nickel and molybdenum of significance to maintaining the balance sheet.

Such an argument is less convincing for phosphorus, and is unsustainable, in both senses of the word, for nitrogen. There is no future in mining the reserves of nitrogen in any soils of the cereal zone; all nitrogen leaving the farm must be replaced, either by fertilizers, legume fixation or the (small) accretions from the atmosphere contained in the rainfall (Table 10.7).

Conclusions

The nature of soil fertility is a particularly complex one depending not only on the chemical and physical characteristics of the soil but also on the biological processes resulting from the many and varied activities of the soil organisms. The soil is a dynamic medium in which chemical, physical and biological activities are influenced by both paddock history and micro-climatic conditions.

Further reading

Arnon, I. (1972). *Crop Production in Dry Regions*, Vol. 1, Leonard Hill, London.

Emerson, W. W., Bond, R. D. and Dexter, A. R. (1978). *Modification of Soil Structure*, J. Wiley & Sons, London, New York.

Graham, R. D. (1983). 'The role of plant nutrition in resistance to pathogenic diseases with special reference to the trace elements', *Advances in Botanical Research*, 10, 221–76.

Graham, R. D. (1984). 'Breeding for nutritional characteristics in cereals', *Advances in Plant Nutrition*, 1, 57–102.

Russell, J. S. and Greacen, E. L. (1977). *Soil Factors in Crop Production in a Semi-arid Environment*, University of Queensland Press, St Lucia, Queensland.

CHAPTER 11

Maintaining the Resource Base

B. R. Roberts

SYNOPSIS In this chapter land degradation processes are explained; the historic background to Australian erosion and salinity is traced; erosion and salinity control methods are described; and the basis for land stewardship and a new land ethic is outlined.

MAJOR CONCEPTS

1 The Australian economy is based on rural production which in turn is based on soil fertility and land stability.

2 Widespread land degradation has become Australia's prime environmental problem.

3 Degradation includes water and wind erosion, salinity, land slip, soil acidity, soil structure decline and decreased microbiological activity.

4 The processes of degradation are generally understood and practical control methods are available for commercial application.

5 Dryland farming systems will be sustainable if economic methods are developed to prevent soil loss, maintain fertility, increase organic matter and limit the accumulation of toxic chemicals in the soil.

6 Effective rural extension based on landholder groups will be the most important element in achieving widespread land care in Australia.

7 The acceptance of a land ethic by the community at large is fundamental to more positive and permanent relations between Australians and their land.

8 Sustainability in dryland farming systems requires not only an ecological basis but economic incentives, education programmes and well-designed regulatory guidelines.

The resource base for dryland agricultural production includes not only the soil but also the vegetation, the water, the wildlife and the landscape as an entity. Because Australia's economy is largely based on primary production, land-based production systems have become known as the 'backbone' of the national economy. In terms of meeting the needs of the human population, Australian agriculture feeds and clothes 15 million of its citizens plus another 15 million in other countries.

Since earliest times the land has formed the basis for Australian societies. For 60 000 years the Aborigines lived as hunters and gatherers, having little effect on the land except through augmenting the natural pattern of lightning fires. However, over the past 200 years of farming, a continuous process of land degradation has occurred.

Attitudes to land in historic perspective

The pioneering era

Going back in history we recognize a familiar pattern common to most settlement eras where the acquisition of land is seen as a new-found security in which the ability of the land to produce wealth is paramount.

The pioneers' relation to their land is one of survival in an environment over which humans have little control. They exploit nature by using the inherent soil fertility, fencing, developing watering points or clearing for cropping or increased animal production. In the pastoral situation there is generally a much greater awareness of humans' dependence on nature, notably their dependence on rainfall and the carrying capacity of native pastures on different soils and in different seasons.

Early in the settlement period landholders became aware of the effects of overstocking on their animals. Unfortunately, the more insidious, but longer-lasting effects on the land, are less easily observed. As such, deterioration of grazing land is generally not recognized until it has reached an advanced stage. This can be called the destruction phase (Figure 11.1).

Recognition or awareness era

The pattern of settlement and consolidation of rural communities generally moves from the pioneering era, in which life is often a battle against nature, to a phase of recognition that production systems need to be in equilibrium with the natural productive levels of the ecosystem concerned, if production is to be permanent. This permanency does not imply stability in the sense of annual yields being constant, but rather stability in the sense that, despite large annual fluctuations in rainfall, the land potential is not reduced through overuse, erosion and salinity.

The recognition phase often does not eventuate until severe economic pressure forces a realization that unless land-use methods (stocking rate, controlled burning, clearing, clean cultivation) are changed, production will continue to decline despite increased inputs through breeding, feeding and fertilizing.

The rate at which awareness of the ecological

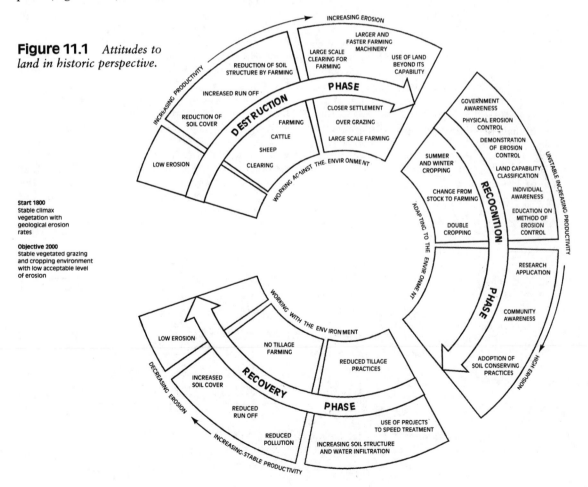

Figure 11.1 *Attitudes to land in historic perspective.*

Start 1800
Stable climax
vegetation with
geological erosion
rates

Objective 2000
Stable vegetated grazing
and cropping environment
with low acceptable level
of erosion

realities becomes an integral aspect of local community values determines largely the possibility of landholders taking early corrective action before soil conservation becomes very expensive, relative to the profitability of current production systems.

Many regions of Australia appear to be entering this awareness era in the 1990s. Though the depth of recognition and acceptance of the problem of land deterioration varies greatly between individuals in the same region, most producers are becoming aware of the need to apply conservation methods and of the growing expectations of the community that rural producers will act as trustees and stewards of the nation's food- and fibre-producing resources.

This awareness phase is likely to continue for several decades and will be greatly influenced by the effectiveness of the nationwide educational programmes.

Conservation or balanced land-use era

There are several parallel activities which require positive political support through funding before the widespread implementation of conservative production methods can be expected.

Firstly, the financial incentives to improve the land will need substantial upgrading before many landholders will be inclined to think further than short-term economic survival. The deteriorating terms of trade which have dominated the rural sector in the recent past have discouraged the development of both land ethics and the stewardship role of the landholder.

As well as financial incentives, there is the need to improve the guidelines to ensure that unacceptable land use, such as the indiscriminate clearing of unsuitable soils for cropping on a vast scale in marginal rainfall zones, does not occur in future. Legislating against certain land uses may be seen as infringing on the freedom of the individual landholder. However, future governments may find it necessary to use land-use controls to meet community pressure to preserve the future productive capacity of the land.

The price of exploitation of the soil must be paid by the community at large, through reduced production and increased product prices. The Australian need today is for a combination of the principles of sociology and ecology in a perspective-making subject which is best termed *social ecology*. Australians can no longer allow the economic pressure of the market place to continue to cause them to live off the country's 'land capital'.

The time has come to recognize the central significance of Routley's three progressive views of humans' relation to their land:
• Man the tyrant (arrogance and exploitation).
• Man the steward (adaptation and dependence).
• Man the co-operator (man/land partnership).
This evolution reflects our movement from a human-centred value system to an eco-centred symbiosis.

Land degradation — what is it?

Land degradation is a collective term that describes the loss of productivity which results from soil erosion, increases in soil salinity, loss of soil structure and so on (Figure 11.2).

Australia's agriculture depends on a thin veneer of fertile soil that is being gradually stripped away. It is estimated that the majority of cropping regions have less than 25 cm of topsoil in which the organic matter is confined. Rain erodes the land even on gradients so gentle that they are almost imperceptible to the eye. The very slow rate of natural soil formation results in soil being regarded as a non-renewable resource, at least in economic terms. Most soils form at the rate of a few centimetres per one thousand years, depending on moisture and temperature levels. Many believe that the current rates of soil degradation or erosion are natural and have occurred since the beginning of time, but this is not strictly so. In localities where the landscape is still in its virgin state, soil erosion occurs very slowly indeed. The small amount of soil that is lost from year to year by the action of wind and rain is usually replaced by the gradual weathering of the underlying rock. Water

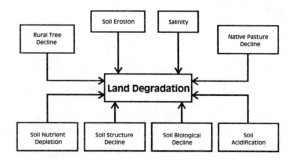

Figure 11.2 *Forms of land degradation.* (SOURCE: Williams, J. 1991. Search for sustainability: agriculture and its place in the natural ecosystem, *Agricultural Science* 4(2), 32–39)

and wind erosion on the scale that is seen today only occur as a result of the actions of humans.

Water erosion is recognized by signs that flowing water has removed or deposited soil. Signs of water erosion are likely to be seen on ploughed slopes or in creeks and waterways. However, water erosion can occur wherever removal of the protective plant cover has left bare soil exposed to the force of falling rain and fast-flowing runoff (Figure 11.3).

Figure 11.3   *Severe erosion of solodic soils.*   (PHOTO: Queensland Department of Primary Industries)

Figure 11.4   *Wind erosion in the western division of NSW.*   (PHOTO: Soil Conservation Service of NSW)

Wind erosion occurs where there is loose sandy soil. It can be seen as the removal of soil from around the bases of plants and the deposition of sand against obstructions such as rocks, bushes and banks. Another sign of wind erosion is the smooth 'swept' appearance of the eroded soil surface and the small parallel depositions of soil against clods and stubble. When severe wind erosion occurs, crops have a sandblasted appearance and may be torn or broken by the impact of grit on the leaves (Figure 11.4).

Salinity usually occurs in areas where sandstone or shale produces soil with a high salt content and where the evaporation rate is high. Runoff carries the salt to the lower lying land and it becomes concentrated by evaporation of the ponded water. The occurrence of salinity is recognized by the death of plants, the accumulation of a white crust on the soil surface and the growth of salt-tolerant plants such as saltbush and bluebush.

Land slip is the movement of large blocks of earth down a slope. Soil which has slipped has numerous scars running more-or-less parallel to

Figure 11.5 *Land slip near Toowoomba, Queensland.* (PHOTO: Queensland Department of Primary Industries)

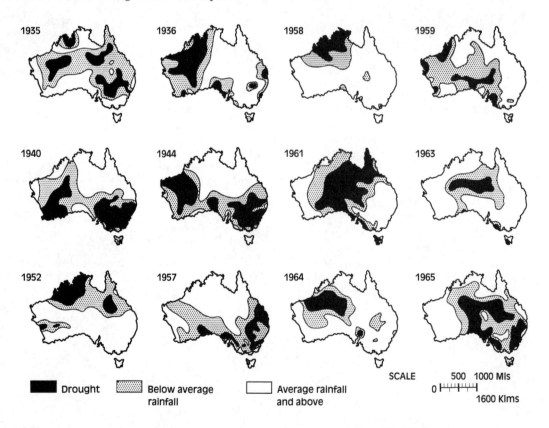

Figure 11.6 *Major droughts rarely affect the whole of Australia at the one time as this series of maps of major drought-affected areas in Australia 1935–65 shows.*

the contour. If the slippage has occurred recently, steps of bare earth will be seen on the upside of the earth which has moved. If the slippage has occurred some time ago, it may be apparent from the angle of the trees and fenceposts. Land slip occurs most commonly in areas where clay subsoil or other impervious subsurface material cause the soil profile to slip when rain saturates the soil (Figure 11.5).

The Australian position

When Australia was being opened up by the early settlers land was plentiful but labour was scarce. The agricultural practices sought maximum yields regardless of the long-term effect on the land. Similar patterns of development were seen in other newly colonized countries. Because action to conserve and stabilize the soil was not taken at the time of settlement, progress towards stabilization has been slower than if no damage had ever occurred. Some soil has been irreversibly affected.

Recurring drought has been an obvious factor shaping Australian agriculture (Figure 11.6) and will certainly continue to be so, but not all drought effects need to be considered as negative. The perception of drought as a national disaster is changing; increasingly it is seen as a recurring natural condition.

Government policies now regard drought as a recurring natural condition, a normal commercial risk that should be included in the management system. Policies encourage self-help and 'drought-proofing' strategies during better seasonal periods.

During economically buoyant periods, settlement advanced into marginal lands. However, when downturns came in the economy, the capital loans that had been invested during the boom still had to be repaid, so the land was exploited even more intensively, e.g. by the ploughing of steeper slopes and drained swamps. In the worst cases, this resulted in complete exhaustion of the soil and failure of the industry. The failure of the wheat industry in the northern lands of South Australia in the 1870s–80s is an example. Not only did the industry fail but there was enormous loss of vegetation and extensive soil erosion. Since then, this region has only supported a pastoral sheep industry. The situation occurred because there was unrealistic optimism as to the reliability of the rainfall in the area north of Goyder's Line (Figure 11.7).

The pattern of declining soil productivity was not the same in all cases, but varied with soil type, climate and topography. On the wheat lands of

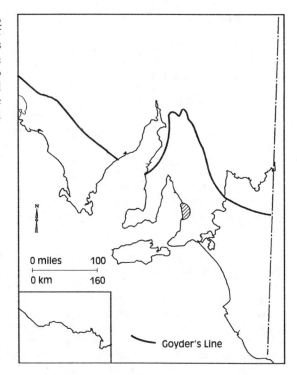

Figure 11.7 *Goyder's Line was defined in 1865 as the northern limit of safe cultivation in South Australia.*

the western slopes of NSW the declining fertility was mainly due to severe water erosion including sheet erosion and gullying. On the wheat lands of South Australia and Victoria degradation was mainly due to wind erosion following short rotations and bare fallowing.

In the 1920s the problem of wind erosion received a great deal of publicity, partly because it was so widespread and partly because its appearance was so spectacular in some areas. Water erosion was actually of greater economic importance. The erosion of the wheat lands of NSW represented a greater economic loss than the combined effects of all other types of erosion in Australia during the first quarter of this century. Over the wheatbelt of north-western and central NSW and in Queensland, summer thunderstorms occur frequently and do great damage. In Victoria, South Australia, and Western Australia the rainfall season extends over winter and the intensity of individual falls is lower. In Queensland, serious water erosion occurs in the intensively cultivated black soils of the Darling Downs and central Queensland. An estimated 45 per cent of that

State's natural grazing lands is also classed as degraded.

In Victoria and South Australia, severe sand drift occurred in the Mallee region as a result of clearing of the vegetative cover for wheat cultivation early this century. The drift in the Mallee has been accentuated by clearing and cultivation of vast tracts of land with few windbreaks, which was encouraged by the relatively high wheat prices and a succession of unusually good seasons after the Second World War.

Strip cropping on the North American pattern had not been found to be adaptable to Victorian conditions due mainly to insufficient rain for summer crops, so windbreaks were planted and various methods of introducing organic matter into the soil, such as the sowing and grazing of oats, were used as the early conservation techniques. By the early 1950s much of the technology which was seen as being urgently needed to achieve better land utilization was being applied to a limited extent. The most significant aspect of this technology was the ability to gain increased production and improved soil fertility in much of the higher rainfall areas of southern Australia by the establishment of improved legume-based pastures with annual top dressings of superphosphate. The economic incentive for pasture improvement was the keen demand for agricultural commodities, particularly for livestock products, which existed and continued for many years after the Second World War. The same incentive stimulated research into new pasture plants, trace elements and ley farming practices. Application of this research has extended opportunities for improved land use by pasture establishment over wider and wider ranges of soils and climate up to the present time.

However, the rate of introduction of conservation practices has never kept pace with the rate at which new land has been brought under the plough. This is still the situation today. In Queensland, since the 1960s more than 60 000 ha have been added annually to the State's new crop land, but conservation measures were planned and implemented on only 50 000 ha each year. The shortfall was added to the backlog of crop land awaiting the attention of the conservation planners. Clearly, the problem is not being solved and is becoming worse.

It is estimated that of the 5 million square kilometres of land used for agricultural and pastoral purposes in Australia about 2.7 million square kilometres, or 51 per cent, require treatment for land degradation. Approximately 22 per cent of the total area needs only the implementation of appropriate management practices, while 29 per cent requires capital works as well.

Erosion processes

Natural erosion is an integral part of the process that has formed the landscape as we know it. The movement of soil from the mountains to the alluvial plains is a slow continual process. When it occurs naturally, this movement has no permanent detrimental effects on the productivity of agricultural land. In contrast, accelerated man-made erosion has a massive ongoing effect on both stability and productivity.

The processes of erosion can best be understood in terms of the factors causing water and wind erosion (Figure 11.8).

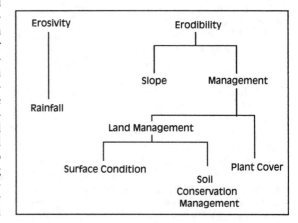

Figure 11.8   *Factors influencing erosion.*

Water erosion

The characteristics of the rainfall and the land are directly responsible for the amount of erosion that occurs in a particular region. The extent to which rainfall is responsible for erosion is called the erosivity of the rainfall. The susceptibility of the soil to erosion is called the erodibility of the soil (Figure 11.8). The erosivity of a storm increases with the intensity of the storm, the size of the drops, and the speed at which the drops hit the soil surface. The erodibility of the soil depends on the soil type, slope, plant cover, moisture content and compaction.

The process of water erosion takes place in three steps:

1 Clods and aggregates are broken down by raindrops striking the surface of the soil and tearing soil particles away from the general soil mass.
2 Detached soil material is transported.
3 The soil mass is deposited. Falling raindrops and surface flow produce widely different effects on the soil. Raindrops act uniformly over the whole surface on which they fall so that they cause smoothing or levelling. Flowing water tends to collect in small rills which, as they converge into streams, cut gullies resulting in the roughening of the surface. Most of the initial erosion damage resulting from rain is caused by soil splash. Only a small amount of soil loss is caused by surface washing alone.

The cross-sectional area of a falling raindrop determines the amount of soil that receives the impact, while the kinetic energy of the same drop determines the force of the impact that must be absorbed at each point on contact. Most of the dispersion of soil particles is accounted for by the energy of the raindrops. Accumulating raindrops falling on the soil surface cause the soil to lose its capacity to absorb water as it becomes fully saturated. This accounts for high runoff during heavy rains. Variation in the size of raindrops, drop velocity, and rainfall intensity alter the splash erosion process. Fine particles of earth in a muddy splash are carried into the air by water rebounding after impact. They may move more than 5 metres horizontally on level surfaces and to heights of up to 1.5 metres; on bare soil, which is highly detachable, more than 50 t per ha may be splashed up by the most intensive rainstorms. When raindrops strike sloping land surfaces, the major portion of the splash moves downhill so that relatively large quantities of soil may be transported by raindrop splash acting alone.

The three types of erosional damage produced by the splash process are puddle erosion, fertility erosion and sheet erosion.

Puddle erosion is the type of erosion which occurs during violent storms when raindrops strike bare earth and shatter clods and soil crumbs, breaking down the soil structure into a puddled condition. The soil's finely broken particles are compacted into an impervious layer of surface mud. The porosity of the surface layer is much reduced by the infiltration of muddy surface materials to the extent that entrance channels to the lower profile are closed.

Fertility erosion occurs when, during the splash process, the finer soil particles are transported away after the coarser material has settled. This type of erosion receives its name from the fact that much of the organic matter and the fertility-bearing elements of the soil may be removed. This may occur even though the land is almost level, though on hills the effect is more pronounced. At the crest of a hill there would be the greatest amount of soil movement with the least amount of energy needed to carry soil. While this erosion can move quantities of soil downhill, it does not carry it away from the base of the slope to any appreciable extent. In the surface flow, some soil is kept in suspension, flowing into rills and then to larger channels and gullies.

Sheet erosion occurs when surface flow moves over land either as a shallow sheet of water across broad surfaces that have no perceptible channels or in rills, gullies, or valley channels. Due to the concentration and velocity of surface flow as it moves downhill it does not affect the entire surface of a paddock equally but acts as a carving agent, roughening the surface by scouring of concentrated flows forming rills. Clear water may not have sufficient detaching capacity to cause much erosion if flowing over well-compacted clay soil. However, if the flow contains highly abrasive soil particles the detaching capacity will be increased and the rate of erosion will be accelerated. Maximum erosion occurs when the surface flow contains just enough abrasive material to detach as much soil as the flow is able to carry.

Increases in velocity produce varying degrees of erosion depending on soil type, and on soil horizon within the same soil. Much of the suspension capacity of water flowing in rills is from the energy of turbulence imparted to the runoff water by raindrops splashing on its surface. These can often increase the transporting capacity by several hundred per cent. The kinetic energy imparted to surface water through raindrops may also give considerable suspension capacity to standing water. Standing water that is churned by splashing raindrops has been found to comprise as much as 20 per cent soil. Raindrop impact can at times move stones as large as 10 mm in diameter when they are partially or wholly submerged in water. The impact can cause them to rise and move some distance downhill.

Rills are formed as a result of flowing water carving grooves into the surface soil (Figure 11.9). They become deeper and broader and eventually develop into gullies. Surface flow tends to produce the largest gullies and the greatest soil losses near the lower ends of long slopes. This is called scour erosion.

Figure 11.9 *Severe rill and gully erosion in cultivation which was unprotected by contour banks, crop or stubble.* (PHOTO: Queensland Department of Primary Industries)

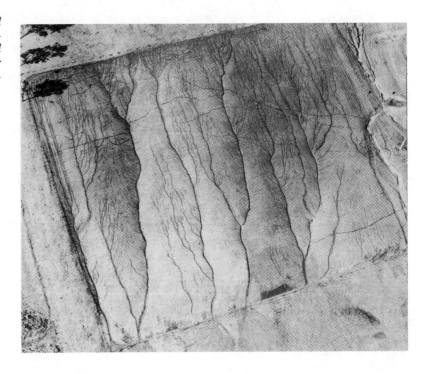

Gully erosion is the most spectacular form of water erosion, but it is only the end-product of the total erosion process. Gullies must be regarded as the products of the sequence of four stages of erosion. First the rill erosion is initiated by relatively small volumes of concentrated runoff into the initial minor channels which precede gullying. As these channels merge lower down the slope, the volume of water may be sufficient to cut a deepening channel which starts to eat back up the slope. This backward movement is caused by waterfall erosion where the runoff enters the gully. At this stage, large blocks of soil slump into the gully and cause a massive increase in soil loss. As the process continues the gully becomes wider and often deeper, and waterfall erosion expands not only at the gully head but wherever large rills deposit water over its vertical walls. The rate and extent of expansion of the gullied area is determined by the volume and speed of runoff entering the gully (Figures 11.3 and 11.10).

Heavier materials, such as coarse sand and gravel which are least transportable, are the first to be deposited by moving water. Fine and light materials are often floated kilometres downstream and deposited in large reservoirs while the less easily transported coarse fractions of eroded soil are deposited upstream. The coarser materials resulting from rill and gully erosion end up in farm dams and small upstream reservoirs located near the bases of sloping paddocks, while the finer and lighter organic matter and clay fractions that are highly transportable are found farther downstream. Debris resulting from erosion may be deposited anywhere along the course of the flow from the point of origin to the sea. Such deposition can increase the fertility of flood plains, but this is not always the case as it can also reduce productivity by placing a layer of sand or gravel on top of rich clay loam or alluvial silt.

Wind erosion

In arid, semi-arid and sub-humid regions, wind erosion is extensive because the land surface is often dry and vegetation is sparse or absent (Figure 11.4). Soil material must be loosened from the soil surface before it can be transported by wind. As only dry soils can be blown, the properties of soil and the effects of land use on the soil play an important part in determining the rate of wind erosion.

Three types of soil movement may occur in wind erosion: saltation, surface creep and suspension.

Saltation is the movement of soil particles resulting from the direct pressure of the wind on the particles. Particles which have been catapulted

Figure 11.10 *A fallow paddock stripped of top-soil to the depth of cultivation. This type of damage exceeded 35 per cent of the surface area of some paddocks.* (PHOTO: Queensland Department of Primary Industries)

into the air by saltation may be maintained in aerial suspension by wind turbulence. Only very fine particles (less than 0.1 mm in diameter) may be affected in this way.

Surface creep occurs when particles of 0.5– 1 mm in diameter which are too heavy to be moved by saltation are pushed along the surface by the impact of particles in saltation. Soil movement is initiated by gusts of wind of over 20 kph which carry the particles in *suspension*. The threshold velocity is determined mainly by soil grain size. Grains of 0.1– 0.15 mm in diameter would require a velocity of 10 kph at 15 cm above the ground. Soil material moved by wind and deposited in dunes has an average diameter range of between 0.16–0.35 mm. Particles in this size range constitute a major part of the total movement. The fine particles of sandy soils moved in suspension usually make up a minor part of the total volume of soil moved in comparison with the amount moved along the surface of the ground.

Salinization

Soil salinization is the accumulation of salts within the root zone of soils to the extent that plant growth is affected. Generally, salinization occurs when high levels of sodium salts and high clay content occur together in the soils of low-lying situ-

ations. The salt usually originates from ground water and surface water coming from geological and soil zones high in salt. In other situations, such as those in Western Australia, salt-laden sea breezes contribute constantly to soil salt levels. These naturally occurring processes produce what is referred to as primary salinity.

Where salinity results from irrigation, clearing or overgrazing, allowing an increase in concentration of salt at the soil surface, it is referred to as secondary salinity. For secondary salinity to occur there must be a source of soluble salts and a mechanism for the transport and further concentration of salt in the soil profile. This concentration usually takes place as a result of evaporation of surface water. Some heavy-textured soils allow water but not salt to leach down through the soil profile. Salt therefore accumulates at or near the surface.

Salinization can occur in irrigated or dryland situations. Saline seepage, often at the foot of hills, can occur in a wide variety of soils, geology, climate and vegetation (Figure 11.11).

When trees are removed from soils which are inherently high in salt, the water table tends to rise and salt accumulates at the surface by a process known as salt effluxion. Through a series of successive wetting and drying cycles, effluxion may lead to levels of salt accumulation high enough to kill plants.

1.

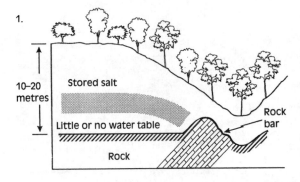

2.

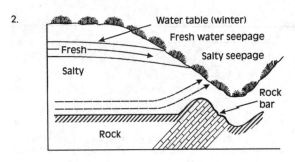

3.

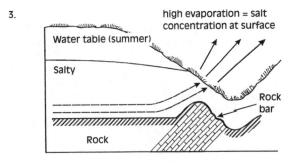

Figure 11.11 *The effect of tree removal on salinity processes in summer and winter after removal of trees.*

Salinity is also associated with the problems of alkalinity and sodicity. When clay with a high sodium content is transported downwards through the soil profile it forms an almost impermeable layer a few centimetres below the surface. Such soils are generally alkaline due to the salt content and are referred to as sodic when the highly dispersed clay concentrates at depth.

Land slippage

The mass movement of soil downhill may start as a shallow form of soil creep in situations where steep slopes have been cleared of trees. Landslip is the term applied to the mass movement of the soil profile (from the surface to bedrock or other impervious layer) as a result of slippage at a cleavage zone lubricated by seepage water. This occurs on steep slopes with deep soils which contain one or more layers of dense clay above which seepage water accumulates, causing a shearing and slipping of the upper layers downslope.

Slippage is more likely to occur during seasons of exceptionally high rainfall. Some regions of very steep topography are naturally prone to land slip, especially where earth tremors occur. This tendency is increased when trees are removed from such slopes, especially in areas where impervious clay layers occur in high rainfall regions. The effects of overgrazing and burning on landslip are complex and not easily predicted, since in some cases a dense growth of long grass may increase the incidence of landslip as a result of the retention of large volumes of water by the dense mat of vegetation.

Other types of degradation

Soil structure decline

Many soils have an inherently weak structure, i.e. the soil particles are loosely aggregated and have weak cohesion. Structure breaks down from an aggregated form to a homogeneous single grain structure as a result of a loss of organic matter (and thus of biological activity) and a mechanical disintegration of the aggregates or crumbs.

A range of physical properties are involved in the breakdown process, but in virtually all cases the symptoms show a clear compaction of the soil, often with a sealed surface layer. In soils of fine texture (e.g. clays and clay loams), continuous exposure of the bared, cultivated soil exposes the surface to the kinetic energy of raindrops which cause splash erosion. This process, in the absence of organic matter, leads to a sorting and a compaction of surface particles which may cause a hard surface layer that prevents infiltration of water and increases runoff and soil loss. The breakdown of surface structure also interferes with soil aeration, since gas exchange is reduced and roots may suffer a lack of oxygen. In addition, the aerobic soil microbial activity will be reduced and essential fertility-enhancing processes such as nitrification will slow down. It has been shown that soils high in fine sand and silt will lose their structure more rapidly under cultivation if organic matter is not retained (Chapter 7).

The exception to compaction in heavy soils is the swelling black clay soils, such as those of the Darling Downs in southern Queensland. These soils are called 'self-mulching' because of the way in which they crack very deeply when dry, leading to a shattering of the profile and crumbling of the surface layer. Part of the surface soil falls into the cracks; when wet, these soils swell and rearrange the aggregates, causing deep aeration and increased biological activity. These soils also have an unusually high moisture-storage capacity as a result of their high colloidal content and the depth of the soil profile.

Reduced biological activity in soils

The level of biological activity in the soil has long been recognized as an important criterion of the health of the soil and is given special significance by organic farmers. When organic matter and its finer derivative (humus) are reduced by excessive cultivation (and thus, oxidation), soil structure collapses in many soils. When this occurs aeration is reduced and microbial activity declines. Similarly, when surface crusting due to a lack of organic matter causes decreased water infiltration, biological activity is slowed down significantly. Such problems will not be overcome by simple addition of chemical fertilizers, since soil moisture and aeration are the limiting factors to crop production in such soil conditions.

Increased soil acidity

When soil pH is below about 5.5 many types of crops will not thrive. The addition of agricultural lime has been the traditional treatment used to increase soil pH. Soil colloids, including humus, are responsible for buffering soils against a decrease in pH. When colloidal matter is removed through over-cultivation, wind erosion or sheet water erosion soil acidity often becomes a problem. This process is exaggerated in coarse sandy soils derived from granite or sandstone.

In recent years the excessive build-up of surface organic matter in improved pastures in the winter rainfall zone has been identified as one cause of widespread acidity problems. Continued use of acidic fertilizers, such as superphosphate, have also been implicated. Acidification in the wheat/sheep production zone has now been recognized as a major problem of dryland agriculture in southern Australia. The application of lime not only raises soil pH but acts as a plant nutrient, particularly for legumes, and has the effect of a soil ameliorant, similar to gypsum. Both lime and gypsum contain calcium which helps soil aggregates to form.

Controlling land degradation

In the farming situation the soil resource can be maintained and enhanced by controlling the degradation processes described in this chapter. Assuming that the farm plan identifies the capability levels correctly, the production techniques to be used on each portion can then be considered. Soil conservation measures, which include the combating not only of soil erosion but also of the other changes referred to, can be considered in two main groups:

1 Mechanical or engineering techniques, including a range of earthworks such as contour banks.
2 Biological or plant-based methods, in which all types of vegetative cover and organic matter are included.

The way in which these two groups of control measures are combined to give an integrated basis for conservation farming depends on the relative importance of problems such as:

• High runoff.
• Soil loss from wind erosion.
• Increase in salinity.
• Soil structure decline, acidity and crusting.

In the case of high runoff, arrangements may have to be made to co-ordinate and control incoming waterflow from neighbouring properties higher up the catchment before a runoff control plan can be finalized for a lower farm.

Control of water erosion

An important early step in any farm plan is the identification of waterflow patterns and the planning of a safe runoff disposal system. As far as possible, runoff water should be kept within natural waterways, but in crop land artificial waterways often need to be constructed to carry runoff from contour banks and stormwater drains upslope. The need for earthworks is largely determined by the extent to which infiltration can be increased through soil surface management, especially stubble retention, and the density of plant cover which can be kept on areas of concentrated flow. Where earthworks are required their maintenance should be an accepted activity of seasonal work, especially the removal of sediment from the channels of banks and constructed waterways.

In principle, water erosion is reduced by reducing both the amount of surface flow and the rate of flow. This is achieved by any techniques which increase infiltration and reduce the unimpeded length of the flow slope. Thus, contour banks break a long sloping paddock into short runoff bays from which water is slowly and safely led across the slope into a well-grassed waterway. Standing crops, dense stubble, deep ripping and cultivation on the contour all reduce runoff and reduce the velocity of flow. The denser the vegetative surface cover, the more rapid the flow which can be safely disposed of.

On crop land with a very small slope (below about 2 per cent) water erosion can often be satisfactorily controlled without banks, using only strip cropping (alternative 25–35 m strips) of summer and winter crops where rainfall is adequate, and minimum tillage and stubble mulching techniques which maximize residual stubble cover.

There are important advantages in the stubble mulching, reduced-till and no-till systems. Apart from the medium and long-term benefits of preserving and stabilizing the topsoil, a number of immediate advantages accrue to the farmer:

- Crop yield is more reliable as a result of improved soil moisture.
- Plant nutrients are retained as a result of less surface wash and runoff.
- Production costs are considerably reduced as a result of lower fuel and machinery costs from fewer operations.
- Contour banks are more easily maintained as a result of less soil being deposited in bank channels.

Control of wind erosion

Wind erosion removes loose surface particles and causes a selective removal of the finer and more fertile particles of soil colloids and humus. This has a particularly detrimental effect on productivity which, when combined with physical damage to young crops caused by sandblasting, results in very low yields.

All wind erosion control techniques in crop land are based on a combination of keeping the soil surface rough, and maintaining a cover of either stubble or a growing crop as long as possible during the windy season. In many regions prone to wind erosion, the establishment of shelterbelts and windbreaks at right angles to the prevailing wind is an effective control.

Control of salinity

In earlier years the relationship between salinity and tree clearing was not understood and many of the first attempts to control salinity treated the symptoms rather than the causes. In more recent times scientists have come to understand the relationship between water use by deep-rooted plants and variations in the height of the groundwater table. Research in Western Australia and Victoria since 1975 has demonstrated that over a wide range of geological conditions rising water tables and the accompanying salinity resulting from surface evaporation can be lowered to acceptable deeper levels. This can be done by growing deep-rooted perennial crops on low-lying crop land, and planting adapted trees in sufficient numbers in the intake bed zones higher up the slopes.

In other cases the construction of crossdrains down to the clay layer over which saline seepage water may move is claimed to collect and drain salt water across the slope from where it can be disposed of in waterways, as in the case of 'Wisalts banks' in Western Australia. In all cases, an understanding of the geological structure of the landscape and the origin of the salt, forms the basis for control techniques. Often the source of salinity is either beyond the farm boundaries or outside the control of the land manager.

Many of the problems of land degradation can be traced to over-zealous clearing of trees.

Clearing trees for new crop land

The Australian wheat industry was built on clearing the land to enable ploughing of the virgin soil. In some cases, reasonably stable farming systems have evolved and flourished on cleared land, provided excessive wind erosion, water erosion or salinity have not reduced yields. However, it is now clear that, had the land-use planners known the limitations of the land, they would not have recommended such large-scale clearing. The problems of unwise clearing arise mainly in the marginal regions such as the eastern wheatbelt of Western Australia, the Eyre Peninsula of South Australia, the Mallee of Victoria, the Western Downs of Queensland and the western division of

NSW. Because the hazards of drought and wind erosion vary greatly with seasons, there are serious difficulties in identifying a particular safe geographical boundary to the cropping zone. The western division of NSW has a long history of enquiries, commissions and policy changes relating to clearing for crop production. This region provides a good example of conditions laid down for so-called clearing licenses and cultivation permits. Such conditions, aimed at minimizing soil loss and soil structure breakdown, are:

- Retention of uncleared and uncultivated areas of native vegetation (usually 20 per cent of the area developed), as perimeter and internal strips and clumps.
- Limit on size of individual cleared/cultivated areas.
- No clearing or cultivation of erodible soils or hazardous sites such as sand hills and steeper slopes.
- No clearing or cultivation adjacent to fences and roads.
- For all soils except clays and those being irrigated, the frequency of cropping is restricted to no more than three crops in nine years, or in more marginal areas, one crop in six years.
- Pastures to be sown at regular intervals alternating with annual crops.
- Stubble not to be burnt except to control shrub or timber regrowth.
- The length of bare fallow restricted in areas where a combination of soil type and unreliable climate predisposes the soil to erosion.
- Cultivation of sloping lands to be carried out on the contour, and in some cases only after soil conservation earthworks have been constructed.
- Should soil erosion occur on any cleared or cultivated areas, remedial measures to be implemented at the lessee's expense.

In practice, the implementation of this scheme has several difficulties arising from the problems of technical decision-making on what is suitable crop land and the political implications of limiting individual freedom.

Reasons for clearing

Several basic questions about tree clearing should be considered: Are the reasons for clearing defensible?; Under what conditions is clearing 'safe'?; How do various clearing methods compare ecologically?; What guidelines should be recommended when clearing is approved?; Is departmental control of clearing necessary?

In virtually all cases, clearing is undertaken to remove trees to create new land for cropping or to remove trees to increase animal carrying capacity. Both these reasons are defensible under suitable conditions where sustainable cropping or sound grazing management ensure a stable landscape with minimum soil loss. Virtually all of the most productive grain farming and intensive animal production regions were built on land originally cleared of *Eucalyptus* and *Acacia*. The argument that clearing necessarily results in land degradation and a loss of productivity cannot be sustained.

Conditions for safe clearing

Sustainable production is dependent on that set of conditions which (a) prevents soil loss; (b) maintains organic matter; (c) retains chemical fertility; (d) protects soil structure; and (e) prevents build-up of toxic chemicals. Any combination of climatic, soil and management conditions which meets these requirements may be regarded as constituting suitable conditions for clearing with safety. In practice, this means limiting clearing operations to those areas where (a) annual rainfall is high enough to produce a sufficiently dense plant cover and yield (grain crops or pastures) and (b) soils are deep enough and fertile enough to produce a good vegetative cover and yield under good management. Add to these basic requirements the need to retain trees on (c) steeper slopes and (d) on erosion-prone and salinity-prone situations, and this provides a useful set of ecologically sound guidelines. Where defined waterways exist, belts of trees should also be retained on both sides of the flowlines to stabilize these areas against excessive waterflow and overgrazing.

Clearing guidelines

The role of trees in a particular landscape must be understood before the consequences of clearing can be predicted. Guidelines must be based on an understanding of the extent to which tree retention is necessary for ecological stability. Thus, where trees and shrubs prevent wind erosion or salinity build-up under natural conditions, the position and extent of retained treebelts may be of special ecological significance in land stability (Chapter 5).

In most cases, a combination of soil erosion control, shelter provision, wildlife protection and aesthetics form the basis for clearing guidelines. This involves decisions on both the proportion and configuration of retained timberbelts. In some

States, a general guideline of 20 per cent of natural vegetation retained in a pattern which interlinks all the habitats is used.

Role of government agencies

The extent to which literate adults should be given the freedom to use technology correctly, has become a social issue: pills, guns, pesticides, machine tools and herbicides all create the same problems for policy makers. Attempts to impose control must balance the requirements of individual freedom and public responsibility.

Where chemicals such as pesticides or herbicides are aerially sprayed, accurate application (without drift) at the recommended rate must be guaranteed if unplanned harmful effects are to be avoided. It must be assumed by the user that the persistence and movement of the correctly applied chemical in the ecosystem has no lasting harmful effects.

Land capability as the basis for farm planning

For several decades, rural land-use planners have stated two fundamental aims in farm planning:
- To use each portion of land to its full potential.
- To protect each portion according to its limitations (Figure 11.2).

The terms 'capability classes' and 'suitability classes' are used to indicate the intensity of use and the level of protection which are appropriate to each section of a planned region, catchment or farm. Generally, production potential and thus land capability is limited by slope and soil depth, but stoniness, wetness or factors such as salinity or erosion proneness may make portions of land less suitable for annual cropping.

The first step in farm planning is thus the mapping of the boundaries of suitable arable (crop) land and the identification of areas in need of special protection (slopes, waterways, saline areas, wildlife habitats). Much of the damage caused by misuse of farmland is the result of ignoring the natural limitations of the land (Figure 11.2). Prime decisions on where annual cropping will be practised must be made on a factual basis otherwise there is often no solution to certain erosion problems further along the decision-making process. As an example, if steep slopes are mistakenly deemed suitable for annual summer cropping in the subtropical zone, no contour bank design or rotation

of annual crops will succeed in stabilizing such slopes under intensive monsoonal storms. Similarly, in the winter rainfall regions of Western Australia and Victoria, if some upper catchment slopes are cleared of trees no choice of cropping systems on the footslopes will counteract the accumulating salinity problem.

Regaining a balance between production and conservation

Recent studies on the decline in yield, breakdown of soil structure and reduction of organic matter, give overwhelming evidence that many Australian cropping systems have been dependent on a continual withdrawal of 'soil capital' or productive potential in the form of mineral nutrients (Chapter 10). The answer to finding the balance between economic levels of farm production and satisfactory levels of conservation of Australia's rural resources lies in the integration of conservation principles into cropping and grazing systems. In the past, conservation was not accepted by many landholders as their responsibility. This has changed considerably since the birth of the Land Care movement in which landholders have taken the initiative and established over 400 Land Care Committees to control land degradation.

However, there are still pressures on politicians to open up more marginal land for cropping, to permit the draining of large swamps, to irrigate soils obviously susceptible to salinity and to clear steep slopes for intensive cropping. Australia must decide on the characteristics of sustainable cropping systems and must plan agricultural development accordingly.

The criteria of sustainability

Since the term 'sustainable' has been used to describe permanent or lasting production systems, many attempts have been made to identify the criteria for sustainability and to estimate satisfactory levels of these factors in quantitative terms. To be both biologically and economically sustainable, farms have to meet five basic requirements of their soil:

1 Maintain or increase soil fertility in terms of plant nutrients.
2 Stabilize the topsoil against erosion and soil losses and so protect the nutrient reservoir,

Figure 11.12 *An outstanding example of a contour bank layout, grassed waterways and stock dams.* (PHOTO: Victorian Department of Conservation, Forests and Land)

waterholding potential and root-zone depth of the crop land.

3 Enhance the organic matter content and microbial activity in the soil as the basis of soil structure, nutrient supply, plant health, water infiltration and erosion resistance.

4 Minimize the build-up of synthetic toxins and harmful natural salts in the soil profile as a means of ensuring rapid plant growth and high quality food crops.

5 Maintain soil pH within an appropriate range.

Other subsets of these basic requirements can be added to this recipe for sustainability, and the acceptable levels of each factor will vary according to local conditions and the cost/price ratio. There is little benefit in attempting to achieve great precision in defining the required levels of these factors. Provided the changes and progress in improving cropping techniques tend to shift the level of these factors in the desired direction, the innovations applied to achieve these should be encouraged within the economic constraints of the time.

Conclusions

It should be emphasized that much is still unknown about Australia's soils and that research is needed to answer many important questions on cropping techniques. At the same time, it should be remembered that the sustainability of cropping systems can be vastly improved simply by applying proven techniques which have been continuously demonstrated to work well for the past few decades. Strip cropping, contour planting, legume rotations, no-till cropping and deep ripping are good examples of such well-tested methods.

Further reading

Russell, J. S. and Isbell, R. F. (eds) (1986). *Australian Soils — The Human Impact*, University of Queensland Press, St Lucia.

Chisholm, A. H. and Dumsday, R. G. (eds) (1987). *Land Degradation—Problems and Policies*. Cambridge University Press, Melbourne.

Roberts, B. R. (1990). *Birth of Land Care — Collected Papers*, University College of Southern Queensland Press, Toowoomba.

Routley, R. V. (1975). *The Fight for the Forests*, Falcon Press, Canberra.

CHAPTER 12

Goals and Priorities for Sustainable Dryland Farming

I. R. Fillery and P. J. Gregory

SYNOPSIS

In this chapter goals and priorities for sustainability of dryland agriculture in Australia are discussed. A research approach is outlined.

MAJOR CONCEPTS

1 Sustainable agriculture is a set of goals or objectives for agricultural systems. The relationship between sustainable agriculture (an end) and farming systems approach as a clearly defined methodology (a means to an end) needs to be appreciated.

2 The inability of agricultural science to deal effectively with issues of sustainability stems from a lack of understanding of the processes actually constraining agricultural systems in the field, including their severity and scale.

3 Sustainable agriculture implies a system that is in balance. The sustainability of agricultural systems should be evaluated in terms of their ability to balance the inputs and outputs of water and nutrients in the long term.

4 Water and nutrients determine the potential for dry matter production in the dryland areas of Australia and are therefore a logical focus for Australian research on sustainability.

Sustainable agriculture is a set of goals or objectives for agricultural systems. It is about managing the land with a healthy ecological balance, a sensitivity to the land's capabilities, using technologies and practices which have minimal impact while maintaining production. Concern about the sustainability of agricultural production has been expressed throughout most of this century, and growing interest in this concept has developed over recent years. Yet, the dialogue has not contributed to a single definition of the term 'sustainable'.

The real issue of sustainable agriculture is not that we should be aiming to develop farming systems that can be repeated over generations without change, but that we must work towards a sustainable future where agriculture is considered within the broader context of sustaining the quality and quantity of life on earth. It is society and the people within it that we want to sustain. The agricultural production system is important to that goal but it should not be considered as an end in itself or independent of other aspects that come together to define quality of living in its broadest sense. Agriculturalists must give primary attention to the total food system — production, processing and distribution.

The current polemic on sustainable agriculture manifests a concern with the thought that modern agriculture, as practised in much of the world today, is non-sustainable. Conventional technologies and strategies have led to an agriculture that uses non-renewable resources at rates which cannot be sustained, and/or which creates a gradual contamination of the environment. Through the use of petroleum-based products for mechanization and chemically enhanced production practices, non-sustainable agriculture as we know it today is, in part, the result of standardized practices over large areas.

The opposite extreme — a perfectly sustainable agriculture — no longer dominates. It would probably be useful to speak of a more sustainable agriculture rather than use the absolute term sustainable. A more sustainable agriculture would rely less on standardized, often chemically enhanced production practices and instead depend on renewable resources and use practices more in tune with local conditions. This implies more diversity in crops produced, changes in rotation practices, the development (or redevelopment) of germplasm well adapted to local environment niches (as opposed to germplasm with broad adaptability), and the necessary accompanying changes in infrastructure. A more sustainable agriculture would be in tune with the local resource base, make maximum use of internal production inputs, and have potential for sustained production and profits further into the future.

Sustainable agriculture and a sustainable future will require some changes to the way we view the world and the manner in which we use resources. Farmers, produce marketeers and consumers would need, however, to adopt energy-conserving, cost-effective food production and delivery systems. Financial and technical assistance should be made available to those wishing to implement more sustainable farming systems. Misunderstandings about the scale and severity of soil conditions thought to undermine the future sustainability of dryland agriculture, ignorance of the processes which determine the chemical, physical and biological integrity of soils, and apprehension about the role of modern technology have precipitated many calls from the public for a return to more 'natural' agricultural systems including organic and low-input farming systems

in the belief that these will solve the problem of sustainability.

Low inputs are not essential to sustainable agriculture, but, under some circumstances, may be an approach towards meeting the goal of sustainable agriculture. If farming systems are to be sustainable in that they minimize the use of external inputs and maximize internal inputs already existing on the farm, a way must be found to reverse the trend towards greater external inputs. Future research, then, will need to balance the needs of greater productivity with those of greater sustainability. Scientists and technologists are charged with improving agriculture by designing new farming systems, selecting the most effective from an array of existing and potential alternatives, and anticipating problems that may arise as agriculture changes.

Sustainable agriculture — a system in balance

Sustainable agriculture implies a system that is in balance. Plants require two soil resources to grow and to reproduce: nutrients and water. Water typically determines the potential for dry matter production in the dryland areas of Australia and is a logical focus for Australian research on sustainability. Therefore, it is logical that the sustainability of agricultural systems should be evaluated in terms of their ability to balance the inputs and outputs of these commodities in the long term; it is illogical to evaluate farming systems only in terms of perceptions of how natural or unnatural a particular input or management strategy is. Moreover, the evaluation of sustainability in terms of the supply of either nutrients or water (or both) provides a useful framework.

Of the major nutrients, nitrogen most frequently limits the growth and productivity of Australian crops. A recent review on the role of legumes and nitrogen fixation in sustainable agriculture concluded that, on average, cropped land in Australia is in nitrogen imbalance equivalent to 12–15 kg N/ha year. Nitrogen inputs, transformations in soil and losses of nitrogen are also closely linked with the processes of acidification and soil hydrophobicity, and soil organic matter is intimately linked with soil structure. Thus, there is a strong argument for the development of research programs which evaluate the management options designed to improve the N status of soils in the short term and to sustain the improved N balances in the long term.

Nitrogen differs from most other nutrients in that there is no long-term storage of mineral N or plant-available N in soil. This is because ammonium N is generally converted to nitrate by bacteria in soil. Any nitrate which is not used by plants is potentially subject to loss by processes including leaching and denitrification. Thus the supply of ammonium and nitrate must be replenished continually. In natural systems this requirement is met by N in precipitation, biological nitrogen fixation, and decomposition of organic materials including soil organic matter, plant and animal residues (Chapter 10).

Strategies adopted to ensure the supply of this nutrient in agriculture include the use of pasture or grain legumes in the farming system, the recycling of crop stubbles in preference to burning, the spreading of animal wastes where animals are housed, the accelerated decomposition of soil organic matter, and, finally, the use of fertilizer nitrogen either alone or in combination with nitrogen supplied from legumes (Chapters 6 and 10). However, the annual negative N balance for Australian crop land, mentioned earlier, implies that N inputs from biological fixation and fertilizers are lower than outputs during the cropping phase of rotations — a feature that cannot be sustained in the long term.

Although there is information about rates of nitrogen fixation by grain legumes, there is only limited information about the rates of nitrogen fixation in pastures. Nutrient deficiencies, acidification of soil, ineffective rhizobium associations and reduced clover performance because of intensified cropping leading to shorter legume phases all contribute to low nitrogen fixation in pastures. It is also possible that large quantities of legume-fixed N are lost from soil because of the cycling of N via animals and the accumulation of nitrates in soil immediately before winter when precipitation exceeds evapotranspiration and the potential for leaching is highest.

Very little is known about the rates of cycling of legume-fixed N in soil including the extent of loss of this nitrogen. Answers are needed to the following questions: How much of the legume-fixed nitrogen is retained in the soil organic N pool or immobilized? What quantity is ammonified, nitrified and subsequently lost from the soil? What is the effect of grazing animals on the turnover of legume-fixed nitrogen? How comparable or dissimilar are the rates of turnover of nitrogen in different rotation systems? In short, what is the effect of crop rotation, pasture and soil management on the N balance over the medium and the long term?

Comparative studies are also needed on the efficiency of utilization of legume and fertilizer nitrogen and their relative importance in the processes of leaching and soil acidification. Leaching of fertilizer nitrogen is often regarded as a major source of nitrogen loss and of acidification. Yet, results from studies in Western Australia indicate that losses of fertilizer N applied to wheat are low even on deep, potentially leachable sands, provided that the N is applied at the correct time. The ability to vary the time and quantity of input of fertilizer N (in other words tailor N inputs to crop requirements), is a management tool which Australian agriculture could exploit more effectively (Chapter 10). At present recommendations about N fertilizer use on cereals are largely based on the average of trials conducted over several sites and several seasons. However, optimum use of fertilizer in a given year must depend on specific site characteristics (e.g. organic matter content) and rainfall patterns. Such strategic use of fertilizer N will only be achieved in practice by the development of models that take account of the interactions between site and seasonal effects. This demands an understanding of the processes involved in N transformations in soils (e.g. the rate of ammonification of legume residues) and the effects of the soil environment and climate (e.g. temperature and rainfall) on them.

The decline in clover in pastures, the susceptibility of clover to water and nutrient stresses and the effect of diseases on growth affect not only the availability of N in soil to crops but also the viability of the meat and wool industries. Management options which improve seed setting and the early growth of annual pasture legumes are needed, along with improved varieties or new species with deeper root systems to enable growth to be sustained through periods when surface soil dries as a consequence of intermittent rainfall. Research also needs to explore the possible role of alternative pasture species as feed for animals and as agents for improving soil structure and soil organic matter content. The feasibility of using grass pastures supplied with N fertilizer in preference to the traditional legume-based pastures warrants examination, particularly if studies confirm that significant quantities of N are being leached under annual legume pastures.

Information on N inputs and outputs and the

cycling of N in soil can be obtained by conducting research within the framework of field experiments which evaluate the effect of rotation and soil management on key nitrogen transformations. Such studies should be made concurrently with those determining the effect of crop rotation on soil acidification, soil non-wetting, and soil physical properties. Studies on management of both legume and non-legume stubbles could be used to assess the effect of retention of stubbles on soil organic N levels, soil structure, the availability of other macro-nutrients, and the effect of stubble retention (with and without herbicides), on the role of macroflora and key microflora in nutrient cycling and structure maintenance.

The high costs associated with running and analysing these experiments make it practicable to establish these experiments on only a few soils. Ideally these soils should be representative of the major soil types within a region. Computer simulation of key N transformations in soil and the ability to predict N budgets in relation to crop, soil and climate is seen as an integral component of extension of results from such studies and a necessary prerequisite to the initiation and evaluation of management and development research. The urgency to find solutions to increasingly intractable soil problems means that time is too short to rely solely on the use of results from long-term trials as guides for management. Computer-based modelling is a powerful tool that will allow the design of improved management practices and the identification of areas of weakness that require further research (Chapter 13).

Soil water

Water is required for crop growth and yields are often proportional to the amount of rainfall received or the amount of water used (evapotranspiration) by the crop. However, water is also important because it is the transporting agent of solids and dissolved materials and it is this indirect role of water that is of great importance when the sustainability of agronomic practices is considered in relation to dryland farming. For example, any management operation that decreases the ability of intense rain to infiltrate the soil surface will increase the likelihood of surface runoff. This has two consequences: less water is available for crop growth (an effect on productivity); and the indirect effect that the probability of soil erosion is increased with soil particles, dissolved nutrients and other salts being moved downslope and, in

extreme conditions, the formation of gullies (effects on sustainability). It is against this background that research objectives and priorities should be viewed. A convenient framework for research is provided by the mass balance equation for soil water in the root zone:

$$\begin{array}{ccc} \text{Additions} & = & \text{Losses of} \\ \text{of water} & & \text{water} \end{array} + \begin{array}{c} \text{Increase in} \\ \text{soil storage} \end{array}$$
$$P + U = E + T + D + R + S$$

where P is precipitation, U is upward capillary flow into the root zone, E is direct evaporation from the soil surface, T is transpiration, D is drainage out of the root zone, R is surface runoff and S is the change of water stored in the root zone. Usually all of the quantities are expressed in terms of a volume of water per unit area (i.e. mm) during the prescribed time interval. Productivity and sustainability are often regarded as incompatible aims but the soil water balance equation makes it clear that, at least in relation to water, they may be complementary aims. If runoff and drainage are minimized to aid sustainability, then more of a given amount of precipitation will be available for evaporation and transpiration and this may, in turn, result in greater productivity. It is clear that without reductions in both runoff and drainage, sustainable agricultural production will not be possible. Research is required into both the magnitude and extent of these processes so that management options appropriate to soil properties and position in the landscape can be found.

Runoff is a major cause of soil erosion resulting in a loss to the farmer of both soil particles and nutrients and the management of soil water is closely linked to management practices that reduce soil loss:

$$\text{Runoff} = \text{Rainfall} - \text{Infiltration}$$

The amount of runoff is determined by both the rate of rainfall and properties of the soil that affect the rate of infiltration (Figure 12.1) and likewise, the amount of soil loss is dependent on properties of both water (rate and quantity of runoff) and soil (soil structure and ability to retain particles) properties.

Practical means of reducing soil loss by controlling runoff are well-developed and commonly applied. Graded banks, conservation tillage to retain stubbles and vegetated waterways all provide means of reducing the quantity and velocity of runoff water and thus the likelihood of soil particle movement. However, less attention has been paid to increasing the infiltration rate of the soil and, in particular, to those features of the soil surface limiting the percolation of water. Questions

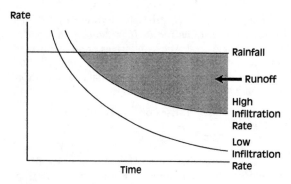

Figure 12.1 *Schematic representation of the relationship between a constant rate of rainfall, infiltration rates and runoff with time.* (SOURCE: So, H. B., 1989. 'Soil water erosion: where and what the unresolved research problems?' in *The Theory and Practice of Soil Management for Sustainable Agriculture*, Department of Primary Industries and Energy, AGPS, Canberra ACT, 12–17)

requiring process-based research are: What is the effect of a surface crust on infiltration? How does a change in soil texture beneath the soil surface affect infiltration? Under what conditions will cultivation to loosen the surface be beneficial?

The physical theory of water infiltration to uniform soils was developed over thirty years ago but its application to field soils is often limited by the presence of cracks, surface crusts and layering within the soil profile. Research is needed to develop a theory of infiltration in non-uniform soils, particularly duplex soils and those in which a surface crust develops. Such a theory could be used to predict likely results of various cultivation practices and, in turn, reduce the number of and necessity for empirical trials.

A particular cause of reduced infiltration resulting in both erosion and reduced crop production arises from the hydrophobic nature of some Australian soils. Non-wetting, siliceous sands are common in South and Western Australia and extend over 2 million hectares. Water repellency induces runoff and erosion of particles by water but more seriously the failure of vegetation to establish in non-wetted areas greatly increases the chances of wind erosion. The chemical nature of the waxes causing the hydrophobic behaviour has been established but the reasons why such materials accumulate in acid sands are not known. It requires research before practical solutions can be developed to overcome the restrictions imposed on sustainability and productivity by this problem.

Once water has entered soil it can either add to the supply of stored water within the root zone or,

if the storage capacity is exceeded, drain out of the root zone. In many parts of the world, drainage is regarded as a desirable part of the water cycle as salts and other substances detrimental to plant growth that might otherwise accumulate are removed from the soil and move via rivers to the sea. However, in many parts of Australia drainage is a largely undesirable process because, as the continent became drier and rivers ceased to flow to the sea, salts have accumulated deep in the soil profile. Water draining through the root zone in elevated positions may result in increased pressure on the ground water causing saline ground water to appear at the surface lower down the slope. Additionally, drainage may raise the ground-water table closer to the soil surface bringing salts with it, thus resulting in saline seepage in low-lying areas. It is widely suggested that major increases in the amount of drainage have been induced by the removal of perennial vegetation and its replacement by annual crops. The deep-root systems of many species of native vegetation (many eucalypts root deeper than 8 m) ensure that little water drains below the rooting zone and the perennial leaf cover means that transpiration can occur year round. In contrast, most crop species only transpire for a limited period of the year (5 months for wheat growing in Western Australia) and have shallow rooting systems (wheat—2 m on deep sands, 0.8 m on duplex soils of Western Australia). Research is needed to establish the components of the soil water balance equation for both arable and pasture crops and perennial vegetation in relation to land surface and catchment hydrology.

Measurements were made on a first-order catchment in a lateritic catchment at North Bannister, Western Australia, using ventilated chambers at regular intervals over one year. The rainfall for the year was 684 mm. Pasture evaporation was 0.57 of the rainfall and evaporation from trees ranged from 2.4 to 3.9 times the rainfall. Such results indicate the potential for judicious plantings of trees at appropriate points on farms to reduce the amount of water moving to ground water. The possibilities of altering agronomic management to increase evapotranspiration from crops also require further attention. Establishing annual crops at the right time at the correct density and with a phenology capable of using all available soil water is a priority for research. Similar research is needed for the development of suitable perennial pastures. Effective crop management combined with effective management of perennial species to minimize drainage from upland regions is probably the only long-term means of sustaining agricultural production in much of the rain-fed wheat-growing area of Australia.

A role for the researcher

Provided that it is well-focused or directed, research can provide most of the information which is needed to either maintain or improve current dryland farming systems. The role of the research process is generally accepted by most agriculturalists though few appreciate that the success of any research programme on dryland farming systems is ultimately linked to the procedures used to define the problem, to assign priorities and finally establish appropriate methodologies.

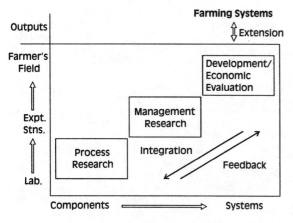

Figure 12.2 *A conceptual model of research and development in an agricultural context.* (SOURCE: Coughlan, K. J. and Webb, A. A., 1987. 'The integration of soils research—a Queensland viewpoint' in *Soil Management for Sustainable Agriculture, Technical Report No. 95*, Division of Resource Management, Western Australian Department of Agriculture)

The research and development process

A conceptual model of research and development in an agricultural context is outlined in Figure 12.2. Three phases of research are depicted: process-based, management, and development/economic evaluation. Process-based research concentrates on explaining what makes a system operate or behave in a particular way. This research invariably deals with components of a system rather than the whole and the subject to be studied must be well defined. Management research, on

the other hand, aims to determine what happens to a system once a set of treatments are applied. Fertilizer response trials are an excellent example of management research since their only aim is to determine the effect on yield of a given input of nutrient. How the effect is manifested is of no consequence—that is the purpose of process-based research.

In the conceptual model, the research process is seen as progressing naturally from process research in the laboratory to management research conducted at experimental stations and finally to development research in farmers' fields. However, in practice the research process rarely follows this logical progression. In order to find quick remedies to problems, agricultural science has relied heavily on management research. Process-based research has, for the most part, been relegated to the laboratory and, at a systems level at least, development research has been ignored. As a consequence, a considerable amount of information has been collected, for example, on the effect of management practices on crop yield over a large number of often poorly characterized sites and growing seasons. Experience has shown that it is difficult to extrapolate this site- and season-specific information from one location to another, particularly in the highly diverse soil and climatic regimes that dominate dryland agriculture in Australia. An understanding of the processes which are responsible for the specific effects does, on the other hand, enable an extrapolation of research findings particularly if these can be incorporated into computer models.

One limitation of the conceptual model outlined in Figure 12.2 is that process-based research is depicted as being laboratory bound. This should not be the case. Indeed, it could be argued that one reason for the inability of agricultural science to deal effectively with issues pertaining to sustainability stems from a lack of understanding of processes actually constraining agricultural systems in the field, including their severity and scale. This fact points to the need to conduct process-based research in the field using soils that are representative of the major soil types and to later complement this with laboratory work.

Research priorities for sustainable agriculture

The following questions need answers:
• Has there been a failure to devise practicable solutions to problems pertaining to sustainability even though adequate information is available on the causes of the problem?

- Is there inadequate knowledge of the scale and severity of the problem?
- Is there an inadequate understanding of the processes which have given rise to the problem, or a need for process-based research?

The broad reasons for decline in soil fertility are understood, but the specific information needed to restore soil fertility is unavailable. There is general agreement about the inadequacy of information on the scale and severity of wind and water erosion, secondary salinity, soil structural decline, soil acidification, soil non-wetting and the decline in soil fertility. Detailed monitoring, probably using a combination of remote-sensing and localized ground survey work, is clearly required along with more field information on the effect of soil properties on processes which contribute to particular forms of soil degradation. The general deficiency in workable solutions to problems pertaining to sustainability reflects the lack of attention given to development research.

There are interactions between different soil constraints, e.g. the lengthening of the pasture ley phase in a cropping rotation might be considered the obvious management strategy to reverse the decline in soil organic matter content. However, this strategy could create other problems namely, increased rates of soil acidification and higher incidences of soil non-wetting. Stubble retention and minimum tillage are other examples of management strategies which could be employed to arrest the decline in soil organic matter, and minimize water and wind erosion. However, the viability of these management strategies may be dependent on good weed control and the use of herbicides. What are the implications of the use of herbicides on the sustainability of dryland soils? In particular, what are the effects of herbicides on the soil macro- and micro-flora and do these have indirect effects on soil physical and biological properties? Resolving these complex interactions will necessitate a more detailed knowledge of processes causing the effects in the field.

Conclusions

There is a need for process-based research to be conducted in the field and for this to be closely integrated with management and developmental research. Considerable reorganization may be needed to mount serious research and extensive effort made to understand sustainable agricultural systems. Nutrient and water balances are the key to both production and sustainability and these must be used to indicate research priorities.

Further reading

Coughlan, K. J. and Webb, A. A. (1989). 'The integration of soils research — a Queensland viewpoint', G. A. Robertson (ed.), *Soil Management for Sustainable Agriculture*, Technical Report 95, Division of Resource Management, WA Department of Agriculture, South Perth, WA.

Greenwood, E. A. N., Klein, L., Beresford, J. D. and Watson, G. D. (1985). 'Differences in annual evaporation between grazed pasture and Eucalyptus species in plantations on a saline farm catchment', *Journal of Hydrology*, 78, 261–78.

Hargrove, W. L. (ed.), (1988). *Cropping Strategies for Efficient Use of Water and Nitrogen*. American Society of Agronomy, Madison, Wisconsin, USA.

'Legumes, nitrogen fixation and sustainable agriculture' (1990). Collection of papers in *Journal of the Australian Institute of Agricultural Science*, 3, May–June 1990.

Robertson, G. A. (ed.), (1989). *Soil Management for Sustainable Agriculture*, Technical Report 95, Division of Resource Management, WA Department of Agriculture, South Perth, WA.

'The theory and practice of soil management for sustainable agriculture', (1989). *Report of a Workshop of the Wheat Council*, Department of Primary Industries and Energy, Australian Government Publishing Service, Canberra.

Analysis Evaluation and Improvement

CHAPTER 13

Analysis and Evaluation of Farming Systems

D. H. White

SYNOPSIS

This chapter introduces modern tools for analysing agricultural systems, i.e. simulation models, mathematical programmes, expert systems and geographic information systems.

MAJOR CONCEPTS

1 Simulation models provide a means of determining the outcome of complex system behaviour. Such predictions depend on the availability of appropriate equations linking environmental variables and system processes, and computers with a capacity for very large numbers of calculations.

2 Reliable simulation models can be used to compare the effects of changes in climate, management, technological innovation and system design. Several useful models are now available for productivity and financial analysis of systems.

3 Mathematical programmes have been designed to estimate the optimal rotation or mix of enterprises on a farm. Since they present some difficulties in representing the often non-linear nature of biological processes and the marked variation between seasons, they must be at least biologically realistic.

4 Expert systems using an 'if . . . then' logic have a role to play as interfaces between models and model users. Using simulation model output, they can identify constraints to production and recommend action.

5 In the future, the linking of simulation models, mathematical programmes, expert systems and geographical information systems will provide the information required to optimize systems for profitability and sustainability.

Dryland farming systems are subjected to great variability in climate, farming intensity, level of farm improvement and management expertise. Interactions between those factors must be taken into account when comparing systems. Assumptions used in farm analysis and management can be incorrect unless the complex relationships of actual farm systems are taken into account. These relationships are constantly changing because of variability in seasonal climate, prices and operational efficiencies. A number of new tools have been developed to assist in this task. Some of these tools are dealt with in more detail in Chapters 15 and 16.

Simulation models

Simulation models are computer programs containing mathematical equations that quantitatively describe the system under study. They are used to simulate the behaviour of a complex system and so allow experimentation of different situations.

Sets of equations can be used to predict changes in pasture growth in response to soil moisture, temperature, leaf area and grazing pressure. Likewise the intake, growth and reproductive rate of the livestock can be mimicked. They are therefore particularly useful for comparing different agricultural systems in the same or different locations over a range of seasons. Because of the very large numbers of calculations involved, computers with adequate capacity are essential to the operation of such procedures. The daily rainfall and other weather data from, say, the past thirty years can be entered to evaluate different management strategies under different climatic conditions. If early autumn rains are followed by dry spells then such models should predict that seedlings have germinated but not survived so that a 'false autumn break' has occurred. Feed shortages at the time of mating, with ewes losing weight, can result in reduced reproductive performance. Feed shortages or adverse weather at lambing can be reflected in heavy losses of lambs and possibly ewes.

Pasture-livestock models have been developed in some State departments of agriculture, universities and the CSIRO to simulate local sheep or cattle production systems. Given the diverse nature of the Australian environment and substantial differences between our grassland ecosystems—annual and perennial improved temperate pastures, semi-arid and arid rangelands, tropical and subtropical grasslands—it was inevitable that quite different models were and are being developed to meet the needs of researchers, farmers and advisers to analyse such systems. This means that users must ensure that they select appropriate models for specific tasks and farming systems.

Pasture-livestock and crop models have been developed, tested and used over a number of years at the Animal Research Institute at Werribee, Victoria, to analyse a wide range of agricultural production systems. These models enable comprehensive productivity and financial analyses to be undertaken in response to varying management strategies and technical inputs. The original model BREW (BREEDING EWES) described either self-replacing Merino or Comeback wool-producing flocks grazing either perennial or annual pastures of ryegrass and subterranean clover in western and northern

Victoria respectively. This model has been used to examine the biological and economic consequences of changing management policies with respect to stocking rate, date of lambing, feeding strategy, sheep genotype and pasture types over a wide range of seasons. It has now evolved to a 10-paddock model, DYNAMOF (DYNAMIC MANAGEment of FEED), which has been used to determine the physical and financial ramifications of using ultrasound scanning for the differential management of single- and twin-bearing Merino ewes (see also Chapter 8). This model has recently been adapted to predict changes in fibre diameter and staple strength as well as fleece weight in response to season and management. These models have been extensively tested using data from the Pasture Research Institute, Hamilton, and from properties in western and northern Victoria. They have also been modified to mimic and analyse prime lamb, beef cattle and pasture-sheep-sheep parasite systems.

The predictions of these models have sometimes appeared counterintuitive, particularly to those who have had little experience in the detailed analysis of agricultural production systems. The rational and logical nature of their predictions often only becomes apparent after detailed examination of these over many simulated years. For example, it is known that, within a breed, heavier ewes at mating will bear more lambs. Therefore, it follows that feeding hay or cereal grains to ewes prior to mating will result in a larger lamb crop. However, it has only recently become apparent that such practices will often be unprofitable. The economic returns from feeding cereal grains to ewes prior to mating were estimated by simulation. The potential advantage of supplemented over unsupplemented ewes was steadily eroded at successive stages of the reproductive cycle and so the higher mortality of multiple embryos resulted in more embryo deaths in the supplemented flock. The unsupplemented flock had a larger number of non-pregnant ewes, but the second and third mating cycles provided these ewes with further chances to become pregnant. By the time the lambs were reared and sold the difference between the supplemented and unsupplemented ewes had been greatly reduced. Because all the supplemented ewes had to be fed to produce only a few extra lambs this strategy, like many other feeding strategies, was found to be unprofitable.

Further examples of the use of models to analyse agricultural production systems are described in Chapter 16.

Recently there has been considerable interest in the development and use of microcomputer programs to analyse agricultural production systems.

The BEEFMAN programs have been developed by scientists at the Queensland Department of Primary Industries and the Gatton College of the University of Queensland as management aids or educational programs for the beef cattle industry of northern Australia. Four programs have been released so far, including the educational games BEEFUP and STOCKUP and the management decision programs, GRASSMAN and STOCKMAN.

BEEFUP and GRASSMAN are designed to help beef producers manage native pastures and timber in the eucalypt woodlands of central Queensland, essentially between Bowen and Charters Towers in the north and Maryborough in the south, east of the 575 mm rainfall isohyet. They are actually simulation models based on years of field experimentation at places such as the Brian Pastures Research Station near Gayndah. The extensive testing that has taken place means that they should be quite reliable for the area for which they are intended.

BEEFUP was originally designed as a game for agricultural students to introduce them to computers and property management through a range of seasons. STOCKUP allows manipulation of the reproductive, culling and mortality levels of the different ages and classes of cattle within a herd. The economic consequences of changing herd structure and adopting various selling policies can be seen quickly, so it is a useful introduction to the factors that influence herd dynamics. STOCKMAN allows the effects of changes in herd management, selling, culling and fertility on herd structure, production and economics to be observed and analysed.

Pastoral properties in the arid and semi-arid rangelands of inland Australia are managed against a background of uncertain markets, and a high risk that climatic downturns will reduce animal productivity and cash flow. Sustainable systems and economic success rest on appropriate long-term property strategies which accept large climatic fluctuations as normal.

The herd economic module of RANGEPACK is a strategy assessment tool that follows a herd or flock through successive years to evaluate the lagged effects of climatic fluctuations on herd numbers, allowing the user to follow the gradual implementation of a new strategy. The outcome of the biological and marketing strategies of a property can therefore be followed through good and bad years and linked to economic returns.

SUMMERPACK is a computer package to aid the management of sheep grazing dry pasture and stubble in Western Australia. It is designed to assist farmers with decisions on stocking rates and feeding regimes over the summer-autumn period.

The CSIRO Division of Plant Industry is producing a number of microcomputer programs which will undoubtedly prove most useful in introducing students and farmers to the power of modelling. These include:
- LAMBALIVE which predicts losses of new-born lambs from exposure in flocks of ewes of different body condition.
- GRAZFEED which predicts the intake on a particular day of different classes of livestock offered specified pasture, with or without supplements. It is a tactical management system for determining the anticipated level of animal production from available pasture, the effect and cost of various supplements at several feeding levels, and the weight of supplement needed for a desired level of production.
- GRASSGRO which predicts pasture growth from temperature and rainfall values, at least for the Southern Tablelands of NSW.
- GRAZCLOCK which displays seasonal patterns of pasture availability, animal requirements, market prices and climatic constraints for a specified locality.
- METACCESS which displays and integrates meteorological data for different locations and time periods, as required.

The release of the SHEEPO package, produced at Werribee, which is already used to analyse a wide range of sheep production systems in south-eastern Australia, puts a particularly versatile management tool in the hands of sheep farmers and their advisers.

Crop-livestock integration

A range of dryland cereal crop models have been developed for different regions of Australia. Crop and livestock enterprises on a farm are mutually interdependent. In addition to competing for land, crop and pasture-livestock enterprises have both beneficial and other competitive effects on one another, particularly within a crop-pasture rotation. Cereal crops deplete the soil of nitrogen whereas leguminous crops and pastures are sources of nitrogen. Many pasture plants are weeds of cereal crops. Legume crops break the life cycles of plant diseases (e.g. Take-all) and pests (e.g. cereal cyst nematode) which are common to grasses and cereal crops. Crop straws, grain and 'hayed off' crops (as in a drought), can all be used to supplement the intake of the livestock.

Analyses of crop-livestock or whole-farm systems must therefore take account of the biological and physical interdependence of the enterprises, particularly where these have major impacts on yield. Likewise, economic analyses of enterprises can also be misleading. It is a common mistake to compare the gross margin of, say, a wheat crop with that of a pasture-livestock system and conclude that the former is more profitable. This is only valid with continuous cropping. Instead one must compare the mean gross margin for the whole crop-pasture rotation with that for a continuous cropping system. This is a more valid comparison and the financial benefits or otherwise of having land in fallow, or of extending the length of the rotation through additional years in crop, may be assessed.

Mathematical programs

Programs such as MIDAS (Chapter 15) are helpful in estimating the optimal mix of enterprises on a farm, particularly if different soil types are involved, and they also help in defining research priorities. For example, a wheat–lupin rotation is extremely robust on the loamy sandplain soils of Western Australia. Because of the profitability of this rotation, it is important not to give priority to conducting pasture research on this soil type, even though it has the highest potential for increased pasture production.

Mathematical programs do present difficulties in adequately representing the often non-linear nature of biological processes and the marked variation between seasons, particularly as these carry over into subsequent years. Considerable effort has been made to ensure MIDAS is biologically realistic; one of the latest versions caters for nine different types of seasons. Mathematical programs have also been used to identify alternative, more profitable pasture-crop rotations in the wheat-sheep zones of Victoria, NSW and South Australia.

Expert systems

Expert systems are computer programs that attempt to solve problems in a manner akin to a human expert in a particular field. Their reasoning is based on 'if . . . then' logic so that they are particularly appropriate where rules or regulations are involved, as in weed and pest control or in diagnosing plant and animal diseases or machinery failures, e.g. if a damage threshold for a particular crop is exceeded by a particular pest then a particular control strategy will be prescribed. The program LUPEST (LUcerne PESTs) developed for the NSW Department of Agriculture is such a system in that it aids in identifying pests in lucerne and giving advice on pest control, thereby avoiding excessive use of pesticides.

Expert systems are not well-suited to evaluating the responsiveness of dynamic biological systems where complex interrelationships govern the responsiveness to management or technological inputs. However, they are of value in providing user-friendly interfaces between model users and the models themselves. They can evaluate model output to identify constraints to production at a particular time and recommend appropriate action, i.e. they make the models a lot easier to use and understand.

Conclusions

Inevitably, the future will bring the emergence of multi-enterprise simulation models in which daily weather data are used to predict both crop and animal production, and the interactions between enterprises in terms of soil fertility, weeds and diseases can be realistically accounted for. The outputs from runs of such models can then be integrated into whole-farm mathematical programmes such as MIDAS for total system evaluation and optimization. When this is achieved, scientists and farmers can really claim that they understand and manage the land within its capability for production and to the best of their ability, thus ensuring productive, sustainable systems for future generations.

Faster computers and geographic information systems allow temporal (time-based) and spatial (land surface) models to be integrated. This will allow large properties and regions to be rapidly monitored and evaluated, taking advantage of the latest in computer- and satellite-based technologies, information on the condition of the land in terms of vegetation and water stress being continually collected by satellites and transmitted for analysis. The computer software and techniques for doing this are being developed rapidly.

Further reading

Morley, F. H. W. and White, D. H. (1985). 'Modelling biological systems', in J. V. Remenyi (ed.) *Agricultural*

Systems Research for Developing Countries, Proc. of Australian Centre for International Agricultural Research, No. 11, 60–9.

Nix, H. A. (1987). 'The role of crop modelling, minimum data sets and geographic information systems in the transfer of agricultural technology', in *Agricultural Environments: Characterization, Classification*, Proc. of Jubilee Symposium of the Australian Institute of Agricultural Science, University of Sydney, Sydney.

Snaydon, R. W. (ed.) (1987). *Managed Grasslands B. Analytical Studies*, Elsevier, Amsterdam, 227–38.

Wilson, J. (1988). *Changing Agriculture. An Introduction to Systems Thinking*, Kangaroo Press, Kenthurst, NSW.

CHAPTER 14

Factors Affecting Farm Profitability

C. M. Boast

SYNOPSIS This chapter identifies the principal factors affecting farm profitability; analyses the strengths and weaknesses in the farming system; explains the key tools for financial and economic analysis and considers the major determinants of cropping and sheep enterprise profitability.

MAJOR CONCEPTS

1 Managers need to plan their farms within the physical and financial constraints that are imposed on the whole farm and for individual paddocks. Constraints on the farm business operate at the national level, the district level, and at the individual paddock level.

2 One of the main objectives of a multi-enterprise mixed cropping and livestock farming system is to combine enterprises that will be complementary in their use of resources, particularly time, labour and machinery.

3 The evaluation of business performance cannot be measured only in profit terms. It is also important that the business avoids damaging its future earning potential.

4 An important element in the success or failure of a farm business is the managerial ability of the farmer. The ultimate test is return on capital.

5 Many businesses fail through having an inadequate level and capital structure. The higher the owner's equity the safer is the business and the less vulnerable it is to a drop in the value of assets or occasional trading loss. The more risky the type of business the higher the level of equity needs to be.

6 Some farmers are content to earn a modest profit while enjoying their lifestyle, others strive to maximize their net income. Distinction should be made between profit and profitability.

7 Enterprise analysis examines crops and livestock and the use of machinery, labour, capital, time and variable inputs in the production process.

8 The farming system must be flexible enough to allow a shift in emphasis from cropping to livestock, or allow a different mix of crops or livestock in the system to take advantage of changing market opportunities without adversely affecting the ecological and biological sustainability of the system.

9 A shift in emphasis between cropping and livestock is likely to have an impact on resource use on the farm, particularly labour, machinery and working capital requirements.

Although farmers and governments are generally unable to influence market prices, producers are in a position to adopt business management strategies and technology that focus on reducing the cost of producing each unit of output. The adoption of new technology, including improved planning and control measures, fertilizers, varieties, chemicals and machines, frequently enables the farmer to turn out more goods at the same cost, or at less cost per unit of output. However, these new technologies may not provide the manager with a long-term sustainable advantage. It is the management of disease, soil structure and fertility in conjunction with new technology that produces the high yields critical in the battle for survival.

Constraints on the farm business operate at the national level, the district level, and at the individual paddock level. Managers need to plan their farms within the constraints that are imposed on the whole farm and for individual paddocks. Within these constraints the farmer has control over the selection of crop and livestock enterprises. Decisions to include some of these enterprise options into the farming system can be made on an annual basis while others may commit the farmer for a longer period of time.

Before it is possible to propose changes to improve enterprise performance, it is necessary to understand the factors which determine enterprise profitability. If managers are not aware of the way their enterprises generate profit and incur cost they run the risk of making irrational strategic decisions.

Identifying strengths and weaknesses in the farming system

Some of the usual objectives of farm business analysis are to:
- Assess the solvency of the business.
- Determine whether the size of the business is economically viable.
- Establish whether resources are efficiently used.
- Determine whether farm profit is satisfactory having regard to weather and market conditions during the year.
- Determine whether variable and fixed costs are excessive.
- Detect signs of inefficient management.
- Gauge whether the combination of enterprises in the farming system is market driven and ecologically and biologically sustainable.
- Provide a detailed system of monitoring the farm business over time as part of the control function of management.

A complete analysis requires the examination of financial and physical data. One without the other would provide an incomplete and even a misleading picture of the farm business.

Forms of analysis

The following discussion concentrates on measures that are meaningful to managers. It is important that the analysis includes the same set of measurements and elements from year to year if useful comparisons are to be made over time. Figure 14.1 gives some useful terminology.

A single net profit or gross margin figure is insufficient for an effective examination of enterprise performance. Further analysis and tests are usually required and details concerning the derivation of both output and costs are necessary.

Figure 14.1 *Glossary of accounting terms*

Current Ratio   Current assets divided by current liabilities.

Debt/Sales Ratio   Determined by expressing annual total debt (interest and capital) repayments as a percentage of gross sales.

Economic Performance   Income and profitability are monitored by calculating net profit, return on capital, proprietor equity and net profit per hectare.

Enterprise Performance   For each farm enterprise, outputs and inputs are measured to monitor the efficiency of resource use, such as land, labour, machinery, feed, breeding, etc.

Financial Analysis   Concentrates on the capital position of the business including solvency, liquidity and net worth.

Gearing   The ratio of equity to debt.

Gross Margin   For any enterprise is gross output less variable costs.

Gross Output   The value of the production of an enterprise or the farm as a whole.

Liquidity Ratio   Liquid assets divided by current liabilities.

Net Profit   Obtained by adjusting the net cash income for total depreciation, net inventory changes and the value of products consumed at home. Net profit is the profit from the year's operation and represents the return to the owner for personal and family labour, management and equity used in the farm business.

Over Trading   A situation where liquid assets fall short of current liabilities.

Return on Capital   Profit as a percentage of invested capital.

Comparative analysis

The technique known as Inter-Farm Comparative Analysis may be quite useful to give guidance to farmers who have few records and minimal information about their farm and its enterprises, but it should never be more than an initial check and guide. It could provide reassurance or warning for those farmers who have little idea how their profitability compares with that of others, provided they are satisfied that the comparative standards are reasonably similar. The main difficulty with this analysis is that farms need to be grouped; there is no ideal way of doing this and farms within a group may be quite different. Variations will occur in soil type, climate, machinery size and building layouts, yields, labour requirements and so on. These would be important considerations in the case of arable farming systems. In addition, farmers within a group may not be typical. The fact that every farm is different limits the application of inter-farm comparisons and it would be optimistic to expect data to be provided to cover the range of differences. The point becomes still more relevant when differences in managerial potential and objectives are considered, particularly on farms of above-average efficiency. However, the comparison may indicate possible reasons for low profits which can be investigated subsequently.

Because of these limitations, emphasis should be placed on comparisons of the farm's current and expected performance against its achievements in previous years or with a previously prepared budget. The farm is compared as if it were only with or within itself. This concept is linked to the idea of obtaining enough information about the farm to replan it as required in the light of its own resources and the farmer's ability and hence establish realistic targets to meet farmers' objectives. Emphasis is therefore placed on what the farmer ought reasonably be able to achieve for a given situation, rather than on comparisons with the results of other farms. Farm planning involves both selecting the right technique and obtaining the necessary data. It is usually the latter that provides problems for advisers. Standard data are available to some extent but rarely in sufficient detail to be applied with confidence to individual farms.

In some cases, assessments can be made of 'practice' against 'theoretical possible limits'. Sometimes the 'theoretical upper limit' may become redundant as new technology and management concepts appear (e.g. the upper limit to lambing percentage may have been considered to be 200 per cent, i.e. twins for every ewe; if ewes lambed three times in two years, a new approach to assessing lambing percentage would be required). The potential yield concept is another example and is discussed fully in Chapter 17.

Reviewing the Balance Sheet — The evaluation of business performance cannot be measured only in profit terms. It is also important that the business avoids damaging its future earning potential. This is best achieved by ensuring that the financial structure of the business is sound. Suggestion of an unsound financial structure can be found in the balance sheet. There are other features that are vital to the strength and ability of the business to survive; these include stability, liquidity, flexibility and growth.

The main objectives in balance sheet analysis are to:

- Establish long-term solvency.
- Establish short-term solvency.
- Establish trends in the financial structure.

Analysis of Financial Structure — An important element in the success or failure of a farm business is the managerial ability of the farmer. However, even a skilled and experienced manager cannot succeed if the financial base of the business is not sound. The most common causes of farm business failure are (a) the ineffective use of finance and (b) poor management.

Long-term solvency depends on sufficient capital reserves to weather unexpected downturns. The prudent farmer will ensure that earnings generated during a 'boom' are reserved within the business rather than dissipated.

Many farm businesses fail through having an inadequate structure and level equity capital. A farm business which relies excessively on borrow-

Table 14.1 *The relationship between capital and borrowings.*

	Farmer A		Farmer B	
	$	%	%	%
Total assets	500 000	100	500 000	100
Liabilities	100 000	20	250 000	50
Net capital	400 000	80	250 000	50
	500 000	100	500 000	100

Farmer A with a 80 per cent equity is clearly in a stronger position to survive a downturn than Farmer B with a 50 per cent equity. Farmer A has less debt to service and the financial position could deteriorate further without becoming insolvent. Indeed there is some spare borrowing capacity. The more risky the type of business, the higher the net capital needs to be.

ings to finance its operations will be vulnerable to booms and busts (Table 14.1).

The higher the level of equity the safer is the business and the less vulnerable it is to a drop in the value of assets or occasional trading loss. Therefore, a farmer who has borrowed 20 per cent of total capital invested is relatively 'low geared', having 80 per cent equity, 20 per cent debt, a ratio of 4:1. On the other hand, the higher the level of gearing, e.g. 80 per cent borrowed (1:4), the greater the possibility of achieving a higher return to owner equity but the chance of insolvency is higher. This is an example of the 'principle of increasing risk'. High gearing should only be considered where consistently good profits are reasonably certain, though low gearing could mean that profitable investment opportunities and increased returns to equity are foregone.

Debt/Sales Ratio — A further measure of long-term solvency is the debt repayment/sales ratio (Figure 14.1). The usefulness of this ratio lies in its relationship to production. The main difficulty with the equity ratio arises with valuing fixed assets—the valuation is frequently subjective and a farmer's equity in the business bears no relationship to the income-generating capacity of the farm.

Generally, a farm business can be considered safe if it has a debt sales ratio of up to 30 per cent. The viability of the farm business will be severely threatened should the ratio exceed 40 per cent. The remedies for at-risk businesses are to increase output through use of efficient sustainable farming practices, good cash flow management and careful financial planning such as restructuring deferred liabilities.

These aspects of solvency relate to the long-term stability of the business. However, the farm business must also survive in the shorter term, weathering temporary stresses and strains. Sometimes farm businesses are not able to continue trading even though their assets exceed liabilities by an acceptable margin. This situation occurs when there are insufficient funds available to meet current debt commitments. To measure the ability of the business to meet short-term debts, management economists calculate the ratio of current assets to current liabilities (Figure 14.1). This is called the current ratio and is an historic measure of the farm's liquidity. Current liabilities are those which have to be met within a short time, usually one year (Table 14.2).

A current ratio (current assets divided by current liabilities) of 2:1 is considered acceptable but this will depend on the type of farm business, the time of year the analysis is conducted and the

Table 14.2 *A comparison between two businesses, each with the same capital but with a different current ratio.*

	Farmer A	Farmer B
	$	$
Current assets	75 000	300 000
Non-current/fixed assets	425 000	200 000
Total assets	500 000	500 000
Current liabilities	150 000	150 000
Deferred liabilities	150 000	150 000
Total liabilities	300 000	300 000
Net capital	200 000	200 000
Current ratio	0.5% (50%)	2.0% (200%)

Farmer A's current ratio is only 0.5 to 1 (or 50 per cent) while Farmer B's is 2 to 1 (or 200 per cent). This is a sign that Farmer A may have invested excessively in *fixed* assets such as land and plant and consequently been left exposed to the risk of not meeting day-to-day commitments. A lending institution may be reluctant to lend money to Farmer A because of doubt whether Farmer A could service the debt. The higher the figure the safer the business in the short run because it is likely to meet demands from creditors by obtaining funds quickly.

composition of current assets. Frequently, this ratio is less than 2:1 because the overdraft is often incorrectly regarded as a medium-term liability. The more liquid the current assets, the lower this ratio can safely be. If the ratio is less than 1 it indicates that the business could face a liquidity problem. While a low ratio can indicate a lack of liquidity a high ratio may suggest inefficient use of resources.

Some current assets are either immediately available or can be made so in a very short time, e.g. liquid assets are cash in hand or in the bank and crops in store. Other assets might be difficult to convert into cash in the short term e.g. growing crops, some debtors and finished trading livestock. If liquid assets are insufficient to cover current liabilities the business could be in a serious financial position if short-term creditors (including the bank overdraft) pressed for payment, even though ultimate solvency may be sound. Therefore a business in this position may have to obtain additional long-term loans so that it can either reduce its current liabilities or increase its liquid assets. The difficulty is to decide which assets are liquid. A rule-of-thumb guide is 1:1, a ratio sometimes referred to as the 'Acid Test'.

Insolvency can occur because of a too rapid expansion of stock holdings and accounts receivable with inadequate cash reserves. The time lag in production means that expenses are incurred without a corresponding increase in sales and if major creditors become doubtful (e.g. if a drought occurs) of the security of their funds, they may demand payment.

Flexibility can be judged in the light of the financial ratios between current liabilities and current assets and also according to the balance of the different types of assets. If too many assets are fixed there may not be sufficient working capital to utilize them effectively. In this case, loans should be obtained on the security of part of the fixed assets to increase the current and medium-term assets or if this is not possible, consideration should be given to selling fixed assets to provide additional working capital.

Farm businesses with regular cash flows, such, as dairying, pig and egg production, are in a better position to meet commitments than is a business that receives income once or twice a year. If the ratios are calculated just before the business receives its annual cheque the position will appear worse than it really is. For this reason a more accurate analysis will be achieved from a cash flow budget which highlights the availability of funds in each period (usually one month) to meet commitments due in that period.

Trend Analysis — The trend statement should not list individual assets and liabilities. A clearer picture is obtained by presenting groups of items. The manager should look for trends which may indicate an improvement or decline of the financial position of the farm business over time. Results are compared with historic values and the objective becomes one of showing improvement over the past results (Table 14.3).

Looking behind Balance Sheet Figures — Calculation of ratios is straight-forward, but the difficulty is to interpret the facts and the associated risk in a logical way. A series of balance sheets will show a good or bad trend and management will be in a position to make some judgement. The initial ratios must be interpreted together with, for example, an investigation of creditors shown in the balance sheet: Are any of them overdue for payment? Are any of them being paid small sums to keep an account turning over? Is there a single very large creditor who may create a financial crisis if a debt is called in? There may be some debtors who are overdue, some debts may be bad or one debtor may be large and failing to pay on time. This situation will adversely affect farm business liquidity. Table 14.4 provides an example of the type of questions that should be asked when trying to interpret the balance sheet.

The ratios have little or no meaning unless the

Table 14.3 *Trend analysis.*

	1985	1986	1987	1988
	$	$	$	$
Current assets	40 000	34 000	36 000	21 000
Breeding livestock	52 000	47 000	66 000	72 000
Land	240 000	250 000	260 000	280 000
Plant and equipment	41 000	53 000	62 000	74 000
Total assets	373 000	384 000	424 000	447 000
Current liabilities	18 000	17 000	21 000	42 000
Liabilities 1 to 3 years	3 000	3 000	30 000	30 000
Liabilities over 3 years	60 000	60 000	60 000	105 000
Total liabilities	81 000	80 000	111 000	177 000
Net capital	292 000	304 000	313 000	270 000
Equity	78.3	79.2	73.8	60.4

The financial structure of the farm business appears sound in that the farmer's equity is still above 60 per cent, however having declined from 79 per cent is of concern. Short-term liabilities are well covered by livestock and current assets, and long-term liabilities are not excessive in relation to long-term assets. However, short and long-term liabilities are rising quite rapidly and signs of financial strain are emerging. To restore the situation some livestock will have to be sold to reduce current liabilities. This situation could have been brought on by excessive drawings or a trading loss as a result of drought or poor commodity prices, leading to a deterioration in both net worth and growth.

Table 14.4 *Looking behind the balance sheet figures.*

Liabilities	Assets
Creditors	**Debtors**
What are the terms of trade?	What are the terms of trade?
Is there a dominant creditor?	Are debtors well spread?
Are creditors' business sound?	Is there a dominant debtor?
	Are debtors considered financially sound?
Financial institutions	**Stock**
What are servicing costs?	Is stock undervalued?
What are the repayment terms?	Is stock value vulnerable to short term market fluctuations?
What is the bank overdraft limit?	Would a forced sale result in a significant reduction in value?
Is the financial institution sound?	What stage of maturity have the animals reached which are valued in the stock figure?
	Is there a demand for the stock?
	Machinery
	What is market value?
	What is the life of the machinery?
	Is machinery obsolete?
	Land and Buildings
	What is the market value?
	Is there likely to be compulsory purchase?
	Are the buildings in good condition?

assets are valued realistically, managers also need to look behind the figures to see what they are composed of. The ratios also need to be looked at in relation to each other so as to give an overall picture of the financial structure of the business. Probably more important than the level of ratios are the trends in these levels over time.

So far we have been looking at historical terms but management must look to the future. It is important to make use of the knowledge of the balance sheet structure to establish the effect management decisions will have on the future structural health of the business. Table 14.5 provides a format for balance sheet analysis.

Table 14.5 *Balance sheet statement*

	Opening balance 1.7.89	Closing balance 30.6.90
Current assets	$	$
Cash		
Debtors	11 120	1 220
Valuation of:		
Produce	800	800
Inputs	700	600
Livestock	150 000	140 000
Sub-total	162 620	142 620
Non-current assets		
Breeding stock	15 000	17 000
Land	400 000	400 000
Plant	56 000	50 000
Sub-total	471 000	467 000
Total assets	633 620	609 620
Current liabilities		
Trade creditors	3 625	4 845
Bank overdraft	97 140	46 690
Sub total	100 765	51 535
Deferred liabilities		
Loans 1–3 yrs		
> 3 yrs	240 000	236 100
Sub total	240 000	236 100
Total external liabilities	340 765	287 635
Net capital	292 855	321 985
Total liabilities	633 620	609 620
Capital account		
OPENING NET CAPITAL	292 855	292 855
Trading profit		21 100
Private receipts		17 988
Less Drawings and tax		9 958
CLOSING NET CAPITAL	292 855	321 985
Ratio analysis		
Proprietor equity %	46	53
Current ratio	1.6	2.8
Net liquidity	61 855	91 085
Return on average capital %		3.4
Return on average owner equity %		6.9

Analysis of economic efficiency

Farm consultants frequently ask clients to list their goals, both business and personal. Perhaps the most common response is: 'to enjoy the way of life', but this enjoyment can be eroded quickly by financial failure. Some farmers are content to earn a modest profit while enjoying their lifestyle, while others strive to maximize net income. For the farm manager, profit is the principal measure of performance. There are many components which contribute to the bottom line. Some components determine output while others affect costs influenced by markets, climatic, physical conditions and management ability.

Figure 14.2 is a trouble-shooting guide for identifying the source of a profitability problem in the farm business. It assumes that profitability will exist. The first step is to look at farm size to see if there is at least the potential for adequate profit or if the farm size is too small. The analysis then proceeds through various measures relating to return on capital and profit. Finally enterprise analysis examines crops and livestock and the use of machinery, labour, capital, time and variable inputs in the production process.

A complete analysis is time-consuming and many of the measures of performance may have satisfactory values. Should the need be to identify and isolate the cause of the profitability problem quickly, a number of steps can be eliminated in the systematic procedure.

Analysis of profit

When analysing profit the distinction should be made between profit and profitability. Profit is a dollar value; profitability is concerned with the size of this profit relative to the size of the business or the value of the assets employed to produce the profit.

Measures of Size — The first step is to analyse farm size to determine if it is large enough to generate adequate net profit under the most efficient methods of production. Historically net profit is correlated to farm size. A low income may be due to the farm being too small.

An income or profitability problem can occur in any year due to poor weather and associated poor yields or due to low product prices or some combination of these factors. If the problem persists in years of average prices and yields it can generally be traced to either farm size or inefficient operations management.

Several measures of farm size can be considered, each with its own advantages or disadvantages. No single measure is entirely satisfactory, so they should not be considered in isolation.
• The value of production is measured in terms of total gross output or gross sales. The value of

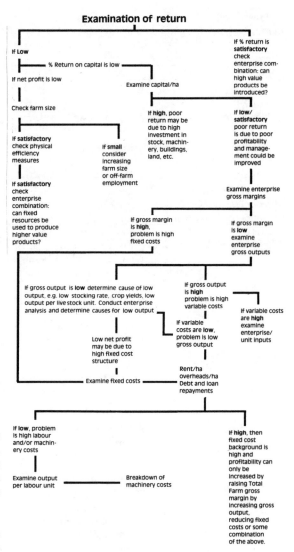

Figure 14.2 *A systematic diagnostic procedure.*

farm production is used because physical units cannot be added together in a meaningful way. This measure combines many different farm products into one measure of size and allows comparison between farms, though its disadvantage is that output and sales are affected by price, yields and managerial efficiency.
• Total capital invested in land, buildings, plant, equipment and livestock with dollar values is measured. This method allows easy comparison between farms.
• Livestock numbers as a measure of size is useful

for comparing size among farms of the same class of livestock.

- Total labour employed is a common measure of farm size regardless of type of farm. Terms such as one-person or two-person farms are often used. This measure of size is affected by the amount of labour-saving technology used and should be carefully interpreted when looking at farm size.
- Total hectares controlled is another common measure of size and can be subdivided into area owned and area rented. The real usefulness of this measure lies in comparing the size of crops for farms of the same general type in an area with similar soil types and rainfall patterns. It is not a good measure where land productivities differ greatly or the major enterprises are not the same. For some comparisons the number of hectares in rotation may be a better measure than total hectares.

If the measure of size indicates that the farm is large enough to provide a better income, the next step is to analyse the efficiency of the operation. The analysis can be broken down into two types: economic efficiency and enterprise efficiency. When analysing crops and grazing livestock, land use has always been a major factor to consider since land is frequently the limiting factor of production. Efficiency of these enterprises has always been on the basis of land use (enterprise yield and gross margin per hectare) while bearing in mind the importance of adequate return on capital. In this case, the return on capital is frequently linked with land use efficiency except in cases of excessive capital investment.

The key measure of economic performance is return on capital. However the analysis can be taken further to assess return to proprietor equity and net profit per hectare. Other measures of economic efficiency include machinery investment per tillable hectare, machine cost per tillable hectare, labour cost per tillable hectare and value of output per labour unit.

Return on Capital — Part of the return from any investment may be in the form of capital gain. However, this does not provide funds for farm development or personal expenses, so it is essential for profit to provide a satisfactory return on capital. This return is calculated by expressing the profit as a percentage of invested capital. This calculation requires two figures: net profit and capital. What figure should be used for capital? — the opening or closing capital, or an arithmetical average of the two? In most instances, the average is appropriate since additional capital may have been

introduced or capital may have been withdrawn during the year. The return of capital should reflect a return on the current market value of all assets. A market-based balance sheet will reflect a more accurate measure of equity and solvency. Using the average of the opening and closing capital provides the calculation in Table 14.6.

Table 14.6 *Return on capital*

	1987	1988	1989
	$	$	$
Opening capital	518 000	524 000	510 000
Closing capital	524 000	510 000	550 000
Average	521 000	517 000	530 000
Net profit	39 075	23 445	54 705
Return on average capital	7.5%	4.6%	10.3%

Return on Proprietor's Equity — Return on capital measures return to both debt and equity and the farmer may be more interested in the return to personal equity in the business. This measure is determined by dividing average proprietor's capital into net profit (less interest payments on borrowed funds) and the result is expressed as a percentage.

Net Profit per Hectare — To facilitate comparisons between properties of different sizes and to compare results within the farm businesses the results can be expressed as net profit per hectare. A way of establishing a goal for net profit per hectare is to establish the return the owner's labour, management and capital could earn in alternative uses, i.e. the opportunity cost of these factors of production become the goal or standard for net profit per hectare. Another goal would be a desired profit/hectare level for management to attain.

Machinery Investment per Tillable Hectare — Dividing the current value of all machinery and equipment by the number of tillable hectares provides a measure of machinery efficiency. This is a useful measure which should be compared with values for farms of the same type and size. Larger farms tend to have a lower machinery investment per hectare because of economies of size. This efficiency measure will indicate either a value too high or too low. A high value indicates that there may be excessive machinery investment, and machinery fixed costs may seriously affect net profit. Low values may indicate that the machinery is too old and unreliable or too small for the area being farmed, and may help to explain poor yields.

There is an important relationship between machinery investment, labour requirements, timeliness of field operations and the amount of contract work hired, and these factors should be evaluated together with regard to farm size.

Machinery Cost per Hectare — This measure differs from machinery investment per tillable hectare in that the total of all annual costs related to machinery is divided by the number of tillable hectares. The total annual machinery costs include all fixed and variable costs plus the cost of contract hired work. Again this measure will vary by type and size of farm and may indicate whether the farm has excessively high or low machinery costs.

Labour Costs per Tillable Hectare — These are determined by dividing total labour cost by the total hectares rotated. The opportunity cost of the farmer may be included in this calculation. Linked to this measure of labour efficiency is machinery size, and type of crops and livestock products produced on the farm. These factors must be taken into account when analysing labour costs per hectare.

Value of Output per Labour Unit — This measures the total value of output per worker unit per year and is affected by the size of the farm business, the combination of enterprises and the investment in machinery and labour-saving devices employed. The higher the value, the greater the efficiency of labour.

Labour-use efficiency can be improved through the use of larger machinery and additional forms of mechanization. However, the principles of the substitution of capital for labour cannot continue indefinitely without a reduction in profit. For this reason it is important to consider machinery costs in conjunction with labour use.

Enterprise analysis

While economic analysis on a whole farm basis may indicate a profitability problem, the source of the problem is often not revealed from the financial statements. Enterprise analysis consists of allocating all outputs and variable costs to individual enterprises for further analysis.

The Gross Margin (GM) Approach — This approach (Figure 14.3) regards the farm business

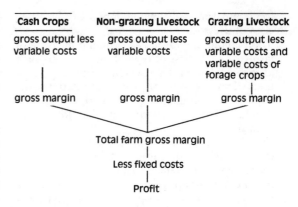

Figure 14.3 *Allocation of costs and output with resultant gross margin and farm profit.*

as a group of productive enterprises, each with its own output and direct and variable costs. These enterprises are centred on the general farm which provides common services for all the enterprises and the necessary co-ordination processes. For the purpose of defining gross margins, a farm is assumed to have a given set of resources: land, labour and capital. The decision to be made is how to utilize these 'fixed' resources over the ensuing production periods. The gross margin of an enterprise is the gross output less variable costs.

Gross output is obtained by using the following formula:

Gross output = (sales + transfers out + closing stock valuation) *less* (purchases + transfer in + purchase of any crop subsequently sold + opening stock valuation).

Output includes off-farm sales, produce transferred-out plus the value of produce consumed in the farmhouse or by workers. If the business purchased some oats which were then fed to stock, the cost would be entered under expenses and would not be included in the gross output statement. However, should the farmer decide to sell the oats purchased, both the costs and the sale price would enter the gross output statements — any revenue included in the procedure is adjusted for debtors and creditors at the start and end of the accounting period.

Gross output can be measured for the whole farm, but for management purposes, when analysing enterprise performance it is necessary to know how each enterprise would perform. For example, a farmer may grow oats which could be sold but instead chooses to feed them to the sheep. In this case, the oats are transferred from one enterprise to another. If the oats are not charged to the sheep enterprise the farmer would have difficulty in obtaining an accurate picture of the profitability of the sheep or the oats enterprise. A realistic figure based on average market values is put on the oats, credited to the oats enterprise and debited to the sheep enterprise. The oats figure is a transfer *out* from the oats enterprise and a transfer *in* to the sheep enterprise. It is a paper calculation for management purposes and the Taxation Commissioner is not interested in this internal transaction because the producer has, in effect, increased output to the oats enterprise by the same amount costs increased to the sheep and the difference in profit is nil. Other examples of interdepartmental transfers include: calves from dairy herd to dairy young stock, calves from dairy herd to beef unit, hoggets from ewe flock to wether flock.

Costs — The main division of costs is, therefore, into variable costs and fixed costs. To be regarded as variable costs in the gross margin sense, costs need to satisfy two criteria. They must (*a*) be specific to a single enterprise and (*b*) vary approximately in proportion to the size of the enterprise. The overhead or fixed costs are those costs that cannot be easily allocated to an enterprise and therefore belong to the farm as a whole (Table 14.7).

Some costs are specific to a single enterprise and are not allocated in calculating gross margins per unit of production because they do not satisfy the second criterion, i.e. they do not vary in proportion to the size of the enterprise, but form a given total sum, at least over a certain range of enterprise size. Examples are machinery repairs and maintenance and depreciation on machinery.

Gross margins are calculated for each enterprise by subtracting the variable costs and can be expressed on a per hectare or per capita basis for analytical purposes, or left as the total for the enterprise as a whole. The fixed costs of the farming business, together with adjustments for stocks on hand when subtracted from the total farm gross margin will give a profit or loss.

The treatment of forage with grazing livestock differs from that of non-grazing livestock because, in addition to the usual variable costs of feed, veterinary and medicine, they must also bear the variable cost of the forage/pasture crops they consume. The livestock gross margin can be regarded as the product of the forage/pasture area and the value of such gross margins can be expressed on a forage-hectare basis to be directly comparable with data for the crop enterprises. Gross margin per forage hectare is the relevant figure to use if

Table 14.7 *Breakdown of costs on the farm enterprise.*

Variable Costs	
Crops	*Livestock*[†]
seed	
fertilizer	bought feed
casual labour	casual labour
contract services	contract services
sprays	sundries (vet etc.)
sundries (twine, etc.)	

Fixed Costs[*]
Regular labour
Unallocated casual labour[**]
Paid management
Machinery repairs and leasing charges
Fuel and electricity
Machinery depreciation (a non-cash cost)
Rent and rates
Depreciation of fixtures and buildings
Sundries (insurance, office, phone, professional fees, etc.)
Financing charges

[*]Because of problems associated with allocation, machinery depreciation, repairs and fuel are included in fixed costs.

[†]Livestock purchases are not regarded as variable costs but are deducted when appropriate enterprises' outputs are calculated.

[**]Casual labour and contract machinery services are employed to do a specific task and therefore it is possible to allocate them directly to enterprises.

the intention is to study the contribution of the forage area to cover the fixed costs and provide a profit on the farm business.

Allocation of forage costs

The dry sheep equivalent (DSE) concept is employed in Australia to enable different types and age groups of livestock to be put on a common basis. Where gross margins per hectare are required for grazing livestock, and more than one type of grazing livestock utilize the same area either at the same time or at different times of the year, a problem of allocation arises. Where there is only one grazing livestock enterprise on a farm, forage costs can be allocated directly to that enterprise. However, the majority of farms have more than one grazing livestock enterprise and the different types of livestock are not normally confined to separate areas of pasture but share the total area. Sheep may follow cows on to a pasture to clean up, and beef animals may eat hay from the same field as the dairy heifers. Since it is not possible to allocate individual fields to particular classes of livestock an attempt must be made to apportion the total farm forage costs between the different grazing enterprises.

The forage requirements for each type of livestock averaged out over the whole year are expressed by reference to the forage needs of a dry sheep. A dry wether is represented by 1 DSE while ewes with lambs and rams represent 1.5 DSE. In order to calculate the forage hectare required to support a given number of livestock of varying types (both grazing and conserved bulk fodder — hay and silage), the total DSE equivalents are first determined (Chapter 8). Forage costs (and forage hectares) may then be allocated to the various grazing enterprises by simple proportion and included in the gross margin worksheet. For example, a simple farm with an annual average number of 420 ewes, 11 rams and 270 wethers may have a total forage cost estimated to be $9000 over a forage/pasture area of 183 ha (Table 14.8).

This method has the practical merit of being simple to use and easy to understand. The stocking rate can be expressed as DSE/ha and forage costs can be expressed as costs/DSE, per head, or per ha.

Of course, the allocation is only approximate and ignores differences in types of forage and the fact that better pastures may be reserved for the dairy herd while sheep and heifers are exiled to the poorer pastures for most of the year. However, in most situations the inaccuracies are small and do not justify the use of a more complicated method.

A characteristic of mixed farming is the large range of potential alternative enterprises, and flexibility in management is substantial. For this reason gross margin analysis can be used to assist in pointing a way to desirable changes. Less profitable enterprises can be either reduced in size, eliminated or reorganized to become more profitable, thereby increasing farm profit. Simple figures of yields of crops per hectare or livestock products per head or per hectare, lambing percentages, and stocking rate are useful. These measures are available, easily understood and in constant use by farmers (Table 14.9).

Such analyses should be regarded as elements of diagnosis, though it is necessary to look further into the reasons for poor performance. Reviewing key factors affecting crop and livestock profitability will point up weaknesses which, if corrected, lead to enhanced profitability and sustainability.

Table 14.8 *Forage allocation.*

Livestock Category	Number	dse/hd	Total dse	% of Total	Share of Forage Cost	Forage ha allocated
Ewes	420	1.5	630	68.7	6 183	125.7
Rams	11	1.5	17	1.9	171	3.4
Wethers	270	1.0	270	29.4	2 646	53.8
Total	701		917	100.0	9 000	183.0

Average stocking rate = 5.0 dse/ha
Forage costs per dse = $9.81
Forage costs per ha = $49.00

Table 14.9 *Enterprise summary.*

ha	% of area	GM	Total GM	% Total gr	Average GM/ha	Average Yield/ha	Potential yield	% of Potential	District average
Wheat	165	36	42 960	47	260	3.0t	3.9	77	2.6t
Barley	107	23	19 400	21	180	2.8t	3.1	90	2.4t
Sheep	189	41	28 413	32	150	36 kg	49 kg	73	42 kg
Total	461	100	90 773	100	197	–	–	–	–

Yield is the most useful measure of efficiency particularly when crops contribute to the bulk of farm profit, actual yield can be measured against potential yield. The total gross margin is affected by yield and price and this is reflected by the combination of enterprises which make up the farming system. The distribution of enterprises in relation to gross margin and area is important as not all enterprises have the same income earning capacity. The average gross margin/hectare in the total row relates to the enterprise mix in the rotation sequence used on the farm.

Table 14.10 *Gross margin budget (livestock).*

Period Enterprise No. head	1990 Merino 454	Quantity/ head	Period Reqd. days	Total quantity	Unit price $	Total $	Per hd. $
Closing valuation				497 hd		15 973	
Sales: Products						23 694	
AWC wool tax				20%		4 739	10.44
Farm gate income						18 955	
Progeny							
Culls				128 hd		1 965	
Transfers out						4 320	
Sub-total (a)						25240	
LESS							
Opening valuation				648 hd		13 560	
Livestock purchases						800	
Transferred in						2 310	
Sub-total (b)						16 670	
Gross output (a–b)						8 570	18.88
LESS Variable costs:							
Feed transferred in							
Type 1 Barley		1 kg	14	6 t	110/t	699	1.54
Type 2		0 kg	14	0 t	0/t	0	0
Type 3		0 kg	30	0 t	0/t	0	0
Purchased feed:							
Sheep costs							
Shearing				454 hd	1.35/h	613	1.35
Crutching				454 hd	0.39/h	177	0.39
Shed hire							
Shearing				454 hd	0.42/h	191	0.42
Crutching				454 hd	0.13/h	59	0.13
Shed labour							
Shearing		1	3	3 md	83/m	249	0.55
Crutching		1	2	2 md	83/m	166	0.37
Wool packs				10 pk	5.3/p	53	0.12
Marking							
Veterinary:							
Type 1 Drench					0.11/h	50	0.11
Type 2 Vaccines						0	0
Type 3 Jet					0.31/h	141	0.31
Transport:							
Stock				128 hd	0.35/h	45	0.35
Wool				10 b	3/b	30	0.07
Stock selling charge				128 hd	0.05%	98	0.22
Yard fees				128 hd	0.3/h	38	0.3
Stock insurance							
VARIABLE COSTS (before forage)						2 609	6
ENTERPRISE GROSS MARGIN (before forage)						5 961	13
FORAGE VARIABLE COSTS						3 028	7
GROSS MARGIN (including forage)						2 933	6
GROSS MARGIN PER HECTARE (ha) 59		49.83					

Table 14.11 *Gross margin budget (crops).*

Enterprise Hectares	Wheat 250	Quantity/ha.	Total quantity	Unit price $	Total $	Per ha $
Closing valuation			0 t	120/t	0	
Sales: Main product	3		634 t	120/t	76 068	304
CBH tolls				14.75/t	9 350	37
Net of CBH toll sales					66 718	
By product	1		120 t	30/t	3 600	14
Transferred out:						
SEED		90 kg	23 t	105.3/t	2 368	
FEED			94 t	105.3/t	9 851	
Sub-total (a)					82 538	
LESS:						
Opening valuation			0 t	120/t	0	
Purchases			0 t	0/t	0	
Transferred in: seed			23 t	140/t	3 150	
Sub-total (b)					3 150	
Gross output (a–b)					79 388	318
LESS Variable costs:						
Fertilizer						
Type 1 Triple		70 kg	17.5 t	325/t	5 688	23
Type 2		0 kg	0.0 t	0/t	0	0
Type 3		0 kg	0.0 t	0/t	0	0
Purchased seed						
Type 1		0 kg	0.0 t	0/t	0	0
Type 2		0 kg	0.0 t	0/t	0	0
Type 3		0 kg	0.0 t	0/t	0	0
Sprays						
Hoegrass		1 L	250 L	18.75/L	4 688	18.75
MCPA		0.35 L	88 L	5.40/L	473	1.89
Ally		6 gm	1500 gm	1.25/g	1 875	7.50
Roundup		1 L	250 L	15.85/L	3 963	15.85
Insurance				5.00/T	413	1.65
Freight						
Fertilizer			18 t	8.00/t	140	0.56
Grain			634 t	3.00/t	1 902	7.61
Other Charges						
Casual labour			5 md	80.00 md	400	1.60
Contract hire					0	0.00
Other					60	0.24
VARIABLE COSTS					19 599	78
ENTERPRISE GROSS MARGIN (GO–VC)					59 788	239

Where a number of alternative strategies are feasible it is essential that selection between alternatives is not haphazard but based on a systematic evaluation of the contribution of each to farm profitability. This will require analysis of the impact changes will have on physical resource use, cash flow, the profit and loss statement, and the balance sheet of the business. Tables 14.10 and 14.11 provide a format for gross margin calculation.

There are five basic strategies to increasing farm profit:

• Increase total farm gross margins using the same

level of fixed costs.
- Increase total farm gross margin with higher fixed costs.
- Maintain the current level of total farm gross margin but reduce fixed costs.
- Lower total farm gross margin with a relatively greater reduction in fixed costs.
- Increase total farm gross margin with lower fixed costs.

Factors affecting crop profitability

Figure 14.4 shows the major determinants of crop profitability can be classified as: yield/ha, price per unit of output, variable costs/ha, and fixed costs (labour and machinery costs/ha).

Yield — While farmers have little or no control over prices received, within limits set by rainfall they can exercise much control over yield per ha through their management of the crop enterprise. Increasing yield is the most likely and most rewarding way to increase profit. Such improvements are not necessarily easy.

Rainfall and climate have a key role in determining yields, land use and stability of the cropping system and must be considered when analysing and developing sustainable farming systems. The potential yield concept (Chapter 17) relates rainfall to potential yield. The yield potential is a good indicator of whether or not the farming system is sustainable and is determined in the following manner for average soil conditions etc.:

- The potential wheat yield is calculated as yield (kg/ha) = (derived April–October rainfall (mm) — 110 x 20).
- The potential barley yield is calculated as yield (kg/ha) = (derived April–October rainfall (mm) — 90 x 20).
- The potential grain legume yield is calculated as yield (kg/ha) = (derived April–October rainfall (mm) — 130 x 15).

As a guide, producers who regularly achieve 70 per cent or more of their potential yield are probably using sustainable husbandry practices. Producers who are operating below 70 per cent but trending upwards are heading in the right direction. Producers who are achieving less than 50 per cent of potential yield or who are trending downwards should analyse their husbandry practices because they may not be economically sustainable in the long term.

Failure of the wheat crop to reach potential yield is usually due to a number of factors which may be limiting. To reach the potential yield requires a long-term planning strategy commencing at least two years before the crop is sown. Once the productivity status of the current farming system is determined an improved farming system can be designed and implemented.

A well-designed and managed rotation sequence combined with a sound tillage and stubble management system is the key to any sustainable farming system. Both these components are controlled by management. The economics of crop rotations are very largely tied up with the effects

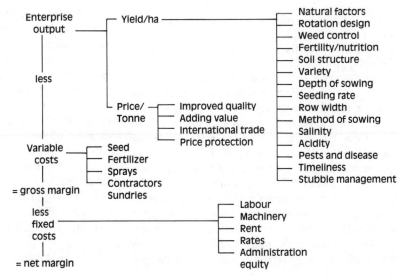

Figure 14.4 *Determinants of crop profitability.*

on yields and the use of resources, and both these factors need to be considered in determining the crop rotation for a particular farm.

The system design must be flexible enough to allow a shift in emphasis from cropping to livestock, or to allow a different mix of crops or livestock in the system to take advantage of changing market opportunities without adversely affecting the ecological and biological sustainability of the system.

The farmer's attitude to work, worry and capital expenditure will also be a major influence, as well as rainfall and the necessity to maintain and enhance the productive capacity of the soil.

A well-designed crop rotation sequence should:
• Combat soil-borne diseases and pests.
• Combat weeds in crops and pastures.
• Maintain/enhance soil structure and soil organic matter.
• Supply some nitrogen to the system.
• Provide feed for livestock.
• Prevent soil erosion.
• Make efficient use of moisture and nutrients.

Most of these points can be illustrated by looking specifically at the effects of introducing a break crop on a farm. The possible financial effects on the farming system are four-fold:
1 The gross margin of the break crop enables farmers to diversify their incomes by growing a variety of crops therefore reducing the impact of price fluctuations.
2 It may lead to a reduction on variable costs for cereal crop production. Fertilizer and spray costs (especially expensive chemicals to control grass weeds) should be reduced for subsequent cereal crops.
3 It will have a beneficial effect on the yields of the ensuing cereal crops.
4 It will effect the fixed cost structure of the farm, relating primarily to the requirements of the break crop for labour and machinery.

The first three effects can be combined so as to measure the likely difference in total gross margin of different rotations. Single enterprise gross margins form the building blocks of farm system design and analysis, which involves evaluating alternative integrated combinations of enterprises in order to determine the most profitable and sustainable mix of enterprises. When comparing rotation systems, the multi-year sum of all gross margins/ha in the rotation cycle divided by the number of years in the rotation cycle, will express the result as an average gross margin/ha per

annum. A comparison of farming systems can be made in this way. The model also assumes that fixed costs will remain the same for the period of the cycle.

It is important to remember that a shift in emphasis between cropping and livestock is likely to have an impact on resource use on the farm (particularly labour, machinery and working capital requirements) and these must be considered in farming system design and analysis. A change in business policy that requires a shift in emphasis between crops and livestock is likely to have a large impact on resource use. On the other hand, a change in the enterprise mix within categories is likely to result in a minor change in the use of resources.

Timeliness in sowing and harvesting can be assisted by use of larger equipment. In the case of sowing, greater management control over timing of operations improves soil moisture utilization efficiency, but this will have to be balanced against the high level of investment in plant and equipment. Early sowing when soils are warm will assist plants escape Cereal Cyst Nematodes (CCN) damage. Weed control and cultivations should not be taking place when sowing should be under way.

In practical terms the opening break and sowing time are in the autumn, however sowing times vary with different crops. It is important to be sufficiently organized as there is some scope to take advantage of different crops and their different sowing times. There will be a cut-off point when the wheat yields will be depressed to the extent that growing wheat may not be as profitable as other alternatives. Other options such as legume grain, pastures and barley could be alternatives to sowing wheat.

The potential yield concept is based on early time of sowing. Any delays in sowing can reduce yields of grain between 80–250 kg/ha/week for every week of delay, depending on the crop and the location of the site. At Roseworthy Agricultural College, South Australia, a field trial in 1989 demonstrated variation in yield according to the time of sowing, with the highest yields of 4.8 tonnes per hectare for crops sown by the first week in May (Table 14.12). By the end of May yields dropped to 3 tonnes per hectare, while those sown by 17 April only yielded 2.5 tonnes per hectare.

With post-emergent herbicides, the time of

Table 14.12 *Influence of sowing date on wheat yield at Roseworthy, South Australia.*

Sowing date	April 17	24	May 1	11	17	24
Yield	2.5	3.5	4.8	3.5	3.2	3.0

spraying can have an enormous effect on crop yield and the return gained from the treatment. Wheat crop yields are set in the first eight weeks or so after sowing. During this period the crop must have no competition from weeds and have free access to nutrients, light and moisture. Late spraying will give a good weed kill but the effect may only be aesthetic. Late spraying may actually reduce crop yields. Applying herbicides at the right time of crop growth and critical weed densities represents management gains without additional cost.

Crops planted by the optimum sowing time are able to make better use of higher soil temperature and therefore germinate more quickly and make better use of soil nitrogen and available moisture. The key to early sowing is the use of knockdown herbicides prior to sowing and direct drilling to keep cultivation to a minimum. This will reduce soil erosion, improve soil structure and increase yields. Alternatively grasses can be controlled the year before in the grain legume phase of the rotation.

Again, the importance of timing arises at harvest. This is particularly important in cropping areas where rainfall at the end of the season will cause losses due to sprouting. In many cases timeliness will reduce losses from bird damage and pod shatter e.g. in the case of lupins.

It should by now be obvious that an important aspect of successful farm management is timing of operations. The organization of resources is a management controllable factor. Machinery breakdowns that hold up sowing and harvesting operations can be very costly in terms of lost yield. A preventive measure would be planned machinery maintenance during non-peak periods.

Price — The second way of improving profit is by obtaining a higher price. Apart from bargaining power and marketing ability, price can be increased by improving product quality, improving the time of marketing, or by adding value. In all cases additional costs will be incurred and the problem is whether the extra returns are greater than the additional costs involved.

There is a need for greater production per hectare, and for higher quality products for export markets. The price of wheat is now based on protein content, and content can vary between 7 and 15 per cent. To upgrade these (protein) values, the farm manager must have an understanding of soil and weather factors in the growing season, the use of nitrogen fertilizer, and crop rotation design. In most instances, inexpensive changes in management, such as selecting a different variety or varying the crop rotation, can lead to worthwhile increases in protein without a sacrifice on grain yield. Managers need to respond to market signals. The Australian Wheat Board has an incentive scheme which offers higher payment for wheat with high protein content.

Wheat grain consists of starch and protein. The grain fills with both at the same time but starch fills faster than protein. In dryland situations moisture usually runs out, photosynthesis ceases and the grain stops filling, leaving the grain with about 90 per cent starch and 10 per cent protein. Protein will not form without nitrogen in good or bad seasons. If there is insufficient nitrogen just prior to grain filling then protein levels will be lower — starch levels continue to rise but protein cannot increase because nitrogen has run out.

Thus, growers have three management practices that they can control to improve quality: (*a*) they have to ensure that adequate nitrogen is available at grain filling; (*b*) they need to plan sowing time; and (*c*) they need to design a good rotation sequence.

Improvement of price by storage and adding value frequently entails processing and other associated costs. For example, grain storage costs include depreciation and interest on plant and equipment, interest on grain stored, loss of weight through drying and handling, and pest control. Instead of charging interest on capital, an alternative approach would be to estimate the additional returns likely (excluding interest charges), and calculate the net return on capital invested.

The two most sensitive factors which influence farm profitability are price and yield. There are many extension programmes to improve yield subject to weather constraints. For price, producers usually rely on last year's price, while the more astute managers would examine critical factors closer to sowing such as futures markets, northern hemisphere crop conditions and exchange rate movements. The critical issue is managing price and the associated exchange rate risk. A range of options is available to growers who wish to manage price risk including: futures and options contracts; forward contracts; fixed price, fixed tonnage contracts; Guaranteed Minimum Price (GMP); cash harvest price; and pool returns.

No organization provides for price insurance at planting time except for the futures and options market and some companies that offer forward contracts. The actual Guaranteed Minimum Price (GMP) is determined during the season and not at planting time. Individual producers are expected to become responsible for their own price risk management. Internationally traded grains are subject to strong down-side price risk as countries respond to a few years of high prices, few production controls and lower prices for other alternative products. Futures and options contracts provide a partial hedge to price vulnerability, largely because there is no requirement to deliver.

Variable Costs — Savings in variable costs may increase gross margins. Such costs frequently constitute between one-third and one-half for cereal production but they are not easy to reduce significantly. This contrasts with pigs and poultry where variable costs which comprise mainly feed are 60–80 per cent of total costs (assuming livestock purchases are deducted from output and therefore not included in costs) and one of the main ways to increase profits is to reduce these costs. However, some opportunities do exist for reduction of variable costs in cropping activities:

- Mixing straight fertilizer instead of buying compounds (though this has difficulties).
- Soil analysis and selecting right choice of fertilizer (e.g. phosphate).
- Timely field operations to avoid cost of more expensive chemical sprays.
- Retaining home-grown seed.
- Reduction or elimination of contractors' charges by buying a machine and doing the job with farm labour. Capital cost/ha can be kept down by buying secondhand equipment or sharing machinery with a neighbour.
- Conservation tillage methods to reduce machinery operating and fuel costs per hectare, but this is usually offset by higher chemical costs per hectare.

Savings in variable costs can be achieved but may not be substantial except for the reduction or elimination of contractors which involves replac-

ing a variable cost with a fixed cost with the intention of reducing total cost. An attempt to reduce variable costs such as seed, fertilizer and sprays will decrease the output more than the costs saved and therefore reduce profit instead of increasing it (assuming that the input level is close to the optimal). For example, soil analysis will avoid wasting fertilizer and the choice of the correct fertilizer can reduce total on-ground costs of the application.

There are three major phosphorus fertilizers for pastures. Single superphosphate has been the main one but changes in pricing of phosphate fertilizers have placed double and triple superphosphate in an economically competitive position (Table 14.13). To assist between choosing the various fertilizers it is essential that phosphorus fertilizer recommendations are expressed as kilograms of phosphorus per hectare. The choice of which fertilizer to use will depend on two factors: the price and the need for sulphur. Where there is enough sulphur in the soil producers have the option of being able to select from the full range of phosphorus fertilizers; where the supply of sulphur is low, the range of fertilizers that contain sulphur is narrowed.

Table 14.13 *Costs of phosphorus fertilizers (example only).*

	Single super	Double super	Triple super
% Phosphorus	9	17.5	20–21
% Sulphur	11	4.4–5	.8–1.5
Rate of fertilizer kg/ha	222	114	100
Tonnes required	133	69	60
Cost of fertilizer ($/t) (ex-works)	140	270	320
Fertilizer cost ($)	18 620	18 630	19 200
Freight cost ($)	2 660	1 380	1 200
Spreading cost ($)	2 797	1 778	1 260
Total cost of fertilizer on ground ($)	24 077	21 788	21 660
Cost/ha ($)	40.13	36.31	36.10

Fixed Cost — Labour and machinery on arable farms together amount to about 40–50 per cent of fixed costs. Therefore cost savings are most likely to come from this area. The efficient organization and integration of people and machines are major factors affecting the profitability of the farm. Striking the optimal level of mechanization requires the correct complement of machinery to allow a high level of timeliness in all operations at an acceptable cost. This theoretical objective becomes very hard to achieve in practice given the options available to farm businesses. For production management to be operated as effectively as possible, detailed planning and thought is required to ensure that the job is done efficiently at the right time with the minimum amount of overtime. A policy of replacing all old machinery with new every four to six years is a luxury policy. It means a fairly high depreciation charge against the business, but the policy should hold repair bills in check. A two-tier strategy is an alternative approach to replacement.

- All 'front-line' machinery, such as a large tractor, harvester and sprayer, should be replaced with new or near new machinery at reasonably regular intervals. These items of equipment can become costly to repair if held too long and unscheduled breakdowns can lead to yield and quality loss.
- 'Second-line' equipment may include everything else on the farm and should be run to the end of useful commercial life and repaired by skilled workers to ensure repair and maintenance costs are at a reasonable level in order to keep breakdowns to a minimum particularly at critical periods.

When high interest rates prevail it is easy to put off until the following year the purchase of machinery, but the equipment is likely to cost more due to inflation. On the other hand, there are tax benefits by using the profit and loss account for the repair component in the purchase of secondhand equipment at a low price. The use of a contractor can enhance productivity in modern farming systems, however the avoidance of a contractor's profit can be achieved through machinery co-operation between farmers. When it is clear that investment in machinery may be necessary it is important that the relative efficiency of alternative machines is assessed. The wrong machine purchase decision and the premature trading-in of a new machine can be an expensive exercise.

When considering capital expenditure on machinery the outlay against the expected return must be considered. In addition, any large expenditure may increase interest charges over a long period. Bank interest is often a factor in reducing farm viability and every dollar spent unwisely will remain as increased borrowing to be financed. For every farm there is a certain viable figure. If interest costs exceed this figure action must be taken and reduction in bank borrowing is likely to be a goal for many farmers.

A 10 per cent saving on these elements of fixed costs can amount to a large saving per hectare. Savings in fixed costs can be achieved by having a seasonal flow of crops that will make use of what should be minimum overhead resources and associated with a spread of enterprises is the cash flow. It is preferable to have a spread of income rather than periods where little cash is coming in.

Economically high and sustainable yields must be achieved from crops and livestock. This will help spread fixed costs over a larger output and achieve a reduction in fixed cost per unit of output.

Pathway to improved system performance

Increasing gross margin by increasing yields per hectare is likely to be the most rewarding way to increase profits. A 10 per cent increase in gross output can increase net farm profit by 20 to 30 per cent. On the other hand, savings in variable costs will not produce substantial increase in profits. Planning fixed costs, especially labour and machinery, is likely to produce cost savings but requires some time to adjust. All current and new investments and their associated fixed costs should be scrutinized.

There is still potential to increase profits and yields further if attention to detail is given to the other management controllable factors that affect yield and price. These include sowing at the optimum time, applying fertilizer correctly and at the appropriate rate, controlling weeds and insect pests and maintaining soil structure through reduced tillage. All of these aspects are essential to the development of a sustainable farming system.

Factors affecting sheep enterprise profitability

Sheep contribute to the economy of the dryland farm by utilizing crop residues and by the technical advantage accruing from mixed grazing in the cereal farming system in southern Australia. Apart from diversifying the farm economy and therefore spreading risk, the sheep enterprise has the added advantage of little tied-up capital. Almost all the capital is invested in the sheep themselves and working capital in every sense of the word is realizable.

Comparisons on a gross margin basis can be misleading. The gross margin fails to illustrate two advantages of the sheep enterprise:

- The extremely low fixed capital requirement, and
- The propensity to fit into the mixed farming operation.

Return on capital invested in the sheep enterprise must always be considered in assessing its contribution to the business.

The factors affecting profit are numerous and are expressed in Figure 14.5. The diagram relates to a breeding flock but can be modified to deal with enterprises that rely on purchasing and reselling sheep.

Once the productivity status of the sheep enterprise in gross margin and in potential production terms is determined, the factors that determine profitability can be reviewed so that improvement strategies can be considered. The determinants of sheep enterprise profitability can be classified as: natural factors; lamb sales; wool sales; replacement cost; stocking rate; variable costs; fixed cost.

Natural Factors — A sheep production model has been developed and relates rainfall to potential stocking rate and level of wool production per hectare.

- The model developed for stocking rate is 1.3 DSE per hectare for every 25 mm of rainfall over 250 mm per annum, i.e.:

Potential stocking rate (dse/ha = (annual rainfall (mm) − 250) x $\frac{1.3}{25}$

- The model for wool is 7.5 kg greasy wool per hectare for every .25 mm over 250 mm per annum i.e.:

Potential wool production (kg/ha) = (annual rainfall (mm) − 250) x $\frac{7.25}{25}$

In the following example where average rainfall is 450 mm, potential stocking rate and potential wool production will be as with the previous formulae:

Potential stocking rate (DSE/ha) = 450 − 250 x $\frac{1.3}{25}$ = 10.4 dse/ha.

Potential wool production (kg/ha) = 450 − 250 x $\frac{7.25}{25}$ = 58 kg/ha.

The following assumptions have been made in the model:

- A dry sheep equivalent is defined as a 50 kg dry sheep grazing under 'normal' conditions where the animal is not gaining or losing weight, or stressed in any way.

Figure 14.5 *Determinants of sheep enterprise profitability.*

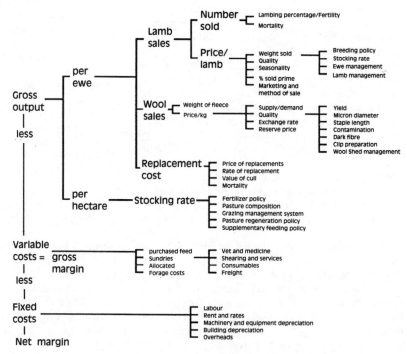

- The 250 mm takes into account rain that falls outside the effective growing season.
- The stocking rate refers to improved pastures and dry crops that are grown to produce hay and grain for livestock feeding.
- The potential stocking rate is the rate that should be achieved during autumn and early winter when there is little or no pasture growth and when land to be cropped is no longer available for grazing.

To maximize output it is imperative growers market even lines of as many fast growing, lean lambs from their flock. Ewe management can influence this objective at mating and during late pregnancy and lactation. Output from the flock consists of wool sales and sales of cast-for-age ewes, cast rams and lambs.

Lamb Sales — Number of lambs per ewe is a function of conception rates, lambing percentage, lambs to weaning, and post-weaning lamb mortality. Lambing percentage can be expressed in several ways:

$$\frac{\text{Number of lambs in the flock at weaning} + \text{Lambs sold before wearing} \times 100}{\text{Number of ewes mated}}$$

This calculation allows for ewe mortality and pre-weaning lamb losses. Lambing percentage is affected by many things, e.g. ewe fertility and rams, nutrition, breeding, stockmanship, ewe mortality, pre-weaning lamb mortality and environment, and season of production.

The number of lambs sold per ewe is determined by the standard of husbandry, management, feeding, genetic factors, disease and weather. Poor results can be related to lack of attention to detail at mating during late pregnancy and after lambing.

The importance of lambing percentage and mortality rates requires emphasis (Table 14.14). In a flock of 1000 ewes it is assumed that 270 lambs will be retained for replacements and provision for mortality. When more lambs are sold, fixed costs

are spread over a larger output resulting in a lower average cost per unit of output.

The price per lamb depends on the target market and it is important to choose a breed or cross that will produce the type of carcass best suited to the target market and produce lambs that provide the carcass weight demanded with the right finish. Market research suggests that lamb consumers require young tender juicy lamb. Consumers want chops with more meat and a fat score of two to three. Virtually all export markets are demanding leaner, meatier lamb carcasses and cuts. Growers will profit from the production of leaner lamb if they produce faster growing lean lambs at a younger age, produce heavier lambs, carry more sheep and increase stocking capacity in some other way. The production of meatier, leaner lambs requires an integrated approach to breeding, management of the ewe and lamb, nutrition and marketing.

The first step to producing leaner lambs is to buy rams that have been bred and selected for fast growth and leanness. Growth rate and leanness are hereditary. Genetically, faster growing leaner rams and ewes breed faster growing leaner lambs. Ewe breeds are chosen for other factors as well as propensity to produce fast-growing lean lambs, such as lambing rate, seasonality of oestrus, wool production and quality, mothering ability, ease of management, availability and cost. In Australia, most slaughtered lambs are produced from first-cross Border Leicester and Merino ewes. Changing breeds is likely to be costly, affecting both cash flow and profitability. This needs to be balanced against an increase in the number and quality of lambs turned off and wool returns.

To produce meatier, lean lambs, growers should purchase rams from studs that emphasize growth rate and leanness, and which uses meat sheep testing (mst) or LAMBPLAN to produce rams of above average growth rate and leanness. These testing schemes offer ram producers and buyers the opportunity to improve the genetic quality of their flocks.

Lamb production in Australia is seasonal and lamb prices tend to follow a predictable seasonal pattern. Most of the lambs are marketed in early spring to early summer. Consequently, prices are at their lowest at this time of the year. As production falls off in the late summer, autumn and winter, prices rise because of supply and demand factors.

Producers have to sell in the market best suited to their production. They must decide whether it is more profitable to go for a higher priced product by lambing out of season and by costly hand feed-

Table 14.14 *The effect of lambing percentage on lambs sold.*

Lambing %	Male Wethers Sold	Ewe Lambs Retained	Ewe Lambs Sold	Total Lambs Sold
100	500	270	230	730
90	450	270	180	630
80	400	270	130	530
70	350	270	80	430

ing to produce a prime lamb, or whether to lamb later and aim to produce a more cheaply fed lamb to market at a lower price. This decision can be made by careful budgeting, taking into account all the variables affecting the individual farm business. The best price is obtained when a high proportion of lambs are finished and marketed before September.

Wool Sales — These are a function of fleece weight and price. These factors are influenced by breeding, nutrition and husbandry management. To produce a clip uniform in fibre diameter, length, strength and colour, attention should be paid to: micron diameter; clip preparation; and wool shed management.

Replacement Costs — With a breeding flock replacement costs entail the cost of new ewes and rams joining the flock set against the cull value of those animals which are no longer suitable and sold. They depend on:

- Rate of replacement. A rigorous policy of culling will increase the depreciation/replacement rate of the flock and must be justified by increased productivity resulting from a shorter genetic interval.
- Cost of replacement. Removal of animals by culling will not result in a reduction of fixed costs. If culling leaves a smaller number of animals in the flock, then each of the remaining animals must carry a larger share of fixed costs. Total flock output must be maintained if the level of productivity is to be maintained or improved and this usually means that total flock numbers must be maintained at farm capacity.
- Value of culls. Animals should be replaced with an animal that will be more productive.

Stocking Rate — The differences in gross margin between the top one-third producers and the average is largely attributed to higher stock rate. The relationship between animal — and per hectare — performance needs to be highlighted and this is related to pasture quality and type, the grazing management system, pasture management and supplementary feeding policy. Vigorous highly productive pastures are a key to high stocking rates.

There is a limit to the increase of the stocking rate. From the analysis of a stocking rate trial at Canberra (Table 14.15), the highest stocking rate yielded the best utilization of pasture but the operation of the law of diminishing returns meant that the point of maximum profit was reached

Table 14.15 *Relationship of stocking rate to gross margin, Peppin Merino ewes.*

Ewes per hectare	Gross margin per hectare	Marginal return
5	84	16
7.5	100	10
10	110	4
12.5	114	−2.0
15	112	−6.0
17.5	106	

(SOURCE: P. F. Byrne, M.Ag.Ec. thesis, University of New England)

before the point of maximum production. The analysis takes into account the cost of feeding in poor seasons, loss of income from reduction in wool cut and lambing percentage as stocking rates of Peppin Merino ewes rise.

Variable Costs — The principle of variable costs is best illustrated in Figure 14.5. Variable costs vary quite widely and again they are dependent on the production system and the type of country the sheep run on. There seems to be no single set of standards to apply.

Variable costs per head may be reduced by increasing efficiency in areas such as disease control, and husbandry methods. The quality of the shearing team and shed organization will improve shearing efficiency.

To make the sheep enterprise more competitive it is necessary to raise both its gross output and its gross margin in relation to the land it uses. This means output per hectare must be raised or a significant reduction made in variable costs. Variable costs are difficult to reduce without risking the income generating capacity of the enterprise. Therefore any increase in gross margin must come from an improvement in gross output per ha. This may be achieved by increasing output per ewe/dse and by increasing numbers per hectare. Figure 14.5 shows many possible ways to increase the gross output per ewe. Each of the factors — improvements in wool sales, reduction in replacement costs, and prices obtained per lamb — will produce improvements. However, there are possibilities for increases from an improvement in lambing percentage, and strategies to achieve this should be explored.

Most producers are preoccupied with per-head

performance. Returns from the sheep enterprise will be highest when stocking rates are high with emphasis on per hectare production and returns, rather than on per head performance. High gross margins will depend on the level of sheep husbandry, grazing management and pasture quality.

Any dramatic improvement in the gross margin per hectare must be obtained from sheep. In the foreseeable future, a rise in stocking rate through improved pasture and grazing management systems and the judicious use of supplementary feeding would seem to hold the key to profitable sheep farming.

Conclusions

In designing a sustainable farming system the manager should select farming practices which place emphasis on long-term profitability rather than short-term gain. To achieve this objective it will be necessary to set yield targets in accordance with long-term sustainable husbandry practices.

One of the main objectives of a multi-enterprise mixed cropping and livestock farming system is to combine enterprises that will be complementary in their use of resources, particularly time, labour and machinery. Combining appropriate crop and livestock enterprises will allow a spread in the use of these resources over the production year.

CHAPTER 15

Modelling to analyse an Integrated Crop–Livestock Farm: MIDAS

D. Morrison, A. Bathgate and J. Barton

SYNOPSIS

This chapter describes MIDAS (Model of an Integrated Dryland Agricultural System), a mathematical programming model designed for agricultural systems analysis. MIDAS selects optimum land-use strategies and other management practices from many alternative activities, while accounting for scarce resources, important interdependencies between enterprises and other biological relationships.

MAJOR CONCEPTS

1 MIDAS, like all models, has limitations which are important for users to understand. Its use is largely complementary to models which simulate single enterprises and represent the biology of these in greater detail.

2 Further extensions from standard MIDAS have involved development of models treating sheep and pasture biology in more detail, models able to address long-term land degradation issues and a model accounting for uncertainty of seasons.

Dryland agriculture based on integrated sheep and cropping farms is important in most mainland States of Australia. In these farms sheep eat the stubble and some grain from crops, while crops and pastures are often grown in rotation, each having important effects on the other. These interdependencies mean that it is important to look at the whole farm rather than just the separate enterprises, so that a systems approach is needed. The complex biological nature of these farms and the importance of economics to the farmer means a multi-disciplinary approach is needed by those carrying out research for and advising farmers.

The requirement that agricultural advisers and researchers take a multi-disciplinary farming systems approach has little meaning unless the farming system can be analysed in a quantitative way. MIDAS has been developed as a tool for quantitative analysis of farms by farm advisers and agricultural researchers servicing the crop-sheep farming areas of Western Australia. One reason it was developed was to allow advisers to investigate how different land uses and other practices compare in terms of overall farm profit, and what implications each has for the management of the rest of the farm. Another reason it was developed was to allow agricultural researchers to consider the impact that different research directions or project outcomes might have on wholefarm profit, and on how the rest of the farm should be managed.

MIDAS

MIDAS was built by economists with modelling skills in combination with the regional advisers and the research staff who were to be the end users of its output. A good model could not be developed without such involvement. In fact, to be highly credible, advisers and researchers needed to be involved in such a way that they viewed it as 'their' model. Building MIDAS required input from soil scientists, agronomists, animal production researchers, economists, farmers and consultants. It also required constant questioning by end users as to the credibility of the output and assumptions, once the model was operational. Feedback from end users has led to constant improvement and updating of MIDAS.

MIDAS is a mathematical programming (MP) model. This means it has some important differences from most models of agricultural systems which are simulation models. Some of these differences are mentioned later in this chapter. Like all MP models, MIDAS can be viewed as having three parts:

1 An objective to be maximized or minimized. In the case of MIDAS, this is maximization of wholefarm profits.

2 Activities which can be viewed as different ways of meeting the objective. In the case of MIDAS, these are different land uses and other farm management practices.

3 Constraints which limit the activities. In the case of MIDAS, they are limited resources of land, finance and labour, and limitations imposed by the biology of the system.

MP models are solved using an optimization program. In each run, the level and combination of activities which maximizes (or minimizes) the objective, subject to constraints, are calculated. In the case of MIDAS this means that such activities as rotations, flock structure and stock numbers which maximize farm profit are calculated while meeting resource and biological constraints.

One kind of MP is linear programming (LP). MIDAS is largely an LP model, which means that it optimizes more efficiently, although it does require that some limitations of LP are addressed in building MIDAS. The following description provides a more detailed explanation of the components of the MIDAS model.

Components of the MIDAS model — activities

Activities are the management options which use the scarce resources of land, labour and capital and directly or indirectly generate income to meet the objective. An activity can be a land-use option (i.e. a rotation), an option which supplies inputs to a rotation such as applying fertilizers, or options which use outputs of the rotations such as selling wheat. Running sheep fits into the last category as they 'use' pasture supplied by a rotation. MIDAS has about 400 activities. The following are the major activities within MIDAS which are concerned with representing alternative management options:

Rotations — Fourteen to twenty different cereal crop/pasture and cereal crop/legume crop rotations are included on each of up to seven different land management units (LMU). The cereal crop in the rotations can be wheat, oats, barley, or triticale. Effectively the number of rotations represented is up to 80 per soil type. In general the rotations are cereal/pasture, cereal/legume and continuous cereal. Any of the above crops may be selected as a component of a rotation. These activities include those costs of pasture and crop management which are determined directly by the rotation, e.g. crop herbicide costs are largely determined by previous land use.

Crop Growth and Crop Fertilizer Response — Within a crop rotation, there is the option to grow one of several cereal crops, and each cereal in each rotation has a different response function to nitrogen fertilizer. Because the response to nitrogen is non-linear, five activities are used to approximate the non-linear response function with five linear segments. This linear approximation of a non-linear function is shown in Figure 15.1. Phosphorus is treated as a single fertilizer requirement. As well as providing grain for sale or feeding, these activities provide stubble for sheep grazing.

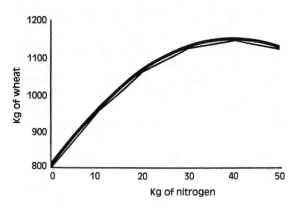

Figure 15.1 *Linear approximation of a nitrogen response curve.*

Cropping Machinery — For any single run of MIDAS, only one complement of machinery is included. This means that the effect of different machinery options cannot be compared within one run but only between runs. Operating costs of sowing are represented in sowing activities. The rate of crop sowing in these activities is limited by the size of machinery. Late sowing incurs a yield penalty. Crop sowing activities represent a progressive reduction in yield for every day sowing is delayed. Thus, while MIDAS could select a very large cropping programme with a small seeding plant, this would be accompanied by a substantial yield penalty.

Sheep and Flock Structure — Up to thirty classes of sheep are represented by individual activities. The classes include different ages of ewes and wethers, from lambs through to cull for age, which is six years for ewes and seven years for wethers. Sheep energy requirements and intake capacity for each class of sheep in each month are represented in these activities. These are calculated as a function of sheep liveweight, liveweight changes, and pregnancy or lactation status. Different selling times of wethers are allowed, so that shipping wethers may be sold at any of four times during the year, and other wethers may be sold at two times during the year.

These activities allow the most profitable flock structure, sheep numbers and stocking rate to be

selected each time MIDAS is run. Sheep costs other than feed are represented in these activities and the returns from the sale of sheep are represented in the sheep selling activities.

Fertilizer Application — Activities are represented which allow different types of nitrogen and phosphorus fertilizers to be applied. They provide nitrogen to the crop fertilizer response activities and to meet phosphate fertilizer requirements. They include the cost of the fertilizer and its application.

Sheep Feeding Strategy — These activities are feed supply options for sheep. They include options of sheep consuming pasture, stubble from crops and grain feeding. The grain is either produced on-farm or bought in.

Feed not consumed in a particular monthly period is passed to the next month by transfer activities. In the case of pasture and stubble transfers, a decline in quantity and quality of feed is represented to account for the biological and physical breakdown of feed over time. The changing quality of stubble and pasture has implications for management, as low quality feed does not have a sufficient energy concentration to meet the energy requirements of sheep without exceeding the limited intake capacity of sheep.

Stubble has three main components: spilt grain, leaf and stem, which differ markedly in digestibility. As high quality components are eaten, the average quality of the stubble is reduced. Several stubble activities represent preferential grazing of higher quality components of stubble by sheep.

Fixed Inputs — The cost of fixed inputs required to support the cropping and sheep enterprises are represented in cropping machinery and sheep overheads activities. Fixed costs accounted for in sheep overheads include the annual equivalent of replacement costs for shearing sheds and equipment, feed storage, water supplies and fencing. Fixed costs for cropping represent the annual equivalent of replacement costs for tillage, sowing and harvesting machinery and for sheds to house this machinery.

Finance — Activities represent the option of borrowing money to overcome a seasonal shortfall of revenue at the cost of interest payments. Conversely, cash flow surpluses receive income from interest.

Selling Output — Grains, wool and sheep which are produced are sold in product selling activities.

Farm-gate prices are represented here.

Other activities have been included, but not routinely. For example, activities allowing comparison of different times of lambing, different cultivation practices, and innovative sheep feeding strategies, have been included in specific runs.

Components of the MIDAS model — constraints

Resource and Physical Constraints
- Soil area. These include representation of the limited area of land and the limited area of each LMU. Each LMU is made up of several different pedological soil types which have similar production characteristics, i.e. similar inputs are required for the same yields.
- Although it could refer to a single paddock, an LMU usually encompasses several paddocks. At present, aggregation is such that no more than seven LMU are represented in any one MIDAS model. Apart from the concern of keeping the model to a manageable size, the available data on crop and pasture yields are not sufficient to allow differentiation of more than seven LMU.
- Credit. As farmers are restricted in the amount of money they are able to borrow on a seasonal basis, this can influence the land use and other practices they can afford to adopt. For this reason, a constraint limiting the amount of credit or overdraft is included.
- Labour. Limited available family labour is represented at the time of peak labour demand. Tasks which exceed the work capacity of family labour incur the additional cost of hiring casual labour.
- Capital. Constraints also represent the limitations that machinery size places on crop-sowing time and the limitations of the fixed inputs of the sheep enterprise.

Logical and Biological Constraints — Some of these constraints represent simple balancing, such as the requirement that no more wool can be sold than is produced, or that the amount of grain fed to sheep, or sold, is not more than that produced or bought. These are:
- A constraint which limits the quantity of wheat sold to the quantity produced.
- A constraint which limits crop area to that specified by the rotation.
- A constraint which limits the quantity of wool sold to that produced.
- Livestock reconciliation. In the short term, a

business will maximize profits if all assets are sold at the end of the term. As MIDAS is a single-year model, profits would be maximised if sheep were sold at the end of the year. To prevent this, balancing constraints are included so that the number of each class of sheep on hand at the end of the year must equal that at the beginning of the year. Constraints of this kind mean that MIDAS is categorized as a single-year equilibrium model.
- Biological and conservation constraints. Other balancing constraints represent biological and conservation limitations in the farm system. These include balancing energy provided by paddock feed and grain feeding with the energy requirements of sheep in each monthly period, and ensuring that the volume of dry matter consumed by the sheep cannot exceed their intake capacity. Also, crop responses to fixed and applied nitrogen are linked to legumes in rotation and fertilizer application activities. Sheep consumption of crop stubble and dry pasture is constrained as a soil conservation measure and to ensure there is enough pasture seed left to germinate pasture the following year.

Components of the MIDAS model — biological relationships

In addition to representing alternative activities, resource and logical constraints, much of MIDAS is taken up representing the biology of the system and biological relationships between different enterprises, which we refer to as interdependencies. To an extent, the treatment of the biology in MIDAS has been described in the sections on activities and constraints, but this section covers some aspects of the biology not previously mentioned and reiterates some that are important.

Interdependencies — Relationships between Enterprises — If the profitability of one farm enterprise was in no way affected by another farm enterprise, then it would be sufficient to look at the enterprises separately. Enterprises, however, do have significant effects on each other, e.g. the profitability of a wheat crop is affected by soil fertility and weed burden, which is partly dependent on the previous land use. The wheat crop, in turn, provides stubble which may affect the profitability of the sheep enterprise. The following interdependencies, some of which have positive and some negative effects, are included in MIDAS:
- The effect of fixation of nitrogen by pasture and crop legume on a subsequent cereal crop.

- The boost in yield of cereals resulting from the disease break provided by crop legumes.
- Reduced pasture production in the first and second years after crop.
- Herbicide costs for crops will vary with previous land use because of its effect on the weed burden.
- The availability of crop stubble for sheep grazing.
- The availability of grain produced on-farm as a sheep feed, which may be cheaper than buying in grain.

Biological Relationships within Enterprises — Some of these have already been mentioned in the description of activities and constraints: diminishing (non-linear) return to additional fertilizer application; lower crop yields with later seeding; energy requirements of sheep and limited feed intake capacity; and the effect of grazing pressure on pasture production. Other biological relationships which have a significant effect on model results are:

- Seasonality of pasture growth. The climate in the Western Australian agricultural areas is Mediterranean. Rainfall mainly occurs during the cooler months (May to October) followed by several dry, warm months, causing a marked seasonal variation in pasture growth and energy concentration. This is represented in different production and energy concentration estimates included for pasture in each month.
- Stocking rate effect on pasture production. Activities represent the reduction in pasture growth resulting from trampling and consumption by grazing sheep. The extent to which growth is reduced is related to sheep numbers. This is included in the model so that realistic stocking rates are selected.
- Sheep liveweights. Seasonal changes in feed availability affect sheep liveweight. Liveweight patterns for each class of sheep were estimated to reflect the likely availability of feed. Representation of changing liveweight is important because if bodyweight loss is not represented, feed requirements would be overestimated when feed is scarce and stocking rates selected would be too low. The effect of liveweight on the selling price of sheep is a further reason to represent seasonal liveweight changes.
- Wool quality. Fibre diameter can have a large influence on price received for wool and it is known that different classes of sheep produce wool of a different fibre diameter. Finer wool from hoggets and ewes, and the higher sale price for fine wool, are represented.

Other biological relationships are included but not routinely, e.g. protein levels in wheat have been represented as a function of various management practices in order to work out whether to adopt a strategy of increasing protein in response to payment for protein.

Model applications

The following are five typical examples of uses to which MIDAS has been put. The first three relate to extension programmes by the Western Australian Department of Agriculture and the second two to research projects.

Optimum rotations

Because MIDAS is an MP model, an optimum solution is obtained for each run and information on the closeness of activities which are not optimal is provided. Tables 15.1 and 15.2 show excerpts from output dealing with optimum rotations on two soil types in the eastern wheatbelt of Western Australia.

The first column of both Tables 15.1 and 15.2 shows the land management unit (denoted by S and a number) and then the rotation type; the second column shows the area of the rotation selected (if it is zero then it is not selected); and the third column shows the 'shadow cost' of the alternative activities, or how many dollars per hectare (at the margin) the rotation is behind the optimum. Note that this kind of information is provided for all of the 400 MIDAS activities.

Table 15.1 *Optimum rotations and closeness of alternatives for yellow-brown loamy sand (S2).*

Rotation	Hectares Selected	Shadow Price ($/ha)
S2PPPP	0	−43.61
S2PPPW	0	−37.14
S2PPW	0	−41.51
S2PWPW	0	−39.31
S2PPWW	0	−30.77
S2PWW	0	−34.62
S2PWWW	0	−30.71
S2WWWW	0	−27.11
S2WWLBEB*	289	0
S2WWL	170	0
S2WLBEB	0	−.20

*BEB—Before the break of season.

On the yellow-brown loamy sand (S2), Table 15.1 shows the model selected to grow 460 ha of a 2 wheat:1 lupin rotation (WWL). For 289 ha of this rotation, lupins are sown before the season break (WWLBEB) to allow extra machinery time for cereal sowing after the break of the season. Actual lupin areas would thus be 460/3 = 153 ha, and 289/3 = 96 ha of those would be sown before the break. Although not selected, the third wheat/lupin rotation is within 21 cents per hectare of being selected. By contrast, the continuous pasture (PPPP) is $43.61 per hectare behind the optimum, and the most profitable rotation not including lupins is $27.11 behind. Based on the shadow costs on this soil type, the optimum land use is a wheat/lupin rotation. However, because of the closeness of wheat/lupin rotations and the uncertainty of data used in the model, it might not be possible to conclude which one is best.

The optimal rotations selected on poorly structured heavy red loam (S6) tell a different story (Table 15.2).

Two dissimilar rotations are selected as optimum — 237 ha of continuous pasture (PPPP) and 223 ha of a 2 wheat:1 field pea rotation (WWF). Lower shadow costs than in Table 15.1 indicate that some of the alternative rotations are not so far behind the optimum as for S2, e.g. even the introduction of a hectare of three pasture:1 crop (PPPW) reduces profit by only $6.07 and five rotations are within $10/ha of the optimum.

In contrast with S2, therefore, the rotational recommendations for this LMU are less clear-cut and would involve a number of choices.

Table 15.2 *Optimum rotations and closeness of alternatives for heavy red loam.*

Rotation	Hectares Selected	Shadow Price ($/ha)
S6PPPP	237	0
S6PPPW	0	−6.07
S6PPW	0	−8.78
S6PWPW	0	−11.35
S6PPWW	0	−6.88
S6PWW	0	−14.63
S6PWWW	0	−13.46
S6WWWW	0	−22.06
S6WF	0	−6.96
S6WWF	223	0
S6WWWF	0	−3.65

P= Pasture, W = Wheat, F = Field peas.

Crop per cent curve

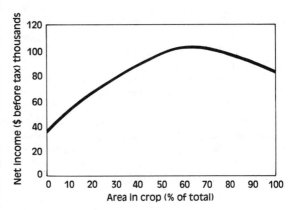

Figure 15.2 *Profit versus per cent of the farm in crop.*

MIDAS can be constrained to different areas of crop and an optimum solution and profit computed. When this is done, profit can be presented as a function of cropping area, (Figure 15.2). This function has a characteristic curved shape; profit initially increases with increased area of crop (but at a decreasing rate) until it 'plateaus off' and declines. The level at which profit plateau is reached varies with the region modelled and relative grain and wool prices, but it usually does not extend below 40 per cent of total area cropped or above 80 per cent of total area cropped. Figure 15.2 shows that profit is maximized at about 60 per cent cropping but that a plateau exists between about 50 and 80 per cent cropping, where less than $5000 of net income separates these points. Extension of results such as these to groups of farmers would not emphasize the optimum level of cropping (60 per cent) but the 50 to 80 per cent range. Reasons for this caution include the continual change in the relative price of grains and wool, farm to farm differences in resources and skills and the degree of accuracy of MIDAS data.

The explanation for the shape of the function is: at low levels of cropping, it is highly profitable to increase cropped area because additional crop can be grown on the soils relatively suited to cropping, the stubbles and grain grown are of high value to the large sheep enterprise, and because cereal crops are grown in longer rotations. As cropping increases, only less suitable soils are available to be sown to additional crop, crop stubble availability no longer limits sheep carrying capacity, and cereal crops are grown in shorter rotations. Above 80 per cent cropping, very tight crop rotations, relatively few sheep to make use of plentiful stubbles, limits on crop-sowing time due to machinery capacity and the cropping of poorer soils, mean lower net income. It is only the legume crops in rotation that prevent net income from crashing well below the levels shown for 90 and 100 per cent cropping.

The reasons listed above for the shape of the curve all relate to the complementarity of sheep and cropping enterprises in the farming system. In fact, it was the advocacy of 100 per cent cropping during the early 1980s, when high wheat prices prevailed, that helped prompt the development of MIDAS and the first issue MIDAS was used to address was the economics of 100 per cent cropping.

Lupins

In recent years some farmers have found lupin yields to be lower than expected and prices have fallen so that the role of lupins has been questioned. Simple analyses that farmers might conduct comparing yield per hectare and then subtracting operating costs, could show wheat to be well in front of lupins. However, simple analyses do not represent such factors as: the very different performance of lupins on different LMUs; the benefits to the subsequent wheat crops provided by nitrogen fixation and other yield boosting effects; lupin stubble availability to sheep; the availability and suitability of lupin grain as a sheep feed, and the flexibility of lupin seeding time.

MIDAS was used for this analysis because it accounted for these factors. The questions addressed in the analysis were:
- Which LMU(s) should be used for lupin production?
- Which are the most profitable rotations on each LMU?
- What is the lowest yield of lupins at which they are still profitable (i.e. the break-even yield)?

Results showed that lupins are profitable in the eastern wheatbelt on at least two of the six LMUs and in each case, the best rotation was 2 cereal crops:1 lupin. The yields used in this case were the expected yields of 0.95 t/ha and 0.8 t/ha on S2 and S3 respectively (S2 is yellow-brown loamy sand and S3 is a gravelly sand).

When MIDAS was run with lupin yields below 0.8 t/ha, lupins became unprofitable on S3 and continuous cereal became the best rotation. On S2, however, the 2 cereal:1 lupins was profitable even when the lupin yield was 0.6 t/ha, an extremely unlikely occurrence given the conservative expect-

ed yield of 0.95 t/ha. The implication of this finding is that lupins are clearly profitable on S2 but only marginally so on S3.

These results strongly suggest that farmers should not reconsider the role of lupins in their farming system unless they are getting yields well below 1 t/ha averaged over the last five years.

Pasture analysis to determine research priorities

Pasture researchers have had to consider on which soil types they should concentrate their research and for which growth characteristics they should be selecting. To help in this process, MIDAS was run firstly to determine at what yield increase would be required for pasture to be profitable on soil types where it is not currently profitable to grow it. Secondly, the value of extra pasture production at different times of the year was estimated to help determine breeding selection criteria.

Results showed that the soils which had the best prospect for the greatest increase in pasture production were not necessarily the best soils on which to concentrate research. The reason for this is the opportunity cost of alternative land uses. Table 15.3 shows that while researchers see S2 as an LMU on which research could increase pasture production substantially, the high profitability of the alternative wheat/lupin rotation means that at least a 60 per cent increase in pasture production is needed. It is very unlikely that this can be achieved.

By contrast, the deep acid sand (S1) has prospects for a smaller increase in pasture growth (see Case Study E), but the low profitability of the cropping alternative means that research here is likely to be of more value to farmers. Soil types S5 and S6 are also better than S2 for further pasture research, not because of prospects for higher yield increases but because of the lower opportunity cost of alternatives.

Table 15.3 *Break-even and expected increases in pasture production on soils in the eastern wheatbelt of Western Australia.*

Soil type	Break-even production increase	Expected production increase through research
S1	Nil	Low
S2	60% +	High — (<60%)
S5	15%	High — (>15%)
S6	Nil	Moderate

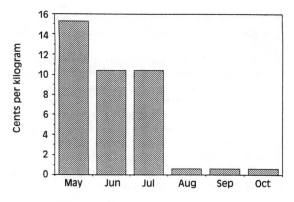

Figure 15.3 *Marginal value of a kilogram of pasture at different times of the growing season (cropped area of 60 per cent).*

Figure 15.3 illustrates monthly difference in the value of additional pasture as it changes during the growing season. The value of pasture reflects the scarcity of sheep feed at that time of the year. This indicates that pasture species which produce extra dry matter early after the break of season are much more valuable, all other things being equal. In fact, selecting for more vigorous early growth could be profitable even if it was at the expense of spring growth.

These results helped researchers decide the soil types on which to direct pasture research and have led to a rethinking of the value of phosphate application to pasture and evaluation of autumn lot-feeding to encourage early pasture growth, oats fodder crops, and trees and shrubs for early forage. Recent and current MIDAS work is part of this evaluation.

Lambing per cent

A large research effort has been and is being undertaken to research increases in lambing percentage. MIDAS was run to investigate the value to eastern wheatbelt farmers of higher lambing percentage. The analysis accounted for the higher price received for the finer wool of young sheep and computed the most profitable use of the extra lambs generated. Although profits were higher with higher lambing percentages in the eastern wheatbelt, they were only of the order of $2 per ewe, minus the cost of inducing the additional pregnancies. The reason for this was found to be that higher profits can only be realized by carrying the extra lambs to at least a saleable age. This means additional costs, either in grain feeding, growing more pasture (and hence less crop), or running fewer adult sheep. The implication of this work is that research should concentrate on low-cost technologies for increasing lambing percentage.

Results also showed that reducing the number of ewes is a strategy to employ to make the most profitable use of a sustained increase in lambing percentage in the eastern wheatbelt. This may seem an unusual finding because it appears that if ewes are more productive then it is better to run fewer of them!

Other uses of MIDAS have been wide-ranging. They vary from investigating the profitability of different lambing times, to examining whether research into new legume crops would be helpful, to considering whether it would be profitable to research methods of treating stubble to make it more palatable, to estimating the contribution of fixed nitrogen and other interdependencies to whole farm profit.

Strengths and limitations

No model is perfect but most have at least some strengths and all have limitations. Model users must understand these strengths and limitations so that they can assess what credibility should be given to results, and to which uses it is most appropriate to put the model.

Optimization and whole farm analysis

MIDAS is strong in terms of the economic analysis it provides through optimization and because it is well suited to whole farm analysis. For an optimization model, it places unusual emphasis on the interdependencies between enterprises and the representation of biological relationships. It has been categorized as a bioeconomic model because it combines the analytical power of MP with representation of biological relationships. These strengths can be seen in the optimum land use and other practices selected by MIDAS runs and also the information that is provided on how close alternatives are to the optimum. The biological information means that the impact of changing biological parameters on whole farm profit can be quantified. The results section best illustrates the strengths of MIDAS.

Biological detail

Although there is a great emphasis on the biology

of the system in MIDAS, the treatment of the biological relationships within enterprises is simpler than in some simulation models of enterprises. For example, the representation of sheep-pasture interactions is more detailed in simulation models concentrating solely on the sheep enterprise, such as GRAZPLAN, while more fundamental levels of the farm system are not represented at all in MIDAS. The focus of MIDAS is at the whole farm level where it is very important to represent enterprises, how those enterprises compete for scarce resources, and their biological interdependencies. It would be possible to model these aspects in more detail, but the costs associated with doing so may outweigh the benefits. As a model becomes more complex, it becomes more costly to build and to run. Results are more difficult to interpret so the chance of errors increases, and the delay before it is useful also increases. Eventually a point is reached where the cost of increased complexity exceeds the benefit.

The policy with MIDAS has been to represent the biological detail insofar as it may have a significant effect on the land uses and other practices selected. Rather than attempt to duplicate enterprise simulation (e.g. sheep-pasture simulation), it is recognized that the models are complementary; MIDAS provides the perspective of the whole farm and enterprise comparison using optimization, while enterprise simulation better represents biological processes. There is some overlap as MIDAS does allow comparison of different ways of running enterprises (e.g. comparison of different flock structures and stocking rates). Similarly enterprise simulation can include comparison of the profitability of some alternatives, but it cannot efficiently find an optimum from many alternatives and provide information on the closeness of alternatives, as shown above.

Single-year equilibrium

Another limitation is the treatment of time. MIDAS is a single-year equilibrium model. This means that it compares, for example, established rotations rather than newly introduced rotations and different flock structures without looking at getting from one flock structure to another. Thus, the financial and biological transition from one state to another is not represented. The reason for not representing the process of change from one state to another is the cost involved in increasing the complexity of models.

As a consequence of the treatment of time, it should be borne in mind by users of MIDAS that:

- It is more relevant to medium-term management decisions than decisions farmers need to make in the short term.
- It may not be financially feasible to move to the optimum practices identified by MIDAS. This is not of concern unless the optimum is radically different from present practices, requiring considerable investment and/or the optimum is only slightly more profitable than present practices.

Probably of more importance is the unsuitability of the single-period MIDAS in addressing alternative practices which have implications for land degradation in the longer term (see below).

Year-to-year variation in season

Uncertainty about yield and rainfall is an important characteristic of the farming system under study. Only one type of season can be considered in a single run of MIDAS, but this effect is reduced by representing the 'expected season' in MIDAS. This means that in estimating a yield or level of nitrogen fixation or a response function for MIDAS, the estimate should be the mean of all previous observations rather than one taken from any particular season. Frequently data were not available to calculate model relationships with such rigour, but all those contributing to the project were aware that data had to be representative of the range of season types experienced, each weighted for its probability of occurrence. A recent development to better represent season uncertainty is mentioned below.

Representation of individual farms

Mostly MIDAS runs are based on one or at most several farms representative of a region, rather than many individual farms. The reason for this is that MIDAS was designed to support the broad extension and research role of the Department of Agriculture, rather than to provide an individual farmer service. Moreover, there are not the resources to operate a MIDAS service for many individual farmers. Individual farm runs are conducted but only to test new versions of MIDAS.

Further developments

The limitations of standard MIDAS, while acceptable for some applications are not acceptable for others and there are some important issues it cannot address. Consequently more sophisticated models have been developed. While these are more powerful than standard MIDAS for some applications, their complexity means that they are more costly to run and fewer people are able to run them and interpret their output. Thus they remain narrowly used research tools. They include:

- Models which examine long-term profit and which are able to incorporate effects of land degradation. Two models have been developed; one is concerned with strategies for gradually reducing salinization of land and the other is concerned with strategies for reducing wind erosion.
- A model which accounts for season uncertainty. This has been developed to test the robustness of answers from standard MIDAS when nine different season types are accounted for.
- A model which represents the biology of sheep and sheep-pasture interactions in much greater detail. In this model the presentation of sheep liveweight and the interaction of sheep and pasture is much closer to that of sheep enterprise simulation models. This has been done to better represent areas in which the sheep enterprise is larger and more important than cropping. This treatment will not be extended as the standard for all MIDAS models unless it is shown that it makes a significant difference to results.

Conclusions

MIDAS makes a particular contribution at the level of whole farm and its strength is economic analysis while accounting for interdependencies between enterprises and biological relationships. There is a dearth of analyses of this type. MIDAS has affected and become an integral part of extension and research planning of some regional centres of the Western Australia crop-sheep farming areas.

Where it is used, it has meant that terms such as 'multi-disciplinary approach' and 'farming systems analysis' have a clear and tangible meaning.

Nevertheless, MIDAS is only an aid to questions of research direction and extension; it complements but is no replacement for the judgement of expert staff. Proper use requires an understanding of the farming system and strengths and limitations of MIDAS.

A prerequisite to success has been the involvement of regional clients at the outset of the project. The degree of success of MIDAS in the different regions is related to the degree of original involvement of regional staff.

Further reading

Dent, J. B., Harrison, S. R. and Woodford, K. B. (1986). *Farm Planning with Linear Programming: Concept and Practice*, Butterworths, Sydney.

Ewing, M. A. and Pannell, D. J. (1987). 'Development of regional pasture research priorities using mathematical programming', in J. L. Wheeler, C. J. Pearson and G. E. Robards (eds), *Temperate Pastures: their Production, Use and Management*, Australian Wool Corporation and CSIRO, Perth, 583–85.

Ewing, M. A., Pannell, D. J. and Morrison, D. A. (1986). 'The place of lupins in the farm rotation: a whole-farm modelling approach'. *Proc. of the International Lupins Conference*, Geraldton, Western Australia, 152–60.

Kingwell, R. S. and Pannell, D. J. (eds)(1987). *MIDAS, A Bioeconomic Model of a Dryland Farm System*, Pudoc, Wageningen.

Morrison, D. A., Kingwell, R. S., Pannell, D. J. and Ewing, M. S. (1986). 'A mathematical programming model of a crop-livestock farm system', *Agricultural Systems*, 20, no. 4, 243–68.

Pannell, D. J. and Falconer, D. A. (1988). 'The relative contributions to profit of fixed and applied nitrogen in a crop-livestock farm system', *Agricultural Systems*, 26, no. 1, 1–17.

CHAPTER 16

Modelling as an Aid to Systems Evaluation

K. G. Rickert and G. M. McKeon

SYNOPSIS In this chapter the criteria used to evaluate farming systems are described. Types of computer models are briefly mentioned and the roles of modelling in the evaluation of farming systems are discussed from the viewpoint of research, extension, farm management and education. Future roles for modelling and the requirements of a good model are mentioned. Examples focus on readily available models that have made a notable contribution to the analysis of dryland farming systems.

MAJOR CONCEPTS

1 To survive in the changing and competitive environment of farming today, operators in the rural industries need to make sound managerial decisions. Computer models can assist in this task.

2 In simple terms, a computer model for a farming system is a series of mathematical equations arranged by a model-builder to mimic the complex processes of the system.

3 A model can be regarded as a collection of hypotheses that express the operation of the system in quantitative terms.

4 Computer modelling is one of several techniques that can be used in systems analysis. Models can evaluate farming systems in terms of productivity, reliability and sustainability depending on the type and purpose of the model.

5 A model can be used to identify new operating principles for better management of the system and to suggest a preferred pathway for new research.

6 Some models can be used to evaluate new technology before its introduction or to evaluate the impact of new technology on existing production systems.

7 Some models have been specifically designed to allow farmers to evaluate alternative management strategies. The range and usefulness of such models is likely to expand in the future.

8 Models can be valuable aids to agricultural education by illustrating the interactions between components in a system and the underlying scientific and managerial principles.

Modelling is a powerful tool for the evaluation of farming systems. A model is merely a simplification of a real-life system that has been designed for a specific purpose. Various types of models exist: physical models such as a toy car have been widely used to represent complex systems, more elaborate physical models are commonly used by engineers to test a structure before it is actually built. In general terms, a properly designed model provides an evaluation of the behaviour of the real-life system at a fraction of the cost of the real-life system.

With farming systems, physical models may take the form of a demonstration farm or a case study. Since the costs associated with setting up a demonstration farm are high, the range of management options that can be effectively tested by such physical models is limited. Moreover, in dryland farming the time required to experience a normal range of climatic conditions is often unacceptably long. To reduce the cost and time factors, computer models can be used in the evaluation of farming systems.

A computer model of a farming system is a series of mathematical equations, arranged by a model-builder, to mimic the complex processes of the system. A computer quickly processes the equations and in doing so transforms input data, such as rainfall, into useful information, such as crop or animal production. Thus, a model can be regarded as a collection of hypotheses that expresses the operation of the system in mathematical terms. A computer program links the various components and provides an objective expression to the whole system. Obviously a model will only provide an accurate estimate of a real-life system if the basic mathematical expressions soundly mimic the various processes in the system and if the input data are accurate. These constraints are commonly known as the GIGO law of modelling: Garbage In, Garbage Out. The model-builder has the responsibility of showing that the model is a valid representation of a system while the model-user is responsible for the accuracy of the input data.

Farming systems and models

Components of systems

A farmer must manage a wide range of interrelating components (Figure 16.1). Physical constraints such as climate, soil type and topography interact with many biological options, such as the selection and husbandry of crops and animals, to determine the type of farm. Mechanical components such as farm machinery and farm infrastructure assist the farmer. All these technical matters contribute to the profitability of the farm along with external economic factors such as input cost and market

prices. Society exerts an influence through Acts of Parliament that determine standards for farm operations and the sale of farm products. The farmer is continually reacting to the various influencing factors through decisions of either a tactical or strategic nature. In addition, the decisions are tempered by the knowledge, goals and attitudes of individual farmers. As a result, a farming system can be regarded as a group of interacting components that operate together for a common purpose and react to external stimuli within a boundary

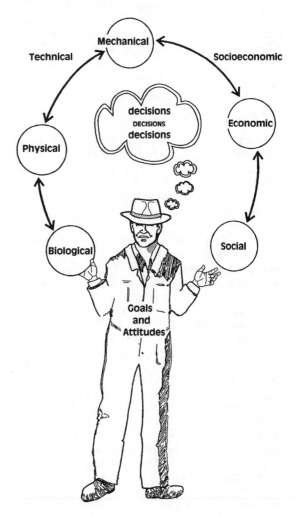

Figure 16.1 *A stylized representation of a farming system. Farmers are continually responding to technical and economic constraints through management decisions which reflect their experiences, goals and attitudes in life. It can be likened to a juggling act because if a decision is grossly in error the whole system will collapse.*

that includes all the significant interactions (Chapter 2). There is a great variety and complexity in the components that constitute a farming system.

Computer modelling tends to focus on the interactions that are expressed in quantitative rather than qualitative terms. Because of this quantitative assessment, modelling is a form of 'hard' systems analysis (Chapter 20). The alternative—'soft' systems analysis—considers qualitative criteria such as prevailing attitudes, value judgements and perceptions in an assessment. Computer modelling is one of several techniques that can be used in systems analysis. It provides information for evaluation by a model-user, and any resulting decision reflects the prevailing goals and attitudes of an individual modeller.

Criteria of evaluation

Models can evaluate farming systems in terms of productivity, reliability and sustainability, depending on the type and purpose of the model. To illustrate these criteria, examples have been selected from a model called PERFECT which evaluates farming techniques in the dryland cropping regions of Queensland.

Productivity—such as bales of wool, tonnes of wheat or cattle sold—is a measure of system output. It is often convenient to express productivity as a form of biological efficiency: level of output per unit of input. Commonly, terms such as wool per sheep (kilograms per sheep), crop yield (tonnes per hectare) and liveweight gain (kilograms per head per year) express biological efficiency. When one term in the ratio is a monetary value, the productivity term becomes a form of economic efficiency, such as gross margin (dollars per hectare). Clearly, the units of productivity vary with the nature of the enterprise and with the nature of the analysis. However, a productivity term is an essential requirement for all systems evaluations, and its magnitude, variability and temporal trend all warrant attention in such an analysis (Figure 16.2).

Reliability indicates the degree to which productivity remains constant in spite of the normal fluctuations in uncontrolled factors such as climate and prices (Figure 16.2). It can be expressed statistically by a coefficient of variation. A system in which average productivity has a relatively low coefficient of variation is considered to be more reliable and more desirable than a system with a high coefficient of variation. Some models that use historical records of climate as inputs allow probability distributions of likely productivity to be

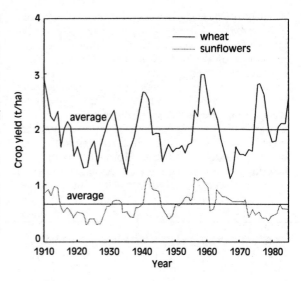

Figure 16.2 *Temporal trends in productivity for wheat and sunflower crops on the Central Highlands in Queensland. Note the wide variation around the average.* (SOURCE: Littleboy *et al.*, 1989. 'PERFECT a computer simulation model of productivity erosion runoff functions to evaluate conservation techniques', *QDPI Bulletin*, QB89005)

generated. In this way a user can gauge the risk associated with achieving a certain level of productivity (Figure 16.3).

Sustainability refers to the ability of a system to maintain the average level of productivity over time. Although sustainability is not simply measured, the temporal trend in productivity is a useful indicator of sustainability, e.g. a decline in crop yield over time suggests that the system is not sustainable because it is being stressed by factors such as increasing soil salinity, soil erosion or a decline in soil nutrients (Figure 16.4). Sustainability also reflects the ability of a system to tolerate abnormal events such as bush fire, flood, early or late frosts, cyclones or hail storms. Some models, often designed as aids to systems management, estimate sustainability in response to management options, such as selection of stocking rate; other models estimate the likelihood of abnormal events through a statistical analysis of historical records. This information can be incorporated into the design of a farming system, e.g. graded banks to reduce soil erosion are designed in relation to the probability of a specified rainfall intensity. These estimates need not be independent, e.g. an increase in stocking rate in a beef production system can increase average liveweight gain per hectare (higher productivity), but this is accompanied by more

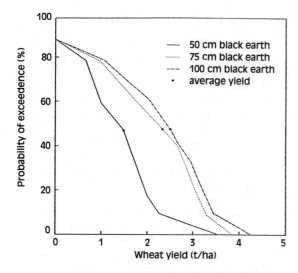

Figure 16.3 *Reliability of a system as expressed through probability distribution of crop yield for different soil depths in the Central Highlands in Queensland.* (SOURCE: Littleboy *et al.*, 1989. 'PERFECT a computer simulation model of productivity erosion runoff functions to evaluate conservation techniques', *QDPI Bulletin*, QB89005)

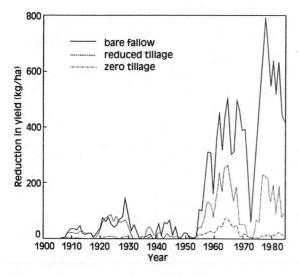

Figure 16.4 *Sustainability of a farming system as reflected by reductions in crop yield due to the predicted soil loss under three management systems for dryland cropping on the Central Highlands in Queensland.* (SOURCE: Littleboy *et al.*, 1989. 'PERFECT a computer simulation model of productivity erosion runoff functions to evaluate conservation techniques', *QDPI Bulletin*, QB89005)

year-by-year variation (less reliability) and overgrazing in the dry season (less sustainability). Thus a farmer must balance the advantage of higher average productivity against a greater risk of low income and land degradation in some years. The final outcome reflects the farmer's attitudes to profit and risk (Figure 16.1). Computer models allow a user to examine these opposing factors in an objective manner.

Types of models

Two classification systems warrant consideration: the first one reflects the mathematical basis to the model, and the second reflects the type of computer software used by the model.

Mathematical Basis — Models can be classified in a two by two matrix: static or dynamic by deterministic or stochastic. In a static model the inputs remain constant for a specific application while in a dynamic model the inputs vary during the application so that the model mimics the performance of a system over time. Deterministic models do not consider the natural variation that exists in a system, but this is considered by a stochastic model. The following examples illustrate the different categories:

- The well-known compound interest law is a static deterministic model.
- The common linear regression model is static and stochastic.
- A monthly cash flow statement is a dynamic deterministic model.
- A model that simulates plant growth from historical records of climate is a dynamic and stochastic model because the end product is a probability of distribution outcome.

Clearly all four categories would be useful in the evaluation of farming systems, but dynamic models are required for the analysis of temporal changes in the system and stochastic models are necessary for estimates of reliability in a system.

Type of Software — Models based on spreadsheets are powerful and convenient tools, well suited for models of static systems. The equations in the model are arranged in a logical sequence and are readily examined and changed while the tabulated output is often supported by a graphic display. This relative simplicity of spreadsheets makes them a useful introduction to modelling and several spreadsheet languages are available. Spreadsheets allow users with a moderate knowledge of computing to develop their own specific models. They are particularly useful for financial analyses and budgeting. However, spreadsheets are limited because they do not readily handle large input files.

Linear program models or LP models have been widely used to determine the optimum combination of factors to achieve a specific goal. Devising the least-cost ration that meets a specified nutrient status from a variety of feeds is a classic role for linear programming. Advanced versions of linear programming, often called mathematical programming, have been used to evaluate farming systems. Output from LP models is usually deterministic and static but this can be changed by linking LP models with dynamic simulation models.

Dynamic simulation models estimate the status of components in the system by processing input data for specified time periods. The models are based on procedural languages such as FORTRAN or PASCAL, and the input files can be very large, such as historical records of daily rainfall. Since the output reflects a response to natural variation in the input data, dynamic simulation models can provide a probabilistic analysis of productivity. Because of these features, dynamic simulation models have been widely used for the evaluation of farming systems.

Expert systems, as the name suggests, attempt to mimic the thinking processes used by a human expert in a topic. When such models focus on crop or pest management they are valuable aids to agricultural extension and farm management. They may be linked to dynamic simulation models to provide a more powerful package for systems evaluation.

In the following sections dynamic simulation models and expert systems will be used to evaluate production systems.

Models as an aid to research

Early computer models of production systems were built by researchers to integrate and extrapolate their research results. This enthusiasm for modelling was not shared by all scientists because the early models were difficult to use, not readily available and of limited practical value. Attitudes changed as more scientists began to regard computer models as tools that were not only a repository for results from past research, but also a precursor for future research.

Repository for research results

This attribute is illustrated by a model that evaluated forage options for beef production in southeast Queensland. The model consisted of two major components (Figure 16.5). One component estimated pasture growth from soil characteristics and climatic records, the second component estimated animal growth from potential liveweight gain and estimates of dry matter intake. In developing and validating this model the authors made use of results from fourteen separate grazing trials reflecting different grazing strategies from a wide range of seasonal conditions. The experiments spanned twenty-five years and included measurements of soil water balance, soil characteristics, pasture yield, pasture composition, pasture growth, pasture nutritive value, growth of steers at different ages for different breeds, and historical records of weather and prices. The model integrated this information from unrelated experiments to form a powerful analytical tool, thereby extending the usefulness of the original results from research. Such a model can be regarded as a facility that not only stores information from research but also applies the information in a wider context.

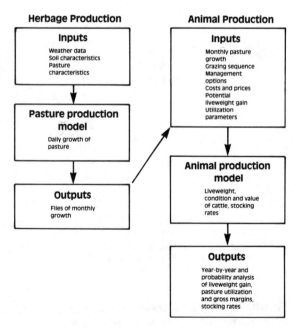

Figure 16.5 *Components of a pasture and animal production model that was developed from research results and was a precursor for new research.*

Precursor for future research

Since a model integrates different components of a system in an objective manner it can be used to:

- Identify new operating principles for better management of the system.
- Suggest a preferred pathway for new research.

For example, when the pasture and beef production models are applied to studies on the interaction between grazing pressure and pasture composition, they show that pasture degradation occurred when more than 30 per cent of pasture growth was utilized by cattle in a relatively dry growing season. Therefore, a safe stocking rate for a specified pasture can be defined as one that does not exceed this critical level of utilization in 70 per cent of years. Mathematically this managerial principle is expressed as:

SSR is < UG/100 ID animals/hectare

where:

SSR is a safe stocking rate;
U is maximum safe utilization in a growing season, (30 per cent);
G is growth of pasture in a growing season (kg/ha);
D is days of grazing in the growing season (e.g. 182 days); and
I is daily feed intake per animal (kg/day/animal).

By highlighting the importance of pasture growth to both pasture condition and animal production, this model has influenced research methods for grazing systems. Pastoral systems are driven by primary production and a lack of understanding by pasture agronomists has led to inappropriate measurements in grazing experiments. In many cases, the only measurement of plant productivity taken was presentation yield which rarely proved useful since it is an expression of the simultaneous processes of plant growth, decay, trampling and consumption. Where measurements of pasture growth have been taken, the interpretation and extrapolation of grazing trial results has been greatly enhanced.

This analysis also led to an experiment to measure pasture growth at different locations in Queensland and to collect essential inputs for the pasture growth model. With this information the pasture growth model can be used to characterize different pasture types at different locations in terms of safe stocking rates. Thus the original model helped to identify a new principle for better pasture management which was subsequently developed and applied through a combination of field experimentation and simulations.

Assessment of proposed research

Funding agencies for agricultural research require estimates of the likely impact on industry of a research proposal. By using suitable biological and economic models, the range of likely results from research can be evaluated in a quantitative manner. The likely benefits from research can be compared against the likely costs for the proposal. Thus funding agencies use the models to help identify cost effective targets for research before allocating funds to a proposal. This form of assessment is increasing though for it to be valid the models must truly reflect the critical biological and economic components of the system.

Regional evaluations by models

Dryland agriculture in Australia must function within constraints imposed by climate, soils and markets. The climatic constraint is not well defined because rainfall, the most important component of climate, varies widely from season to season and year by year. Whenever a traditional farming system is changed at a regional level, the evaluation of the new system is confounded by the wide variation in climate, location and management options. Computer models can overcome this problem by allowing historical records of rainfall to be used in a regional evaluation of a new farming system under different management options. Two examples follow.

Expansion of cropping in Queensland

Wheat cropping expanded in the Maranoa region of south-western Queensland during the 1970s and early 1980s in response to low prices for wool and beef. Although this was an attractive diversification for the traditional grazing industry there was little information on the level and reliability of wheat production for the different centres in the region. Estimates of this information were supplied by a computer model that simulated crop growth and yield from long-term weather records. A decline in mean crop yield along an east-west transect was accompanied by increased variability in crop yield, a higher likelihood of planting failures, and a reduced likelihood of achieving a specified yield such as 1000 kg/ha (Table 16.1). Clearly, wheat growing in the western Maranoa is less productive and less reliable than the traditional eastern wheat growing centres. These

Table 16.1 *Predicted yields of wheat from 1889 to 1981 along east-west transect in the Maranoa region of south-western Queensland*

Centre	Mean Yield (kg/ha)	Coefficient of variation %	Probability of 1000 kg/ha %	Probability of failed Planting %
East-west transect				
Dalby	1685	41	86	5
Miles	1474	47	73	8
Roma	1192	56	64	12
Mitchell	1018	57	52	16
Charleville	1005	65	46	19

(SOURCE: Hammer, G. L., Woodruff, D. R. and Robinson, J. B., 1984. 'Reliability of Wheat Production', D. A. K. McNee (ed.), 'Cropping in the Maranoa and Warrego' *QDPI Information Series*, QI84012, 16–28)

responses reflect trends in the amount and distribution of rainfall.

The analysis also showed that the conditions experienced in any period of twenty years may not be a reliable indicator of long-term conditions, e.g. from 1961 to 1980 average crop yields were 115 per cent and 89 per cent of the long-term average for Dalby and Charleville respectively. Thus, a farmer's perception of the level and reliability of production, even after twenty years of farming, might be a distortion of the long-term values because the range of conditions experienced is limited. The model overcame this problem and gave a more complete evaluation including objective estimates of the risks associated with cropping in a marginal region. When linked to an economic model the risk associated with marginal cropping can be interpreted in economic terms, e.g. wheat growing could be profitable in the Maranoa region of Queensland, assuming a specified cost structure and soil type.

However, although these analyses give a valuable insight into the productivity of the system, they are a simplification of the real world. Issues such as the integration of crops with livestock and techniques to minimize soil erosion after converting grassland to crop land were not considered. The model proved to be a useful but limited analytical tool.

In the above example, simulation modelling based on historical rainfall allowed a re-examination of appropriate land use. Historically, large changes in rainfall distribution in eastern Australia occur on long time scales that span a decade or a generation, and in response, the type of farm enterprise may need to change. For example, during a dry time period grazing perennial pastures might be a more appropriate land use than annual cropping. As the cause of climatic variability is better understood and future projections of climate change become more certain (e.g. the Greenhouse effect), simulation models will allow alternative land uses to be compared.

Evaluation of history of grazing

This example is concerned with the analysis of trends in stocking rate for beef producing regions in northern Queensland. A model of grass production was used to estimate safe stocking rates for beef cattle in the Dalrymple Shire. A stocking rate was deemed safe when the animals consumed 30 per cent of the pasture that grew in summer. Pasture growth was estimated by a pasture growth model using historical records of rainfall with additional adjustments for the effects of trees and land use in the shire. Safe stocking rates were then estimated and compared with the actual stocking rate, calculated as the ratio of total number of beef cattle to the area of the shire. After 1965 the actual stocking rates for the shire exceeded the estimated safe stocking rates, and from 1982 to 1988, a period of below average rainfall, the actual stocking rates were about double the safe stocking rates (Figure 16.6). The increase in stock numbers after 1965 coincided with improved cattle husbandry, the introduction of better adapted breeds of cattle, a dramatic fall in the price of export beef, and widespread land degradation. Prior to 1965 stocking rates were 'safe' and botanical composition of the pasture was relatively stable. Thus, the model provided an analysis of regional trends in stocking rates that was based on criteria that foster sustainable production from pastures. The analysis linked observed land degradation to over-

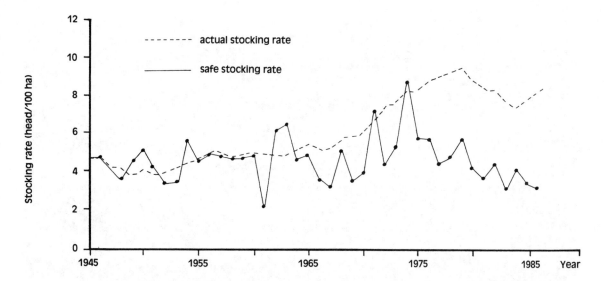

Figure 16.6 *Comparison of actual stocking rate (- - -) and predicted safe stocking rate (—) for the Dalrymple Shire in North Queensland. Safe stocking rate = pasture growth (kg/ha Dec.–May) x safe utilization in summer (30%) x shire index (effect of trees and other land uses)/animal intake for 6 months (1800 kg per cattle equivalent). Pasture growth was calculated using a pasture growth model. The shire index (0.3) was calculated for 1945–63 (a period of pasture stability) as the ratio of the actual shire stocking rate (4.54 cattle equivalents per 100 ha) to the calculated safe stocking rate for the cleared pasture (14.92 cattle equivalents per 100 ha). The actual shire stocking rate was calculated as the ratio of total beef cattle to the shire area.*

grazing and suggested a stocking rate that can be sustained in the region.

The above example also illustrates the problems that can arise in dryland farming systems when the cattle component of the system changes through the introduction of new technology, without an appreciation of the impact of this change on the interrelated pasture component. A model can be used to evaluate new technology before its introduction.

Evaluation of new technology

Some models have been used to analyse the impact of new technology on an existing production system. Two examples follow that clearly illustrate the benefits from such an analysis.

Evaluation of reliability

The Mitchell grass plains of north-western Queensland have a gently undulating topography and fertile, cracking clay soils. The annual rainfall of about 400 mm is strongly summer-dominant and highly variable. Traditionally the grasslands

are grazed by sheep and cattle but production is curtailed by drought and by poor quality of pasture in the dry season. The topography, soil type and rainfall distribution suggest that runoff in summer could be collected in a shallow dam. This water could then be used to irrigate grain and forage crops that would supplement the pastures during periods of deficiency (Figure 16.7). The concept was demonstrated at Richmond from 1968 to 1978 when grain sorghum crops were grown in 70 per cent of years on an irrigated farm adjacent to the dam. Forage sorghum was also sown in the bed of the dam as the water level receded. Local graziers were attracted to the technology but because it was a dramatic departure from the traditional system a computer model was built to estimate the long-term reliability of the concept. Runoff and crop yields were estimated from sixty years of climate records using a range of dam designs and catchment characteristics. The resulting analyses showed that, even with the design characteristics at an optimum, the system was not viable because the runoff was too unreliable (Table 16.2). The successful field study was conducted in a series of better-than-average years. By using a model to place the new technology in an historical context a more realistic evalua-

Figure 16.7   *Demonstration of a shallow storage irrigation system on the Mitchell grasslands of north-western Queensland*   (PHOTO J. Clewett)

tion was obtained. The model discouraged adoption of the unreliable technology.

Integration of technology

The second example evaluates a new management practice for a self-replacing flock of Merino ewes grazing annual pastures in western Victoria. A computer model that estimated pasture and animal production for this type of enterprise was modified and refined to estimate the economic benefits gained by using ultrasonic scanning to identify ewes with multiple lambs. Such identification allows twin-bearing ewes to receive special feeding to improve the survival and growth of their lambs. Thus, ultrasonic scanning should enhance the real-

ization of opportunities for flock improvement through selection for higher fecundity, reduced reproductive wastage and increased wool production.

The model permitted a range of management strategies to be compared in a variety of seasonal and market conditions. Economic benefits from scanning were greatest when pregnancy and lactation corresponded to periods of feed shortage that arose from either lambing in autumn or drought. The financial benefits from scanning were also shown to increase as the proportion of twin-bearing ewes in the flock increased from 5 to 30 per cent, thereby indicating that the technology was most beneficial when there was already a high level of fecundity in the flock and a high level of management on the farm. Relatively small benefits

Table 16.2 *Performance by decades for a shallow storage irrigation system in north-western Queensland. The simulation spanned six decades from October 1918 to September 1978. The catchment area was 1660 ha; the dam size was 400 ML and irrigation area was 40 ha.*

	1918–28	1928–38	1938–48	1948–58	1958–68	1968–78
Mean annual runoff (mm)	14.1	4.2	28.6	67.6	12.0	80.0
Years with no runoff	4	3	3	3	4	2
Number of crops sown and irrigated	3	5	3	7	4	7
Years with sufficient water to irrigate 40 ha	2	1	3	7	3	7
Average yield of irrigated grain sorghum, crops tonne/ha	3.3	2.8	3.4	3.5	2.7	3.2

from scanning occurred when wool producing flocks had low levels of fecundity. Since this situation exists in many Merino flocks in Australia the evaluation concluded that ultrasonic scanning has a rather limited and special application in the Australian wool industry.

Improved management through systems evaluations

Farmers mainly use computers for letter writing, record keeping, farm budgeting, farm accounting and other tasks that centre on the financial analysis and management of a farm. A range of commercial software is available for this purpose as well as custom-built spreadsheet models. All are static and deterministic models. An element of natural variability is achieved by comparing average, worst and best-case scenarios. Experience has shown that, compared with traditional methods of farm record keeping and accounting, the computerized system requires a similar involvement of time but it provides far more information for decision making. In recent times, a new generation of simulation models, or expert system models, has become available as aids for decision making. They consider the interactions between the dynamic components of the system. Case studies

from some of these models provide examples of the range and application of the new software which is also expanding the role of the microcomputer in farm management.

Dryland cropping systems

WHEATMAN is a model to help wheat farmers in Queensland choose a preferred management strategy for each crop in each paddock on a farm. It predicts the probability of yields and frost damage with different combinations of variety, soil type, planting time, initial status of soil moisture and nutrition, and fertilizer history. WHEATMAN was developed from a larger wheat production model that used 100 years of daily climate data for each of seven wheat growing regions centred around Goondiwindi, Miles, Roma, St George, Emerald, Biloela and Dalby. Subsequently the effects of frost, and the influence of soil nutrient status and fertilizer applications on grain yield and protein content were incorporated in the model. It was tested by 200 farmers before its commercial release.

Time of planting can have a large effect on wheat yields in the eastern Darling Downs (Table 16.3). The highest crop yields were obtained with the earliest date of sowing but this was also associated with a very high risk of frost at anthesis. Sowing wheat early in June is recommended for

Table 16.3 *Information from the* WHEATMAN *model for a Waco clay with 85 mm of available water at sowing at Dalby in Queensland. It shows the interactions between wheat varieties, sowing time, the likelihood of frost damage and crop yields.*

| | Sowing time | | | |
Varieties	1 May	1 Jun	1 Jul	1 Aug
	Date of anthesis			
Suneca	9 Sep	25 Sep	7 Oct	29 Oct
Kite	24 Aug	18 Sep	1 Oct	16 Oct
Hartog	15 Aug	12 Sep	28 Sep	13 Oct
	% risk of frost at crop head height ($-3.2°C$)			
Suneca	24	5	2	0
Kite	58	10	4	0
Hartog	78	18	4	0
	Average (long-term) crop yield (t/ha)			
Suneca	1.75	1.52	1.25	1.00
Kite	1.48	1.57	1.28	1.09
Hartog	1.16	1.79	1.39	1.17

the eastern Darling Downs because the frost risk during anthesis is acceptable. The reduction in yield for late plantings is due to the increased temperature and pan evaporation around anthesis compared with early sowing.

Similar analyses of the interaction between planting time, frost risk and variety selection have been popular with farmers using the WHEATMAN model. Survey results have indicated that after running the model, 30 per cent of farmers changed their views on planting time and selection of varieties. Thus the WHEATMAN model gave greater insight into the operation of their dryland farming system.

Woodland management for beef production

GRASSMAN is a management aid for farmers that graze eucalypt woodlands in eastern Queensland from latitudes 20 to 26°S and with an average annual rainfall exceeding 575 mm. It is an expert system that allows a user to insert key performance criteria, and is based on research by pasture scientists. It is designed to test and identify preferred management strategies with a focus on interactions between tree management, seasonal conditions, grazing management, beef production and pasture production and condition. A user may select one of twenty-two different eucalypt com-

munities with a specified tree density and then develop a management scenario for the next fifteen years by indicating a choice of stocking rates, seasonal conditions, frequency of burning and timber control measures. The model estimates regrowth of trees and pasture and animal production for the proposed scenario (Table 16.4). An economic evaluation is also available. In this manner the model can be used to identify preferred management strategies for land in a major beef producing region of Queensland.

Herd structure and management

The third example of a model for farm management is called RANGEPACK HerdEcon. It is a dynamic herd and flock model that has been linked to cash flows for properties in the semi-arid rangelands of Australia, and it enables a manager to develop and evaluate strategies. In particular, the model links herd biology and marketing options with economic returns, copes with climatic variation, dynamically follows herd structure through successive years, accommodates single or mixed enterprises of cattle, sheep or goats, and can have new prices or strategies introduced by the user. Thus, the model can mimic a real property in terms of herd structure, receipts and expenditure in response to marketing strategies and a specified sequence in seasonal conditions. For example, con-

Table 16.4   *Output from the* GRASSMAN *model.*

Community: Narrow leaf ironbark (*Eucalyptus crebra*) on duplex soil with no understory.
Management: Mature trees with a basal area of 15 m²/ha killed by stem injection in year 1; stocking rates adjusted to utilize 30% of pasture growth in summer; pasture burnt every four years.

Selected Scenario

	Seasonal Conditions	Timber Treatment	Basal area m²/ha	Height of Regrowth m	Stocking Rate ha/adult	Yield in May kg/ha
1	average	stem inject	0.3	1.0	4.3	1340
2	poor	—	0.9	1.4	4.9	1650
3	average	—	1.1	1.8	3.1	2254
4	good	—	1.4	2.2	2.4	2530
5	average	burn	0.9	1.5	3.3	1450
6	average	—	1.1	1.9	2.9	2210
7	very poor	—	1.4	2.3	8.9	1670
8	very good	—	1.7	2.7	2.0	2460
9	poor	burn	1.6	2.4	5.5	870
10	poor	—	1.9	2.8	4.9	1500
11	average	—	2.3	3.2	3.9	1980
12	good	—	2.8	3.6	2.6	2330
13	average	burn	2.8	3.4	4.1	1180
14	good	—	3.4	3.8	2.7	2090
15	average	—	4.0	4.2	4.0	2010

Table 16.5   *Evaluations with* RANGEPACK *of management strategies in response to drought for two consecutive years in a sequence of ten years for a beef property in central Australia.*

Herd: 3000 breeding cows plus followers.
Options: A — Do nothing
         B — Sell old steers at start of drought
         C — Sell all steers and heifers at start of drought
         D — Sell old steers and old cows at start of drought
         E — Feed breeding cows a supplement ($1 per day)

Option:	A	B	C	D	E
Accumulated cash surplus $'000	426	664	625	549	466
Years with cash deficits	5	4	5	4	3
Maximum annual cash surplus $'000	250	310	600	290	250
Maximum annual cash deficit $'000	−200	−280	−260	−340	−350

sider a beef cattle herd near Alice Springs with 3000 breeding cows and a total of 6000 cattle. What strategy should be used to minimize the financial effects of two poor years (years 3 and 4) in a sequence of ten years? Of five strategies evaluated, selling marketable cattle early in the dry period proved to be most beneficial (Table 16.5). It is also a simple task to extend the sequence of poor years in the model and thereby estimate the sustainability of the property to a prolonged drought, in terms of positive cash flow and herd structure. In a similar manner, a wide range of season-by-management scenarios can be evaluated for properties in the extensive grazing lands of Australia.

Education through systems evaluation

Because a computer model is a repository for information pertaining to a production system, models can be used to access and present this information for educational purposes. Models used for education will emphasize interactions between various components in the system, since this task is well handled through modelling and is difficult to teach in a more traditional manner. Educational models may have up to three modes of operation:
• Interactive operation.
• Automatic simulation.
• Problem solving.
   In the interactive mode a model may mimic a

farm while the student pretends to be a farmer. If the model accesses historical records of weather and prices on a predetermined time step, such as season-by-season, the student must respond to the prevailing conditions. The student makes managerial decisions and soon learns that a specific decision not only depends on past decisions but also influences future decisions. The object of the exercise is to manage the farm in a profitable and sustainable manner. Models that operate in this interactive mode allow students to experience some of the tribulations and satisfactions of farming, and their role in agricultural education is likely to expand with improvements in computer software and hardware. The flight simulator, used in the education of airline pilots, is a sophisticated combination of a physical model and a computer model operating in an interactive manner. Perhaps one day, students will be using a farm simulator.

Secondly, students may use computer models of production systems to conduct simulation experiments where the treatments represent different management strategies. Results from an experiment may appear as a graphic or tabular representation of a response surface. If the model is so designed, students may alter a parameter that governs a key component in the system and thereby assess the influence of that parameter on the system (Figure 16.3). This activity can be expanded into a sensitivity analysis whereby the most influential components of the system are identified. Thus, models operating in the simulation mode may be used to evaluate management strategies and to teach the nature and operation of components in the system.

BEEFUP is an example of an educational model that operates either in the interactive or automatic modes. A student must manage 1000 ha of eucalypt woodland in one of five locations in Queensland. Trees can be cleared and cattle can be purchased and sold at will. The model processes fifteen years of weather and prices and a student aims to create a profitable enterprise without degrading the pasture. By using the interactive mode a novice to the industry experiences the interaction between seasonal conditions, forage supply and animal performance, and how these interactions are influenced by stocking policy, land clearing and geographic location. Tactical responses to drought and price fluctuations can also be assessed. Alternatively, the model can be switched to the automatic simulation mode to test broad management strategies over all seasons. Also, the scientific basis of the model can be demonstrated by applying the model to specific situations and analysing the results in detail. In this way scientific principles and their application are taught in an active manner.

Models that function as expert systems to solve problems are the third type and have a valuable role in education. They provide a rapid evaluation of a system that may involve a complex mix of biological and technical information. Again, by planning a series of questions a student can use these models to evaluate managerial strategies, such as the information from the WHEATMAN model in Table 16.3.

In the future, farming system models might be linked to computer-assisted learning packages to provide students with a self-directed learning and assessment facility.

Conclusions

This chapter has emphasized the evaluation of farming systems by a range of computer models that are either available commercially or well documented in scientific literature. Only a small selection of the models available for systems analysis were mentioned. Indeed, the authors were biased towards applications in northern Australia, but the principles outlined in the chapter have general application.

Analytical models of farming systems originated as aids to research but in the last decade the role of models has widened to include evaluation of new technology, identification of preferred management tactics and strategies, and agricultural education. The rapid increase in the power and availability of micro-computers has helped stimulate the development of new software to meet the needs of extension officers, farmers and educators. All model-users learn from the ability of a model to integrate, in an objective manner, the various components of a system into a whole. This is the unique and most valuable attribute of computer models of farming systems.

The complexity of the technical and economic components of farming is steadily increasing and frequently changing. To survive in this dynamic and competitive environment, operators in the rural industries need to make sound managerial decisions. Computer models can assist in this task. For these reasons, and because of the increasing awareness and availability of computers, the range and sophistication of models will increase in the future, particularly those that assist farm managers. Such models will probably target specific problems for specific rural industries. The new models will have a more advanced analytical ability because of better computer programming and because of an expansion in the basic knowledge dealing with components in the system.

The models used in this chapter were developed through contributions from many different disciplines including soil, plant and animal scientists, extension officers, economists, computer programmers and potential users. Thus, a team approach was essential for developing the models, but experience has shown that the team must work towards a well-defined objective. In this way model development provides a clear and beneficial focus for the team members and the industry they serve.

Further reading

Muchow, R. C. and Bellamy, J. A. (eds) (1991). *Climatic Risk in Crop Production: Models and Management for the Semi-arid Tropics and Sub-tropics*. C. A. B., London (in press).

Remenji, J. V. (ed.) (1985). 'Agricultural systems research for developing countries', ACIAR *Proc.*, no. 11, Canberra.

Van Keulen, H. and Wolf, J. (eds) (1986). *Modelling of Agricultural Production: weather, soils and crops*. Pudoc, Wageningen.

Wilson, B. (1984). *Systems: Concepts, Methodologies and Applications*. John Wiley and Sons, Sydney.

CHAPTER 17

Monitoring the Functioning of Dryland Farming Systems

R. J. French

SYNOPSIS This chapter defines potential productivity and discusses the factors which contribute to it. It examines reasons why the potential has not often been reached on the average farm and suggests ways in which the farmer might implement a monitoring system to better appreciate changes which are occurring.

MAJOR CONCEPTS

1 The climate determines which crops and pastures can be grown and their potential yield.

2 The soil properties and management skills of the farmer determine the degree of success in approaching the potential.

3 Farmers have in the past relied on memory to develop their farming programmes but this is no longer good enough. Good records are essential.

4 Farming today is a multi-factor business and this requires an upgraded set of criteria to help in decision making.

5 Farmers should monitor trends especially in climate, nutrient flows, soil losses, organic matter status, pest and disease levels and any changes in the soil's physical status. An essential feature in monitoring farming systems is to record the nutrient lost from each paddock via harvested products.

In essence, agriculture involves harvesting the energy of the sun to produce food and fibre for human populations. Growth of vegetation is necessary on farms to:

- Produce yields of crops and pastures and provide an income for the farmer.
- Provide cover for controlling erosion. Grazing and tillage operations need to be carried out with care to maintain this cover.
- Provide organic matter from plant roots and residues, and thereby maintain the soil structure and microbial biomass.

The climate determines which crops and pastures can be grown, and their potential yield, and the soil properties and management skills of the farmer determine the degree of success in approaching the potential yield. The production of crops per hectare from the average farm has not increased very much during the last forty to fifty years despite the investment of large sums of money into research. The average yield of wheat over the last thirty years has only increased by 4–5 kg/ha/year (Table 17.1). This rate does not compare favourably with the average yearly gains of about 40 kg/ha/year in the USA and 80 kg/ha/year in the UK.

Table 17.1 *Average yield of wheat for individual States and for Australia in different decades (tonnes/ha).*

	Australia	WA	SA	Vic	NSW	Qld
1950	1.22	0.95	1.18	1.36	1.11	1.38
1960	1.23	0.97	1.18	1.44	1.34	1.34
1970	1.30	1.08	1.19	1.63	1.22	1.36
1980	1.36	1.17	1.26	1.76	1.49	1.46

Similarly, the productivity of livestock systems has not improved to anywhere near the levels predicted several decades ago. In 1964 when sheep numbers in Australia were 165 million it was predicted that the numbers should reach 265 million by 1990. In fact, by 1990 the numbers were only 170 million.

There is a need to constantly monitor for discernible trends in productivity and soil reserves. In the past farmers have relied on memory to develop their farming programmes. Today farming is a multi-factor business and this requires an upgraded set of criteria to help in decision making. Farmers should monitor trends especially in climate and productivity per millimetre of water use,

Table 17.2 *The average climatic values for five locations in Australia's dryland cropping areas.*

Merredin (WA) — Latitude 31° 31' S

	J	F	M	A	M	J	J	A	S	O	N	D	Year
Rainfall (mm)	12	14	23	21	38	52	49	37	23	18	13	12	312
Evaporation (mm)	386	328	295	171	99	67	68	79	113	206	275	371	2458
Daily max. air temp (°C)	32	32	31	27	24	21	20	20	22	24	27	29	26
Daily min. air temp (°C)	18	19	18	15	13	11	9	9	9	11	14	16	14

Georgetown (SA) —Latitude 33° 22' S

	J	F	M	A	M	J	J	A	S	O	N	D	Year
Rainfall (mm)	20	21	19	37	52	57	56	57	53	46	31	24	473
Evaporation (mm)	340	300	229	155	101	75	75	98	145	210	280	332	2340
Daily max. air temp (°C)	32	31	28	23	18	15	14	16	19	22	26	29	23
Daily min. air temp (°C)	14	14	11	8	6	4	3	4	5	7	10	12	8

Ararat (Vic.) — Latitude 37° 20' S

	J	F	M	A	M	J	J	A	S	O	N	D	Year
Rainfall (mm)	42	31	42	51	61	47	66	71	69	66	47	26	625
Evaporation (mm)	210	190	125	72	50	35	38	50	80	120	150	190	1360
Daily max. air temp (°C)	26	27	23	19	15	12	11	13	14	18	21	24	19
Daily min. air temp (°C)	11	11	10	7	6	4	3	4	5	6	8	9	7

Wagga Wagga (NSW) — Latitude 35° 07' S

	J	F	M	A	M	J	J	A	S	O	N	D	Year
Rainfall (mm)	43	38	43	44	54	45	54	54	50	61	43	41	570
Evaporation (mm)	241	169	158	95	55	33	33	47	72	102	150	222	1380
Daily max. air temp (°C)	31	31	28	22	17	14	12	14	17	21	25	29	22
Daily min. air temp (°C)	16	16	14	9	6	4	3	4	5	8	11	14	9

Toowoomba (Qld) — Latitude 27° 33' S

	J	F	M	A	M	J	J	A	S	O	N	D	Year
Rainfall (mm)	138	123	99	62	56	61	55	42	49	74	89	121	969
Evaporation (mm)	200	160	150	125	90	65	80	90	110	165	185	215	1635
Daily max. air temp (°C)	28	27	25	23	20	17	16	18	21	23	26	27	23
Daily min. air temp (°C)	17	17	16	13	9	6	5	6	9	11	14	16	11

nutrient flows, soil losses, organic matter status, changes in the soil's physical status and pest and disease levels. An essential feature in monitoring farming systems is to record the nutrient lost from each paddock via harvested products.

If the productivity is to be improved then a programme of action is needed for individual farms. This will involve:

- A farm resource data base of the climate, land form, and soil types.
- Selection of the appropriate crops and pastures that can give the best economic return in the environment.
- Monitoring long-term trends in a range of basic soil properties.
- Monitoring seasonal soil and plant factors to identify limitations to growth, and then to devise plans to correct them.
- Measuring the success of the year's management practices by expressing yield as a percentage of the potential yield, and calculating the net income per hectare.

Evaluation of the farm resource base

The basic resources of a farm that need to be compiled are climate, landscape and potential yield.

Climate

The key climatic factors are solar radiation, rainfall, evaporation, humidity and maximum and minimum air temperatures. The amount of solar radiation sets the potential for growth because it provides the energy for photosynthesis. This factor is not measured at many locations in the farming areas, but an estimate of the incoming energy can be obtained from the evaporation of water from a class A pan. (In this chapter 'evaporation' refers to evaporation from a class A pan with a bird guard.) An average relationship between sunlight energy and pan evaporation is:

$$\frac{\text{Sunlight Energy (KJ cm}^2)}{\text{Pan evaporation (mm)}} = 0.4\text{--}0.5$$

Evaporation can also indicate the degree of stress on plants. Hence the growth will be influenced by the ratio of the water supply to evaporation during the growing season. The development of plants can also be related to the accumulated day degrees of temperature from sowing to germination.

Rainfall, evaporation and maximum and minimum air temperatures are recorded at many of the official meteorological recording stations. The values for five different locations in dryland cropping areas in Australia are shown in Table 17.2. In some areas, additional climatic data, e.g. vapour pressure deficit, humidity and soil temperatures are recorded, and these values can increase the precision of climatic assessment on farms that are close to the recording site. Farmers generally record the rainfall on their own farms, but some innovative farmers now have their own small on-farm weather stations to record many of these factors thereby providing a better basis for decision making during the growing season.

The major climatic variation is rainfall and the extent of the past variation can be gauged from a set of decile tables for the town of Georgetown (South Australia) (Table 17.3). The data show decile rainfall values for each month of the year, combinations of months in the growing season, April to October, and the full year. Using the data set (Table 17.3) a decile trend graph can be developed for the growing season. The trends for Georgetown for the 1987 growing season are shown in Figure 17.1. The rainfall for April is 15 mm, then this value is added to each of the decile 9, 5 and 1 values for May to October, and the four values are plotted on a graph. Similarly the combined value for April and May is added to each of the decile 9, 5 and 1 values for June to October, and these four values are plotted. It is possible to estimate what the decile value for the combined April to October growing season is likely to be by noting the degree of convergence of the lines on the graph.

The decile trend helps in making decisions in the growing season. Thus if by the end of July, a decile 8 growing season is indicated, higher yields could be obtained by applying more nitrogen. If the trend is heading towards a decile 3 the farmer must consider buying more fodder or reducing sheep numbers. The growing season at Georgetown has been defined as April to October, although the growing season can vary from May to October, or from April to November in different areas. In such cases, it is possible to recast the decile columns to cover these different months.

While most decile tables have been derived from data collected at meteorological stations it is possible to develop such tables for a farmer's own property, provided accurate monthly records have been kept for at least forty years. The method is shown in Figure 17.2.

A further factor in understanding the effect of climate is to assess trends in annual rainfall. One

Table 17.3 *Rainfall distribution patterns for Georgetown (SA).*

Rainfall deciles (millimetres) calculated for 1874–1985

		J	F	M	A	M	J	J	A	S	O	N	D	Year
Highest on record		112	130	88	150	231	165	131	134	172	195	110	115	750
	9	49	57	44	86	95	104	97	88	90	91	72	54	639
	8	31	40	31	54	77	86	84	82	78	76	49	37	577
	7	24	29	23	46	67	71	70	69	67	61	38	29	536
	6	18	18	18	36	57	66	60	64	60	50	30	24	481
Decile	5	12	10	14	32	42	52	52	57	50	36	25	19	462
	4	8	4	11	23	39	44	45	51	42	28	19	14	429
	3	5	2	7	17	30	35	37	44	32	23	13	10	400
	2	3	1	4	10	23	25	32	33	23	17	9	6	367
	1	1	0	1	4	11	14	22	24	17	9	4	3	340
Lowest on record		0	0	0	0	2	1	5	2	8	3	0	0	229
Average		20	21	19	37	52	57	56	57	53	46	31	24	471

Combinations of groups of months

		Apr	May –Oct	Apr –May	June –Oct	Apr May June	July –Oct	Apr –July	Aug –Oct	Apr –Aug	Sept Oct	Apr –Sept	Oct	Apr –Oct
Highest on record		150	546	317	477	398	393	433	327	503	264	568	195	627
	9	86	463	148	394	232	304	314	244	373	167	450	91	503
	8	54		119		199		277		342		395		455
	7	46		107		178		240		300		363		410
	6	36		90		151		203		272		339		384
Decile	5	32	315	78	268	138	205	192	149	247	89	305	36	352
	4	23		71		121		170		222		272		319
	3	17		60		105		152		207		251		291
	2	10		51		93		134		188		222		261
	1	4	209	39	163	66	123	109	83	151	44	199	9	223
Lowest on record		0	110	6	80	26	73	51	41	89	22	106	3	128
Average		37	321	89	269	146	212	202	156	259	99	312	46	358

Decile 1 means that in one year in ten, the rainfall will be less than the amount listed in the column for a specific month or group of months. Decile 5 values are approximately equal to the average rainfall. Decile 9 means that in nine years out of ten, the rainfall will be less than the amount listed in the column or conversely, in only one year in ten will the rainfall exceed that value.
The lower half of the table shows a combination of decile values of rainfall for progressive months during the growing season.

Figure 17.1 *A decile trend graph showing the probability of rainfall in the growing season and estimates of wheat yield for George-town, South Australia. Using the data set (insert) the cumulative monthly rainfall is plotted (□——□) for the year 1987.*

	A	M	J	J	A	S	O
mm	15	76	59	67	63	42	20
Cumulative rain (mm)	15	91	150	217	280	322	342

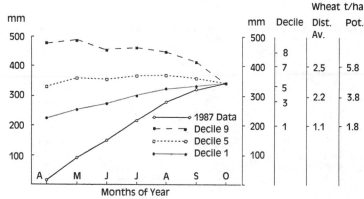

Figure 17.2 *Derivation of farmers' decile rainfall tables from their own forty years of rainfall records.*

The steps to be taken are:

1 To obtain decile values of individual months: rank each month's rainfall in ascending numerical order. Then, as the records are for forty years, the decile intervals are for four years. Thus count up four values from the lowest recording and that value becomes decile 1. Then count up another four values and that value is decile 2. Continue this process up to the highest recording.

2 To obtain decile values of groups of months: add up each year's rainfall for each of the groups of months listed in the bottom half of Table 17.3, viz, for April and May, April and May and June, through to April–October; and for May–October, June–October, etc. Rank the combined values for each of the forty years in ascending numerical order and derive the decile values.

such method of assessment is to calculate a cumulative sum of the difference between each individual year's rainfall and the long-term average. The running total has the capacity to define groups of years in which the rainfall has mainly been below the long-term average, and other years when the rainfall has been above average, in perhaps 70 per cent of the years.

The variations in average annual rainfall at six locations in South Australia in different groups of years are given in Table 17.4. The groups of years were determined by the change in direction of the cumulative rainfall trend. Variations in the decile rankings of growing season rainfall and the variation of rainfall in different groups of years need to be taken into account in assessing the effectiveness of research projects and the trend in district and farm yields.

Landscape

Variations in slope and soil type occur on most farms and existing individual paddocks often have variations in the depth and type of soil. If farm productivity is to increase, then it is first necessary to classify the basic land and soil resources and then manage them according to their own individual benefits or limitations for plant growth. This approach leads to the development of a Farm Plan. The basic features of a Farm Plan are:

1 Mapping of land classes. The main criteria are:
Arable land

Class I   Less than 2 per cent slope; no erosion hazard.

Class II   2–4 per cent slope — slight to moderate erosion hazard may require contour working lines.

Class III   4–10 per cent slope — moderate to severe erosion hazards, requires contour banking.

Class IV   Mainly sown to pastures but may grow an occasional crop.

Non-arable land

Class V   Requires contour furrowing and improved pastures.

Class VI   Capable of little improvement in production.

Class VII   Little agricultural value — main value is in tree planting.

2 Location of different soil types. The key factors in deciding the suitability of soils for plant growth are:
• The relatively unchanging soil physical properties, e.g. the depth of the root zone, the available water holding capacity, the probability of water logging, etc.
• The presence of chemical factors such as salinity, soil acidity, nutrient toxicity e.g. from boron or aluminium.
• The degree of degradation by erosion or salinity.

Table 17.4 *Variations in the average annual rainfall in different groups of years at farming locations in SA.*

Years	Kimba	Booleroo Centre	Georgetown	Mt Barker	Sedan	Lameroo
			(mm)			
1903–26	380	407	510	840	332	429
1927–45	308	340	407	730	259	337
1946–56	403	440	536	809	332	426
1957–67	317	329	407	683	270	347
1968–81	387	445	559	772	324	404
1982–9*	281	374	463	678	263	367

*The downward trend is still continuing.

The different soil types can be marked on to a clear plastic sheet which is then overlaid on to the map of the different land classes. Interactions between land classes and soil types determine the locations of future fence lines and the optimum use of this land — whether it be cropping, sown pastures, natural pastures or trees or native vegetation.

Potential yield for the property

The climate of the district has been identified as a key factor in the resources of the farm. The productivity should be calculated from the climate to provide a measure of the potential yields of crops and pasture that can be grown so that these values can be used as criteria to measure the success of the farming operations (Chapter 14).

Research has established relations between the yield of crops and pastures and the amount of water used during the growing season. Many of the relations are expressed as complex mathematical equations. However, where farmers want to calculate the relations on their own farms, simpler relations can give a good estimate of yield, e.g. the ratio of water use (i.e. stored soil moisture and rainfall) to evaporation can be used as a measure of the length of the growing season. To promote germination the ratio needs to exceed 0.3; best growth occurs when the ratio is 0.9. Values of 1.2 or higher usually indicate water logging and

reduced yields, and growth ceases when the ratio falls to 0.2.

The types of crops and pastures that can be grown in a district are also influenced by climatic factors. To obtain high yields, the plants need to complete flowering before the week with an average maximum temperature that is species specific. Thus temperate cereals need to complete flowering by the week which averages a daily maximum air temperature of 23°C. For temperate grain legumes and annual legume pastures the critical temperature is 20°C. Flowering after the onset of these temperatures will result in a reduced number of flowers and the filling of the grain. To be successful these species need assured moisture from sowing to flowering and an absence of frost at flowering time.

The potential yield of any crop or pasture is determined by the amount of incoming solar radiation that is intercepted. While these measurements are sometimes made on research stations, a satisfactory estimate can be made by using the class A pan evaporation data.

Potential yields can also be related to the amount of water used by a crop from sowing to harvest. The water used is obtained by measuring the soil water at sowing, adding the rainfall and measuring the remaining soil water at harvest. Only a few farmers measure soil water, but an estimate of water use was made so that the majority of South Australian farmers could assess the potential yield. This estimate was the 'derived' April to October rainfall. The derived April rainfall includes any water in the soil on 1 April due to long fallowing practices or to late summer rains.

Figure 17.3 shows the relation between potential yield of wheat and the 'derived' April to

Figure 17.3 *The relation between grain yield of wheat and the derived April–October rainfall for selected experimental sites and farmers' paddocks in SA. The sloping line indicates the potential yield. The curved line (—·—) shows the district yields in SA. These are only about half the potential at 250 mm and one-third the potential at 400 mm. The responses to different treatments are shown by lettered lines linking points. Yield increases were obtained by the application of nitrogen (points linked by a B line), phosphorus (C line), copper (D line) and control of eelworms (F line). Yield reductions occurred because of delayed time of sowing (A line), effects of weeds (E line) and waterlogging (G line).*

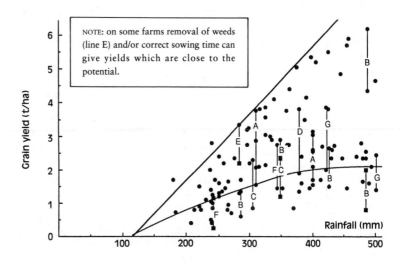

NOTE: on some farms removal of weeds (line E) and/or correct sowing time can give yields which are close to the potential.

October rainfall. The estimate of water use can be divided into:

- Rainfall that is lost by direct evaporation from the soil and plant leaves.
- The amount of water (rain and subsoil moisture) transpired by the crop to produce yield.

The evaporation figure is equal to the point on the X axis where the sloping line starts — as shown, it is 110 mm. This value will vary somewhat depending on soil types, e.g. on hard setting soils with poor infiltration the value may reach 170 mm. In low rainfall years, e.g. with less than 150 mm rainfall from April to October, the evaporation will be about 60 per cent of the rainfall. In average to good seasons, on most soils the evaporation loss ranges from about 25–40 per cent of the growing season rainfall.

The sloping line in Figure 17.3 represents the potential amount of grain that can be produced for every millimetre of water that is used (transpired) by the plant. For wheat, the potential production is 20 kg/ha/mm. Similar data are available for grain legumes and annual legume pastures. The average evaporation loss with grain legumes is 130 mm rainfall and potential grain yield is 15 kg/ha/mm transpired. For annual legume pastures, the average evaporation loss is 70 mm, and the potential dry matter production is 45 kg/ha/mm transpired. These relationships can be summarized as follows where water use denotes the change in soil moisture plus the growing season rainfall.

Wheat   Grain (kg/ha) = [water use(mm) −110] x 20

Grain Legumes   Grain (kg/ha) = [water use(mm) −130] x 15

Legume Pasture   Dry Matter (kg/ha) = [water use(mm) − 70] x 45

Other estimates of the potential yield can be obtained from a formula which includes a constant for each crop, water use and average daily pan evaporation in the growing season:

$$\text{Wheat} \quad \text{Grain (kg/ha)} = \frac{40 \times \text{Water Use (mm)}}{\text{Average daily evaporation from sowing to harvest (mm)}}$$

$$\text{Grain Legumes} \quad \text{Grain (kg/ha)} = \frac{35 \times \text{Water Use (mm)}}{\text{Average daily evaporation from sowing to harvest (mm)}}$$

$$\text{Legume Pasture} \quad \text{Dry Matter (kg/ha)} = \frac{70 \times \text{Water Use (mm)}}{\text{Average daily evaporation from sowing to maturity (mm)}}$$

While climatic factors are the basis for calculating the potential yield there are nevertheless soil types with specific physical limitations which can prevent this yield being obtained, e.g. the lack of adequate depth of root zone, low waterholding capacity. These factors seriously limit the potential yield of a specific crop or pasture, and it is necessary to reassess whether alternative land uses should be adopted.

Monitoring the long-term trends on farms

The landscape and the soils are the basic farm resources for defining management practices. As a result of tillage operations, the growth of crops and pastures and removal of products, fertilizer programmes and grazing practices, various changes in soil properties can develop.

It is most important to monitor the level of the soil properties using updated techniques and to

Table 17.5 *Optimum and lower values of organic carbon and nitrogen contents in top soil (0–100 mm) of different textures.*

Texture	Organic C+ — %		Total N+ — %	
	Optimum	Lower Limit	Optimum	Lower Limit
Sandy	0.6	0.5	0.06	0.05
Sandy loam	1.0	0.7	0.10	0.07
Loam	1.5	1.1	0.15	0.11
Clay	2.2	1.6	0.22	0.16

*Lower limit is the level at which degradation may occur.
+The tests for organic carbon and nitrogen are available from commercial firms.

define the direction of trends. Analysis should be done every four to five years, or at the end of a rotation. It is also important to assess whether the farming system is operating efficiently and to find out what is necessary to overcome the limiting factors. Assessment should be based on:

1 The incidence of erosion hazard. In general, most farmers can only record the year and a descriptive account of the loss of soil due to erosion.

2 The levels of organic carbon and nitrogen and their relation to the texture of the soil. Optimum values and safe lower limits for organic carbon and nitrogen determined for different soil textures are summarized in Table 17.5. The lower limit represents the stage at which the condition of the soil surface seriously limits growth and leads to soil degradation.

3 Soil structure. This defines the aggregation of soil particles and their influence on the intake of water and air into the soil, and the effect on root growth. Aggregation tests are available from commercial firms and these record the percentage of aggregates greater than 2 mm, and those less than 0.25 mm. Such tests can predict the need for lime or gypsum to improve the structure.

However, farmers can also carry out simple tests which broadly indicate the state of the soil structure. These tests include:

— An infiltration test in which an open cylinder is placed on the soil surface and 50 mm of water is added. The time taken for the water to infiltrate into the soil is a measure of the structure. The depth to which the water penetrates is a measure of the presence or absence of any compacted layer.

— A slaking test in which an aggregate of air-dry soil is placed on a piece of fly wire and put into a jar of water. The time taken for the aggregate to collapse relates to the amount of organic or inorganic binding material, e.g. organic matter and clay. A quick collapse indicates the need for a change in farming practices to restore the organic matter content.

— A dispersion test. This can be carried out as part of the slaking test. Due to a lack of bonding, clay particles diffuse slowly from the soil aggregate and produce a spreading cloud. The bonds between the clay particles are reduced because of high levels of sodium and magnesium and from cultivating the soil when it is too wet. In the field the diffused clay blocks the soil pores that transmit and store water. The addition of gypsum is needed to help solve this problem.

4 Soil compaction. This is a measure of the density of the soil. Compacted subsoils can develop due to excessive cultivation or excessive traffic from tractors and machinery. Measurement consists of taking a core of soil from the selected layer, weighing and drying it and relating these values to the volume of the core. Where the value exceeds about 1.35 g/cc, limitations to root growth and yield are probable because the pore space is greatly reduced. Changing farming practices are therefore warranted.

5 Soil pH. Over a period of time, there is a tendency for soils to become more acid due to the nitrification process and to the removal of calcium magnesium and potassium in the products of legume crops and pastures. Measurements of soil pH can be made either in water or calcium chloride. It is important to record which method was used because pH in calcium chloride can give a value 0.5 or more of a unit lower than that recorded in water.

Soil acidity reduces the availability of molybdenum, phosphorus and sulphur but increases the probability of toxic levels of manganese and aluminium. Application of lime is recommended when the pH (water) falls below 5.5. The lime required to balance the removal in products varies from 50–60 kg for a tonne of lucerne and clover hay, 22 kg per tonne of cereal hay, and 3 kg per tonne of cereal grain.

6 Soil salinity. An increase in the salinity of soils can occur due to the clearing of native vegetation on sloping country. Agricultural crops and pastures use less water than the native vegetation. This extra water infiltrates through the soil layers, dissolves the accumulated salt and then carries it to the valleys where salt rises to the surface with the water table. The effect of soil salinity on plant growth varies with the soil texture. Low soil salinity tests range from about 0.3 milli-siemens/cm (m s/cm) on sands and loams, to 0.7 m s/cm on clays. Practically all legumes (crop and pasture) require low salinity levels. High salinity tests range from 0.9–1.2 m s/cm for sands and loams, and up to 2.5 m s/cm for clays. Only canola and barley can grow satisfactorily beyond these values. Most other crops and pastures can be grown within the above values.

7 Soil nutrient status. Changes in the soil nutrient status can occur from a decline in organic matter content, a reduction in fertilizer application, the removal of nutrients in saleable products and the erosion of soil. For efficient plant growth, it is necessary to operate a balance sheet of the nutrient status of each paddock. A check

on the accuracy of the balance sheet can be made by measuring the level of available nutrients in the soil at the end of each rotation cycle.

8 Water repellence. The inability of soils, particularly sands, to absorb rainfall can be due to the organic materials which form a coating on the soil granules. Applying a cylinder of water to the soil (see above) can decide if such a problem exists.

9 Soil organisms. These play an important part in cycling organic matter and nutrients in soil and thereby contributing to soil fertility. Earthworms are the most obvious organisms and these can be assessed by sampling soils in July–August in winter rainfall regions. Under favourable soil management practices counts of between 300–400/m^2 soil, 10 cm deep, have been recorded. The effects of other soil organisms such as free-living nitrogen fixing bacteria can also be determined, but usually only in special laboratories.

Monitoring all the above factors provides an effective assessment over time of the changes occurring in the farm landscape. Aerial photographs (whether colour or infra-red) provide an effective means of monitoring changes in the landscape over time. The photographs should be taken at the same time of the year so that appropriate comparisons can be made.

Monitoring of paddocks during the growing season

Farmers are faced with the problem of managing the variability of costs and prices and the uncertainty of the growing season rainfall. They also need to understand the implications of research findings and determine if they are relevant to increasing the productivity on their paddocks. The yield of crops and pastures in any one season depends on the interaction of many biological factors, some promoting yield, others restricting it. To understand these interactions, it is important for the farmers to monitor and measure many factors in their paddocks during the growing season.

For best results, the monitoring should be carried out at four times during the growing season viz:

• Before the opening rains — this provides the opportunity to assess the stubble handling operations needed before the crops are sown, to assess the seed supply of self-regenerating pastures, the possibility of root diseases and the soil water status.

• At the tillering to boot stage — here the monitoring aims to identify and measure the factors limiting growth both above and below ground and to find ways of overcoming the problems.

• In the early spring time — to assess the incidence of root and leaf diseases on crops, and to assess the possibility of grassy pastures carrying over disease problems into the following year's crops.

• At harvest time — to assess the 'harvest index' of crops (i.e. the ratio of grain produced to total dry matter) and to record the grain protein levels from each paddock.

The factors to be monitored can vary according to whether the paddock is in a crop or pasture.

Crop paddocks

The principal factors to be monitored are:

• The amount of stubble cover that exists prior to sowing and whether it can be handled by mulching, discing or burning.

• The soil moisture content at or just prior to sowing. Further moisture measurements should be made at the tillering stage, and at the final harvest (Table 17.6). Measurements can be made by taking soil cores and drying the sample or by the use of neutron moisture meters recording the moisture in permanent tubes (Figure 17.4).

The incidence of water logging can also be assessed. Excess water in the root zone reduces the levels of soil oxygen and the nutrient uptake, particularly nitrogen. The reduction in yield is greatest when water logging occurs in the few weeks before flowering (anthesis).

• The incidence of weeds, both pre-seeding and post-seeding, and the most economical way of controlling them. Weeds will greatly reduce yield if there are more than 50 weeds/m^2 at seven weeks after sowing.

• Time of sowing can have a big impact on yield, e.g. in the majority of districts in South Australia best yields are obtained by sowing early in May. Some farmers may even take a risk and sow dry at this time for yields are reduced by 200–300 kg/ha for each week's delay in sowing beyond this date.

The impact of time of sowing in a mid-north district in South Australia is shown in Table 17.7. The data highlight the ratio of water use/evaporation in two growth intervals, and the effect these ratios have on dry matter pro-

Table 17.6 Variation in yield of wheat due to changes in the water use/evaporation ratio.

High Yield

	Rain (mm)	Water Use (mm)	Evap. (mm)	Water/ evap.	Dry Matter kg/ha	Grain
Sowing-tillering	111	93	128	0.72	770	
Tillering-anthesis	117	185	229	0.81	7990	
Anthesis-soft dough	30	81	160	0.51	2730	
Soft dough-harvest	34	32	223	0.14	−330	
Total	292	391	740	0.53	11 160	3900

Low Yield

	Rain (mm)	Water Use (mm)	Evap. (mm)	Water/ evap.	Dry Matter kg/ha	Grain
Sowing-tillering	83	73	183	0.40	520	
Tillering-anthesis	42	93	232	0.40	2330	
Anthesis-soft dough	21	49	178	0.28	1350	
Soft dough-harvest	7	21	195	0.11	−300	
Total	153	236	788	0.30	3900	1060

Table 17.7 *Effects of wheat sowing time on water use/evaporation ratios and yield in the mid-north district of South Australia.*

Date of Sowing	6 June	19 July	3 Aug	21 Aug
Sowing-Anthesis				
No. days	133	105	91	79
Water use (mm)	310	274	257	216
Evaporation (mm)	356	368	372	390
Ratio water use/evap.	0.87	0.74	0.69	0.55
Dry matter (t/ha)	9.32	6.92	4.29	3.50
Anthesis-Harvest				
No. days	44	35	41	37
Water use (mm)	90	60	57	49
Evaporation (mm)	316	297	306	292
Ratio water use/evap.	0.28	0.20	0.19	0.17
Dry matter (t/ha)	2.83	1.99	2.46	1.13
Harvest				
Total dry matter (t/ha)	12.15	8.91	6.75	4.63
No. grains/m²	12 910	10 150	7210	5960
Grain yield (t/ha)	3.76	2.86	2.08	1.55
Kg/ha/mm WU	9.4	8.6	6.6	5.8

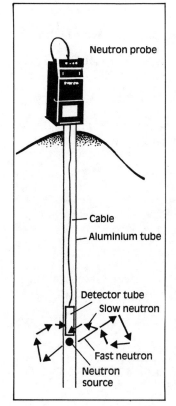

Figure 17.4 *How a neutron moisture metre probe measures soil moisture.*

Neutron probe

Cable
Aluminium tube

Detector tube
Slow neutron

Fast neutron

Neutron source

duction, grain yield and grain production per millimetre of water use. Highest yields are obtained when the ratio from sowing to anthesis exceeds 0.80 and the ratio from anthesis to harvest approaches 0.3.

• Sowing depth and seedling establishment. The depth of sowing should not be greater than the coleoptile length. High yields of wheat require the successful establishment of about 200 plants/m², giving 600–800 tillers by the tillering to boot stage.

• Root diseases and nematodes. These organisms cause major reductions in yields of crops and pastures. The main ones are rhizoctonia, Take-all ('hay die') and Cereal Cyst Nematodes.

Measures to assess the likely effect on yield consist of:

— Soil sampling the paddocks several months

before sowing and having incubated tests, conducted by commercial firms, to assess the potential level of infection.

— Examining the roots of plants at about six weeks after sowing. About 1 kg blocks of soil are taken from the paddock and the soil washed off and the roots placed in plastic dishes and the incidence of diseases assessed.

All the following factors influence the yield of crops and need to be managed so the best use is made of the sunlight energy.

• Nutrient uptake. The success of previous fertilizing practices can influence the growth of crops and pastures in the current growing season. Whether or not the nutrient supply is adequate can be judged by deficiency symptoms or by plant analyses. The analysis can be conducted either on the youngest expanded blade, or the younger open blade or on samples of the whole plant above ground.

Table 17.8 shows the relation between climatic factors and the accumulated nutrient uptake for optimum growth for amounts of dry matter increasing from 1 to 8 tonnes/ha. The climatic factors show the accumulated amounts of water use (soil moisture change and rainfall), evaporation and day-degrees of maximum and minimum air temperatures. While these values may vary somewhat due to localized climatic factors, soil types, varieties and perhaps a luxury consumption of nutrients in some situations, they do provide a basis for monitoring the factors limiting growth and water-use efficiency in a season. Any major deviations from these optimum values will reduce the yield of dry matter.

The data highlight the different amounts of nutrients needed by three plant species. Lupins and medic pastures need higher uptakes of nitrogen, calcium, magnesium and zinc than wheat. The medic pasture also has a higher need for both phosphate and zinc compared with wheat. In the early growth stage, e.g. to produce the first 1 to 2 tonnes/ha of dry matter, the medic pasture has to take up about twice as much phosphorus as wheat and 50 per cent more nitrogen.

In determining a balance sheet approach for nutrient levels, a key factor is the amount of nutrients removed from the paddocks by the products. Table 17.9 shows the amount of nutrients removed in a tonne of wheat and lupin grain, and in a tonne of medic hay. Variations occur in the amounts of nitrogen, potash, calcium magnesium and zinc. If the medic pasture is grazed the loss of nutrients is much lower than

Table 17.8 *Climatic factors and nutrient uptake in plant tops for optimum growth of wheat, lupins and medic pasture in South Australia where dry matter increases from 1–8 t/ha during the growing season.*

Wheat

Dry Matter (t/ha)	1	2	3	4	6	8
Water use (mm)	60	85	100	120	160	200
Evaporation (mm)	100	140	165	190	230	275
Max Temp (°C)	800*	1100	1250	1400	1750	2100
Min Temp (°C)	300*	430	520	620	800	950

Nutrient Uptake

		1	2	3	4	6	8
P	kg/ha	3.5	5.7	7.5	10.5	15.5	18.0
N	kg/ha	43	60	77	95	125	145
K	kg/ha	40	70	100	140	160	150
S	kg/ha	3.0	5.0	6.0	7.0	9.5	12.0
Ca	kg/ha	3.7	5.0	6.0	8.0	9.0	10.0
Mg	kg/ha	2.0	3.0	4.0	5.0	7.5	10.0
Cu	g/ha	15	22	30	35	45	60
Zn	g/ha	30	48	58	70	90	115
Mn	g/ha	80	105	120	145	185	220

Lupins

Dry Matter (t/ha)	1	2	3	4	6	8
Water use (mm)	80	130	170	210	275	310
Evaporation (mm)	100	150	185	240	300	370
Max Temp (°C)	1150	1400	1600	1800	2200	2600
Min Temp (°C)	500	700	850	1000	1200	1400

Nutrient Uptake

		1	2	3	4	6	8
P	kg/ha	3.3	6.0	8.5	10.5	14.0	16.0
N	kg/ha	45	80	110	140	185	215
K	kg/ha	27	48	67	85	120	148
S	kg/ha	2.7	5.0	7.1	9.0	13.0	16.8
Ca	kg/ha	20	35	50	65	85	100
Mg	kg/ha	6	11	15.4	20.1	28.3	35.0
Cu	g/ha	10	20	28	33	42	48
Zn	g/ha	35	75	110	140	170	200
Mn	g/ha	220	410	550	680	870	1000

Medic Pasture

Dry Matter (t/ha)	1	2	3	4	6	8
Water use (mm)	40	70	100	130	190	250
Evaporation (mm)	50	80	125	170	250	340
Max Temp (°C)	400	700	1000	1300	1900	2200
Min Temp (°C)	200	300	370	470	700	900

Nutrient Uptake

		1	2	3	4	6	8
P	kg/ha	7	10.5	13.5	17.0	23.5	27.5
N	kg/ha	65	100	125	150	200	240
K	kg/ha	55	80	100	120	165	190
S	kg/ha	4	7	10	12	17	18.5
Ca	kg/ha	15	25	35	45	65	75
Mg	kg/ha	4.5	8	11	13.5	20.5	23.5
Cu	g/ha	15	25	33	38	58	70
Zn	g/ha	67	95	115	135	165	185
Mn	g/ha	70	110	140	160	220	240

* Cumulative maximum and minimum temperatures

if removed as hay. Grazing animals return from 60–80 per cent of the nutrients in the feed they consume. Nevertheless, nitrogen losses can be high due to volatilization from the urine patches.

- Damage from insects. Crop yields can be reduced by damage arising from infestations of insects, such as red-legged earth mite and lucerne flea. It is important to inspect crops from 6–8 weeks after sowing and assess the need for control measures. While many crops can recover from this damage, yields are nevertheless reduced.
- Leaf diseases. A range of leaf diseases can, in certain seasons, reduce yield and quality of the grain. Diseases such as septoria blotch, stripe rust, yellow leaf spot and barley yellow dwarf virus are not uncommon and have distinctive symptoms. These are best recognized from photographs in booklets which are available to farmers. It is important also to record which varieties have a resistance to a specific leaf disease.

Again, actual photographs (colour or infra-red) can illustrate the degree of success of farming practices in each season. Photos taken just before the heading stage of crops can identify uneven growth due to soil compaction, diseases, nutrient deficiencies, and indicate where improvements in management practices are needed to increase productivity.

Pasture paddocks

A similar set of soil and plant factors need to be monitored during the growing season to ensure high production from pastures. There are also some additional factors:

- The amount of seed in the paddock before the opening rains. For instance, a good medic pasture needs a minimum establishment of 200 plants/m^2, with a preferred figure as high as 1000 plants/m^2. After allowing for about 10 per cent of seeds germinating, and only 5 per cent becoming established, the seed reserves need to be at least 200 kg/ha. Seed testing kits are available to measure the amount of seed available. Stock may have to be removed from a paddock if the seed supply is minimal to prevent consumption of seed as they tend to eat the larger pods with greater seed content.
- When resowing legume species, it is essential to inoculate them with the appropriate rhizobial strain. Some recently selected strains for medics are now able to persist in slightly acid soils.
- Insect control. Insects can indirectly reduce the seedling numbers in autumn, and directly reduce the seed yields in spring. For varieties that are susceptible to lucerne flea and red-legged earth mite, it is advisable to spray with the appropriate chemicals 10 days before the opening rains, and again 4–6 weeks after germination. Pastures should also be monitored in late August to check on the incidence of spotted alfalfa-aphids, blue-green aphids and pea aphids. Some medic varieties now have resistance to aphids. These include Harbinger and Parabinga for 250–400 mm rainfall and Paraggio and Sephi for 375–500 mm rainfall.
- Viability of pastures seeds after grazing. Often the success of a particular legume species is due to the percentage of seeds that remain viable after passing through sheep. Thus, from 35–40 per cent of the seeds of Persian clover, Balansa clover and Cluster clover remain viable and 20 per cent for burr medic compared with only 1–7

Table 17.9 *Nutrients removed in each tonne of grain for wheat and lupins and each tonne of pasture hay as well as from 5 kg wool and 50 kg meat.*

Nutrient Removed		Wheat (tonne)	Lupins (tonne)	Medic Pasture (hay) (tonne)	Wool (5 kg)	Meat (50 kg)
P	(kg)	3	4.5	3.2	0.02	0.4
N	(kg)	23	57	30	1.0	3.0
K	(kg)	5	10	28	0.1	0.1
S	(kg)	2	4	2.8	0.2	0.4
Ca	(kg)	0.4	2.5	10.1		
Mg	(kg)	1.5	2.8	3.3		
Cn	(g)	7	8	9		
Zn	(g)	16	32	23		
Mn	(g)	40	16	30		

per cent of the range of all other temperate legume pasture species.

- Persistence of grasses. Both perennial and annual species contribute to production in farms. The perennial pastures are best suited to the high rainfall regions and consist of species such as perennial rye grass and phalaris. Annual grasses are common in the cropping areas and include annual rye grass, barley, brome and silver grass. They usually provide good early feed for livestock but they also become hosts for building up root diseases which reduce crop yields in the following year.

A key monitoring role in pastures is to follow the growth of grasses. If the paddock is being cropped the following year, the grasses should be removed by chemical spraying in mid-July. If the spraying is delayed until mid-September the incidence of most diseases is not reduced. Problems are occurring now with some grasses that have developed resistance to chemicals.

Where saline areas exist on farms, these need to be set aside and sown with salt–tolerant grasses such as *Puccinellia* and *Agropyron*.

- Nutrient supply. The approach in developing a balance sheet of nutrients for crops, also applies to pasture paddocks. It is also important to carry out soil tests for available nutrients because the soil levels are influenced by the nutrients returned under grazing and by the concentration of nutrient in dung patches and sheep camps. Plant tests similar to those for crops can also be carried out on pastures to assess the nutrient status.

- Measurement of pasture availability. The management of pastures under grazing requires a continual evaluation. Variations in growth at different times of the year, the need to prevent lower quality fodders from dominating the stand and the need for a minimum plant cover for erosion control are all affected by grazing pressures. Some of the criteria that form the basis of monitoring grazing pastures are:

— Active vegetative cover. If the pasture height is more than 12–15 cm, production is wasted. Grazing should be increased to reduce the height to 3–5 cm to let in light for the developing tillers.

— Minimum cover to prevent erosion. The amount of dry matter needed varies from 600 kg/ha for flat land to 1000 kg/ha for undulating land, to 1500 kg/ha for steep land.

— The optimum levels of pasture production at which livestock should be removed to prevent a loss in animal production:

Sheep	— ewes	500 kg/ha
	— ewes in lamb	700 kg/ha
	— ewes and lambs	1400 kg/ha
Cows	— dry	600 kg/ha
	— in milk	1600 kg/ha
Beef Cattle	— in calf	1000 kg/ha
	— with calf	1600 kg/ha
	— post weaning	800 kg/ha

Careful attention should be paid to the dry matter level over summer because of the quality of the feed, the daily loss of dry feed (due to disintegration) of 20–40 kg/ha/day, the potential loss of pods containing pasture seeds, and the erosion hazard.

The above data identify the need for farmers to estimate the amount of pasture. This can be assessed by the use of quadrats and cages. The existing level of feed can be assessed by cutting quadrats of $0.5 \, m^2$, drying the feed and weighing and calculating the production per hectare. The growth during the grazing period can be measured by the use of cages that protect the pasture from the grazing animals. The assessment is the same as for the quadrat cuts.

With the experience gained from making these pasture cuts over several seasons, it is possible for farmers to make visual assessments of their pasture production and to adjust their grazing pressures accordingly. The information can lead to the need for portable electric fences to increase stocking rates during the spring flush and the need for small holding paddocks or feedlots to handle the sheep over the summer.

In general, the dry matter intake for livestock is:

Adult dry sheep	10 kg/head/week
Ewes in lamb	18 kg/head/week
Ewes with lamb	25 kg/head/week
Cows — dry	42 kg/head/week
— early lactation	105 kg/head/week

Similarly adequate water supplies for livestock range from:

	ewe	*cow*
January	3.7 litres/day	53 litres/day
July	0.1 litres/day	22 litres/day

- Stock health. The general health of stock over the year needs to be monitored for any incidence of worms or other parasites. Measurements of liveweight trends are helpful.

Measures of success

Monitoring on farms provides the basis for production, profitability and sustainable agriculture.

The first measure of success is to evaluate the yield obtained and this is best done by expressing the yield as a percentage of the potential yield that is possible with the growing season water supply. Figure 17.3 shows the relation between wheat yield and an estimate of growing season water supply, viz. the 'derived' April–October rainfall. It also shows that district farm yields are only about half the potential that is possible for the water supply, despite the technological inputs into wheat growing. By plotting their own paddock yields on this graph the farmers can obtain an estimate of the efficiency of their farming practises. The monitoring programmes during the year should be able to define the factors that are limiting the yields in each paddock. In most cases, there are several factors that are limiting yield at any one time and hence there is a need for developing multi-factorial demonstrations on farmer's paddocks to identify the main ones. For example, in a cropping rotation trial at Tarlee, South Australia, annual monitoring of grain yield showed that grain yield per mm of water use declined in the first four years from about 5 to 2 kg/ha/mm, but in the following six years the yields rose to over 9 kg/ha grain per mm. It is probable that any changes in farm management will take a number of years to show their impact on improved productivity.

In another field experiment on a farming property in the lower north of South Australia the yields of different crops in a continuous cropping programme have shown a steady rise towards the potential yield. The farmer has consistently monitored the amount of stubble that has to be handled before sowing, the type of tillage and fertilizer programmes, and the need for weed control in order to promote growth and improve the soil structure of the red-brown soil. The pattern of yields is shown in Table 17.10.

A similar potential yield establishes the relation between the dry matter yield of annual legume pastures and the water use to the date when growth ceases (Figure 17.5). In general, this occurs a few weeks after flowering when the daily pan evaporation exceeds 4.5 mm, but late spring rains can delay the onset of this evaporation rate. For South Australia, the date will usually vary from mid-September to mid-November in different districts depending on the climatic inputs, and these dates will therefore determine the calculations to estimate the water use. The graph shows that on-farm production is again appreciably below the potential.

The feed produced in the growing season must (in general) meet the needs of animals for the whole year. Hence although the production can be related to the seasonal rainfall, the carrying capacity should be related to the annual rainfall. As a rule of thumb, in cereal and high rainfall districts in South Australia, after deducting 250 mm from

Table 17.10 *The yields under continuous cropping and their percentage of the potential yield on a farming property in the lower north of South Australia.*

Year	Crop	Rainfall* (derived Apr–Oct) (mm)	Yield t/ha	% of Potential Yield
1978	Wheat	489	2.2	29
1979	Peas	439	1.2	22
1980	Barley	451	2.2	31
1981	Barley	389	2.7	45
1982	Peas	334	0.9	29
1983	Barley	445	2.8	39
1984	Beans	407	1.8	43
1985	Wheat	408	3.8	64
1986	Peas	390	2.2	56
1987	Barley	344	3.3	66
1988	Beans	378	1.8	47
1989	Wheat	390	5.2	81

*See text for explanation

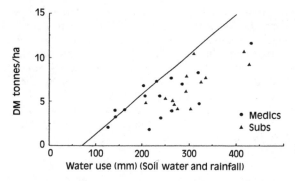

Figure 17.5 *The relationship between potential dry matter production of an annual legume pasture and the amount of available soil moisture (soil water + rainfall) during the growing season in southern Australia. Evaporation accounts for 4.5 mm/day (on average).*

the annual rainfall the remaining rainfall should provide enough feed to carry 1.5 sheep/ha for each additional 25 mm of rain.

Other measures of success from improved farming practices are rising trends in the level of soil fertility, a reduction in degradation from erosion and salinity, reduced incidence of root diseases and an increase in the biomass. Above all, however, the net returns per hectare are the final factors in determining whether or not the farmer can sustain a livelihood.

Future plans

Farming still has a very important role to play in the world's economy, for most of the food in the 21st century will come from rain-fed areas. It is therefore important that a strong link be established between agricultural science and the application of these findings, so that increased productivity can be obtained on a wide range of farms with differing soils and climates.

While much degradation has occurred in our farming lands in past eras (Chapter 11), farmers should not be blamed entirely for the problem. Future successful farming requires that they be encouraged and assisted to change their management practices. Neither 'sustainable agriculture' nor 'land care' are possible unless farmers can maintain their livelihood.

Farming is a business which has specific components relating to the climate and soils on the property, the need for monitoring a range of factors in each paddock, the need for scientific advice which is relevant to each farm and the production of financial budgets. Dryland farming systems involve crops, pastures and livestock. All of these involve biological processes, and management attempts to integrate the various factors and generate farm production and profit. Monitoring is a vital part of management.

Farming depends on growth of plants and a main aim in the future is to use the water supply more efficiently. There are several prerequisites:

- Increasing the water supply by fallowing (mechanical or chemical), cultivation on the contour, weed control, preventing soil compaction and water repellence.
- Reducing the evaporation by minimum tillage, surface mulching of stubble, early time of sowing, growing varieties with early vigour and good leaf area.
- Increasing the transpiration efficiency by early sowing, control of weeds and root diseases, control of insects and leaf diseases and optimum fertilizer application, both in the amount of nutrients available and the ratio between them to promote maximum growth.

In most cases, to achieve higher yields it is necessary to conduct paddock monitoring of weeds, root diseases, and tillage practices in the year before the crop is sown. There is no one single method of farming for sustainable agriculture in all soils and climates. Thus farming programmes may involve:

- Ley farming, which usually consists of alternate years of crop and pasture.
- Continuous cropping, which can be either continuous cereals or a rotation of cereals and grain legumes.
- Phase farming, in which the paddock may be cropped for 4 to 8 years, and then changed to legume pastures for 2 to 3 years.

In the future farming is going to require an increasing technological data bank of a whole range of climatic, soil, crop and pasture data relating to the requirements for optimum growth. Decision making will rely on this data bank and on the levels of various factors identified by monitoring biological processes in each paddock. The data bank is beyond the memory capacity of farmers, and future farmers will therefore need skills to operate home computer programs. Records can then be kept of annual rainfall patterns, the yield of crops and pastures and stock numbers in different paddocks in past years, and of the tillage programmes. Decile rainfall trends can be developed for each season and there needs to be a monthly list of factors which must be monitored in each paddock to contribute to the decision-making process. As well, the computer will also be the basis for financial budgeting whereby all costs, prices and returns can be tabled for each month of the year.

Other factors which can assist the decision making of farmers are:

- A programme to conduct experiments in individual paddocks to establish what factors limit production.
- On-farm weather stations that can measure the specific values of climatic factors. These values can contribute to a better understanding of growth, seasonal rainfall, the likelihood of leaf diseases in crops, erosion hazards etc.
- The use of remote sensing, such as near infrared photographs, which can readily identify variable growth in paddocks, and recording equipment that can detect nutrient deficiencies, crop diseases, digestibility of livestock feed and grain protein content (Figure 17.6).

Future farming practices are therefore going to depend very much on the identification of the soil and climatic factors on the farm, the monitoring of soil, plant and disease, the undertaking of research projects that can identify the factors limiting growth on individual paddocks, and the development of multi-factor integrated farming systems to increase productivity and sustain the farmer's livelihood.

Conclusions

The production per hectare of the average farm has not increased very much during the past forty to fifty years, despite the investment of large sums into agricultural research. Questions arise as to whether agricultural research has been relevant to farming systems, whether there has been too much emphasis on single factor research, or whether the extension of the research findings to the farm has been adequate.

Farming is becoming increasingly complicated because of the interactions between ecological, agronomic, social and economic factors. The

Figure 17.6   *Aerial photographs can be powerful tools in any monitoring programme.*

farmer is faced with the problems of managing variability of inputs and prices and the uncertainty of seasons, and is being asked to practice more sustainable methods. The economic viability of a farmer is one of the major factors influencing farm management practices.

In the past, farmers have relied on memory to develop their farming programmes, today farming is a multi-factor business and this requires an upgraded set of criteria to help in decision making. Farmers should monitor trends, especially in climate, nutrient flows, soil losses, organic matter status, pest and disease levels and any changes in the soils physical status. An essential feature in monitoring farming systems is to record the nutrient lost from each paddock via harvested products.

Further reading

Carter, E. D. (1987). 'Establishment and natural regeneration of annual pastures', *Temperate Pastures: Their Production, Use and Management*, J. L. Wheeler, C. J. Pearson and G. E. Robards (eds.), Australian Wool Corporation/CSIRO, 35–51.

Carter, E. D., Le Leu, Kylie, M. and Baldwin, G. T. (1989). 'Predicting the Emergence of Annual Pasture Legumes: Commercialisation of a Soil-coring Technique', *Proc. 5th Aust. Agron. Conf. Perth*, p. 438.

French, R. J. and Schultz, J. E. (1984). 'Water Use Efficiency of Wheat in a Mediterranean Environment'. I. 'The Relation Between Yield, Water Use and Climate', *Aust. J. Agric. Res.* 35: 734–64, II. 'Some Limitations to Efficiency', *Aust. J. Agric. Res.* 35, 765–75.

Morbey, A. S. C. and Ashton, B. L. (1990). *Lot Feeding of Sheep During the 1988 Drought—Technical Report No. 155*, Department of Agriculture, South Australia.

Rovira, A. D. (1984). 'Countering Soil Borne Cereal Diseases', *Rural Research*, 123, 12–16.

Rovira, A. D. (1990). 'Ecology, epidemiology and control of Take-all, Rhizoctonia Bare Patch and Cereal Cyst Nematode in wheat' in *Aust. J. Plant Pathology*, 19, 101–111.

Tisdall, J. M. and Oades, J. M. (1982). 'Organic matter and water stable aggregates in soils', *J. Soil Science*, 33, 141–163.

Tisdall, J. M. (1988). 'Soil managed as an ideal medium for roots leads to high productivity', *Proc. Wimmera Conservation Farming*, Victoria, 5–16.

Wegener, P. F., McDowall, C. J. and Frensham, A. B. (1989). 'Monitoring farming systems: effects of cropping practices; limitations to wheat yields', *Technical Paper No. 23*, Department of Agriculture, South Australia.

Development in Dryland Farming Systems

CHAPTER 18

Research in a Farming Systems Framework

R. L. McCown

SYNOPSIS This chapter explains the new pressure on agricultural research to be efficient in providing knowledge and technology to improve the performance of production systems, and the need for a systems approach to achieve this efficiency. It presents a systems analytic framework for research which combines Farming Systems Research and Agroecosystem Analysis concepts.

MAJOR CONCEPTS

1 The Farming Systems Research (FSR) approach was developed to aid research priority setting in a developing country context. Heavy reliance is placed on the study of individual farms and participation by individual farmers.

2 Conway's Agroecosystem Analysis and Development scheme was developed in a developing country context as a means of rapidly appraising the problems and opportunities of agriculture at the scale of a specific land system or watershed.

3 The broadening of FSR by incorporation of agroecosystem performance concepts and relaxation of the original farm scale provides a flexible framework for considering how to use research resources most effectively to benefit a given agricultural system.

4 Operations research is an approach for studying large, complex systems with the aim of improving their management. It depends heavily on simulating system performance using numerical models of the system.

Agricultural research has played an important role in the development of modern agriculture. However, it has been subjected to a critical reappraisal of its cost-effectiveness and the way it is funded. There has been a shift from a situation where research priorities have been set mainly by scientists who conducted the research using public money, to one in which users of the research pay a significant share of the research bill and also have a significant voice in how funds are spent.

In the early days of the application of science to the traditional art of agriculture, research was conducted in close proximity to production systems and the managers of these systems. Most of those who trained in agricultural science were from the land. There was the general expectation that agricultural science would efficiently show the way to improved agricultural practice. For several decades the rate at which it did this was quite spectacular. However, further progress required increased scientific specialization. The early farm-oriented scientist was replaced by specialists in a wide range of disciplines. While this reductionist approach enabled certain technological advances, the corporate capability of researchers to see the production system in perspective declined together with the incentive to do so. Agricultural scientists were rewarded less and less for research that efficiently improved agricultural practice and more and more for excellence in their discipline. They continued, however, to set their own research priorities.

The setting of priorities for research is central to this chapter. In recent times, agricultural scientists have done a much better job of conducting research than in planning and evaluating it. There has been an emphasis on scientific originality (has it been done before?) and excellence (will it impress scientific peers?) at the expense of a general problem-solving approach which gives adequate weight to whether research objectives are relevant to the needs of client managers of agricultural systems. Post-graduate students learn much about how to conduct research to gain new knowledge, but little about how to recognize what is worth finding out when resources are scarce.

Research is now increasingly dependent on funding from managers of production systems, i.e. farmers, (via producer organizations) and researchers are accountable to them within agreed projects. Financial resources for research are scarce and allocations can be expected to be made on the basis of perceived benefits to the production system concerned. Scientists face a new competitive environment in which the judging of research proposals is done with a new rigour and, to an important degree, by clients. Many scientists accept these changes as logical and overdue but there is concern that the new arrangement may merely substitute new biases for old, with different but equally serious inadequacies. To complicate matters, there are other clients with heightened concern about non-production performance aspects of agricultural systems, e.g. land degradation, effects on the environment beyond the farm, and product safety.

A process that can improve the ability to identify research priorities in terms of improving or sustaining system performance and which draws upon the respective comparative advantages of researchers and clients in doing so is required.

Setting priorities for research to improve production systems

An historically important example of scientists getting the priorities wrong occurred in the 1960s when it became painfully apparent that farmers in Third World countries were not adopting the new technology that expatriate agricultural experts were offering them. Not surprisingly, the initial interpretation was that farmers in these circumstances are rather irrationally conservative, but, in fact it was mainly because the researchers were solving problems that were not really important to those farmers. Farming Systems Research (FSR) methodology is a scheme for thinking about farming systems, their problems, and their improvement, and for aiding the setting of research priorities.

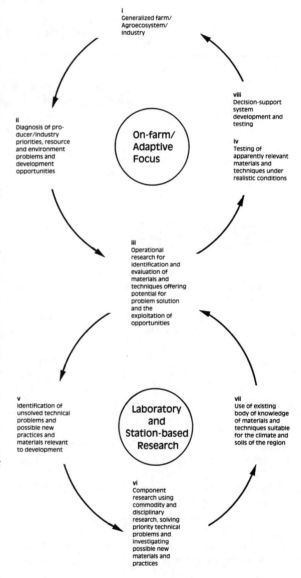

Figure 18.1 *The flow of activities in Farming Systems Research (FSR).*

Figure 18.2 *The structure of a farm system showing both biotechnical and social factors important in scientific analysis of economic decision making by a manager.* (SOURCE: adapted from figure by Norman (1980) and cited by Anderson *et. al.* in Remenyi (ed.) (ASS)).

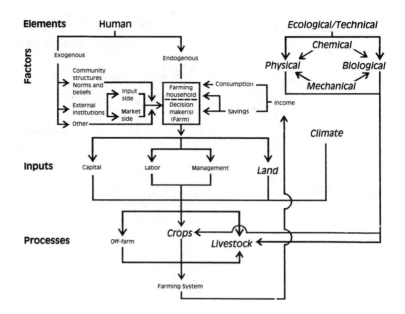

The objective of FSR is to improve the performance of the farm system and the welfare of the farm family/firm (Figure 18.1, i). To maximize the relevance of research activities, emphasis is placed on diagnosis of problems and opportunities on-farm (Figure 18.1, ii) and the testing of apparently relevant materials and methods on the farm in conjunction with the farmer (Figure 18.1, iv). Where they are judged to be potentially adequate (Figure 18.1, iii), materials and techniques are drawn from the existing body of knowledge (Figure 18.1, vii). Where unsolved technical problems and possible new relevant practices and materials are identified, appropriate research is initiated, and the outcome contributes alternatives for evaluation (Figure 18.1, iii, iv) and adds to the store of knowledge (Figure 18.1, vii).

In attempting to diagnose the performance of a farming system, one encounters formidable complexity. However, a hierarchial classification of the elements of the system, as in Figure 18.2, helps make a methodical approach seem feasible. In terms of the farmer's objective, the farming system is an economic system with an important sub-system which is ecological and technical in nature (italicized in Figure 18.2). Clearly, effective analysis requires a multidisciplinary team which includes social, biological and physical scientists.

The production ecology of a farming system can be analysed in terms of three properties: productivity, stability, and sustainability. These properties provide the framework for diagnosing problems and for setting objectives for research. A research project in northern Australia provides a case of agricultural research conducted using this approach. The objective was to assess the feasibility of dryland grain cropping in areas of arable soils in the 'top end' of the Northern Territory and in north Queensland. At the time the study was initiated there were virtually no farmers there, though there had been several expensive attempts at cropping previously and another was soon to begin. The main question was '*should* farmers be here?' Figure 18.3 outlines the analyses of the problems and an evaluation of prospects for various alternative technologies that has constituted this research programme.

Initial work focused on filling important gaps in understanding of the key problems of nitrogen supply and crop establishment. Research in an earlier era had shown that on the red earth soils in this tropical wet-dry climate, seedbeds often slaked and crusted before seedlings emerged and leaching loss of nitrate was often high. New studies showed that high surface soil temperature was an even greater problem than crusting and that both were due to the high rate of drying of the soil surface. Identification of the most promising approach to alleviating problems of both seedbed conditions and nitrogen supply stemmed from two assumptions. The first assumption was that integration of cropping with the existing beef cattle grazing system would be economically beneficial. The second assumption was that crop planting using no-tillage technology, which leaves a mulch on the soil surface, would be as beneficial in reducing erosion as

Figure 18.3 *The structure of a research project in northern Australia.*

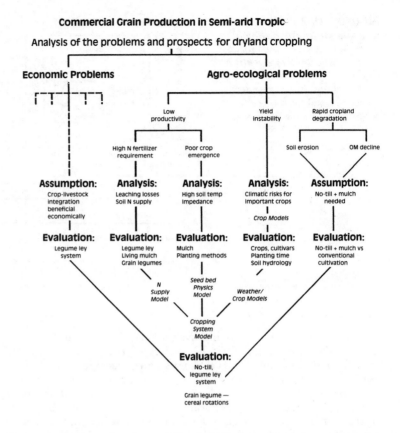

it had been found to be in other areas of the tropics. The second research phase focused on evaluation of legume leys, including the value of the nitrogen and mulch they leave behind. Leys were evaluated in terms of contribution of nitrogen to coarse grain crops, and also for their effect on cattle performance and cattle on ley performance. The efficacy of the amount and type of mulch and of planting method in alleviating problems of high soil temperature and strength were evaluated.

One of the important features of this region, characteristic of the semi-arid tropics, is high year-to-year variability in crop yield. There are several important climatic risks. One is the risk of inadequate duration of favourable planting conditions (neither too dry or too wet) sufficiently early in the season. Another is the risk of having inadequate mulch at planting. A third is the risk of failure of rainfall during the growing season. In none of these is quantification of problem frequency feasible using experimentation, especially since the question needs to be answered for many different places. Similarly, difficulties exist in the evaluation of the efficacy of conservation tillage (no-till + mulch) in reducing soil erosion. A different approach for evaluation of aspects of systems such

as these that vary importantly and unpredictably, from year-to-year and among locations, was needed. An appropriate approach is explored in the next section, followed by its application to this particular example.

Operations research concepts and methods in farming systems research

Operational (or operations) research is the application of scientific methods to complex problems which arise in managing large man-made systems. The need for special methods arises from the dilemma that, while improvement in the management of such systems in industry, the military, and government would benefit from knowledge gained from experimentation with the system, it cannot usually be carried out. This difficulty is overcome by the construction of a model of the system, upon which experiments can be conducted (simulations).

The aim of operations research is to improve the performance of the system in terms of the system objective. This is accomplished by improving the structure of the system, e.g. adoption of improved technology, or by improved managerial decision making. A major aspect of the latter is the reduction of uncertainties and improved means of managing risk. Because of the diversity of functions within large complex systems, operations research is conducted by multi-disciplinary teams whose composition relates to the nature of the system in question. Since its origin in the late 1930s operations research has grown into an important discipline of science, with indistinct boundaries with systems engineering and information technology.

Agricultural research is concerned with improving the performance of large, complex systems, yet there has been little recognition of operations research, as such, in agricultural research. The reasons for this seem to lie in the distinctive nature of agricultural systems and particular historical role of agricultural research. Agricultural systems can be viewed as either a man-organized system with important uncontrolled variables or a natural system with purposeful intervention (Figure 18.2). Not surprisingly, the system is seen as a man-organized one by agricultural economists, and it is they who have recognized the relevance of operations research to agriculture. On the other hand, agricultural scientists have seen the system largely as an ecosystem in which technology influences physical or biological efficiencies. Their contributions have largely been in developing new relevant technologies, which the farmer has the opportunity to adopt. Economics deals with the 'whole' system and its methods enable this. Much of the historical contribution of agricultural research in improving specific technologies has depended on the ability of scientists to diagnose and interpret needs and opportunities in farming systems. This has resulted from expert judgement and largely without the aid of scientific methods for dealing with the 'whole' system. This situation has changed gradually since the advent in the 1960s of the 'systems approach' in agricultural research aided by simulation modelling.

In certain respects, the systems approach in agriculture has been operations research by another name. The systems approach is a methodology for dealing objectively and, as often as practicable, scientifically, with the complexity of systems. A mathematical model of the system is constructed which mimics the behaviour of the real system, allowing study of the response of the system to various factors to be made using the model. Once an adequate model exists, simulation experiments can be conducted cheaply at any location for which there are data on weather and soil and in as many years for which there are weather records or

for as many years as are needed to make estimates of year-to-year variability. (Requirements for long sequences of historical weather data are declining with the development of weather simulators.) Two of the most important attributes of agroecosystems are the variability and unpredictability of yields among growing seasons, and sustainability of average yields in the long term. Models provide the most practical way for research to efficiently explore the range of possible management strategies for addressing this variability for different climates, soils, and farmer goals and socio-economic circumstances.

Models of relevant components of farming systems have existed for over fifteen years, so one might ask why there has not been more achievement from research involving models. In the first place, adequate simulation of yield response over a sufficiently wide range of conditions has been found to be more difficult to achieve than expected. Secondly, only a minority of modelling efforts had the improvement of a farming system as a concrete objective. The interests of research groups with the required skills for biological modelling have usually been in model building. Less often this interest extends to model evaluation, and even less frequently does it include model application. Furthermore, systems simulation does not share the explicit utilitarian orientation of FSR and Agroecosystems Analysis. Even when the expressed purpose is improvement of farming systems, there is no intrinsic means to consider important socio-economic factors.

In spite of past shortfalls and limitations, the potential of these methods in agricultural research remains high, and there is a resurgence of interest in them. One reason for this is that substantial progress has been made in modelling strategies and in the development of models of key land and crop processes; a large portion of the development costs of useful production system models has been paid. Another reason is that, as a result of the personal computing revolution, there is a new degree of receptivity to this computing-intensive methodology. Also, a new interest of agricultural research lies in helping producers to make better management decisions and recognition of the potential of simulation modelling to aid development of decision-support tools. Operations research has as its main aim and indicator of performance the improvement of management decision making. Agricultural systems modelling has always qualified as operations research but it has often failed to go beyond development of the model. With the new premium in agricultural research on utility in improving producer decision making, it seems appropriate for agricultural research to embrace and exploit the well-developed discipline of operations research for this important research activity.

Since they stem from common ancestors, the concepts in the *systems approach* and *operations research* have many similarities with FSR. However, at the applications level they are very different. Would their merger result in an aid of greater value to research planning in countries with developed agriculture and agricultural research institutions? The notion of integrating the science of using models of systems with FSR is not new, but there is little evidence of it happening.

The key interface of concepts and methods occurs in activity Figure 18.1, iii. This is the area where decisions are made; it is often referred to as the planning stage. These interrelated objectives can be distinguished. The first is identification of those possible changes to the system that might be relevant and technically and economically feasible. The second is the evaluation of selected potential innovations in terms of efficacy and feasibility prior to either testing them on farms or committing research resources to them. Effective operations research in this context requires (*a*) a perspective of agriculture systems that is in keeping with both FSR focus on relevance, together with a scientific systems approach to research and (*b*) a good understanding of the existing farming system or agroecosystem. The breadth of understanding needed may require a multi-disciplinary team, but it is essential that the team includes generalists — individuals with special talent in evaluating and integrating information and interpreting outcomes in terms of overall system performance. In most FSR to date, the quality of decisions in the planning stage has depended on the experience and good judgement of the scientists involved. While it is hard to imagine the value of these human ingredients declining, the planning process can be greatly aided by computer-based tools. The combination of an appropriate physical/biological simulation model of a relevant system or enterprise and a compatible method for economic analysis provides a powerful aid to evaluate system performance and the effects of possible changes or innovations in terms of both short- and long-term ecology and economics.

The operations research activity is not exclusive to the planning stage of FSR. Field research needed for developing and testing models and testing the operational feasibility of complex technical strategies overlaps with field research on components (Figure 18.1, vi). Adequate description of certain problems such as climatic risk (Figure 18.1, ii) requires the use of models. Development of decision-support systems for producers in variable climates (part of Figure 18.1, iv) requires use of models to generate relevant probability distributions of outcomes. On-farm testing can be conducted in only a few places and years, and generalization of results depend heavily on use of the simulation models driven by weather from other years and locations and specified for other soils.

Operations research approach — dryland cropping in Australian tropics

The research project described in Figure 18.3 is searching for the best methods for overcoming various problems of the farming system, but it is also concerned with the question 'will the best be good enough?' The important phenomena being researched are sensitive to the weather being too extreme. A method for assessing the impact of weather effects on technological alternatives that reflects the frequency of occurrence of outcomes is needed.

One of the main constraints to a potential grains industry is yield instability (Figure 18.3). Crop models provide the basic tool that is needed to integrate the effects of various weather factors on various processes affecting performance and to use the information that exists at a given location as historical daily weather records. Figure 18.4 shows maize and sorghum yields simulated for 100 years at Katherine, (NT) with sufficient nitrogen fertilizer to produce 85 per cent of maximum yield. The respective models have been tested and adapted to predict accurately for the climate and relevant cultivars. Clearly, high maize yield can be achieved with gratifying frequency, but the risk of low yield is also impressive. These contrasts are less marked with sorghum. Quantification of frequencies is simplified by plotting yields as cumulative distribution frequencies (CDFs) (Figure 18.5).

Variation in the performance of maize and sorghum is shown after several adjustments to the data to better approximate and compare on-farm performance. Yields have been discounted by 25 per cent to approximate the effects of miscellaneous yield losses not dealt with by the model, e.g. pests. Output has been adjusted for differences in the value of nitrogen used to produce 85 per cent of maximum for the two crops (150 kg/ha for maize; 100 kg for sorghum) and differences in current grain prices ($120 for maize; $100 for sorghum). In the best 40 per cent of years, net returns from maize exceed those from sorghum. In the remainder of years, returns are higher with

Figure 18.4 *Simulated maize and sorghum grain yields for Katherine, Northern Territory.*

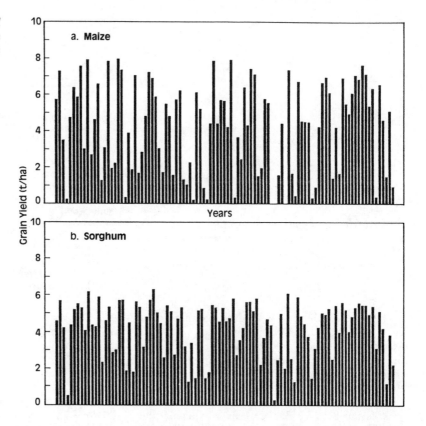

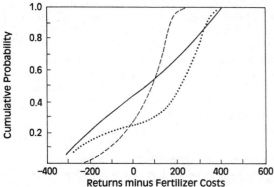

Figure 18.5 *Cumulative probabilities of cash returns net fertilizer costs for maize based on simulations in the period 1888–1987 (—), sorghum for the same period (– – – –), and maize 1978–87 at Katherine, Northern Territory (....).*

sorghum. Although median incomes are similar, variation in income is much smaller with sorghum.

When the research project began there was considerable local interest in maize. Simulated returns for the first ten years of the project, during which time data for the maize model were collected, are shown in Figure 18.5. Median returns for this period was >$200/ha while that for the full 100 year period was about $60/ha. This illustrates the value of an operations research approach in utilizing direct experimentation (in developing and testing models) to guard against drawing wrong conclusions from experimental results from a non-representative period or location.

Another operations research objective of the project is the evaluation of various cultivars or physiological attributes. Figure 18.6 shows the yield implication of having a maize cultivar with a longer vegetative period (twenty-three *vs* nineteen leaves) and a 5 per cent longer grain filling time, using DeKalb var. XL82 as a standard, at Katherine (900 mm mean annual rainfall) and Douglas Daly (1200 mm). In a comparable 21-year period at the wetter location, the later cultivar is about 1 t/ha better in 70 per cent of seasons (never worse), and clearly the preferred type. At the drier site, later maturity is appreciably better in 40 per cent of years, and somewhat worse in another 20 per cent. Superiority here is less clear and analysis for a larger sample of years is warranted.

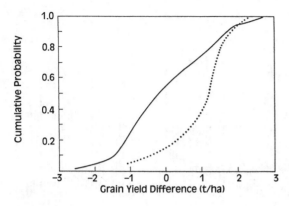

Figure 18.6 *Cumulative probability of differences in simulated maize grain yield between DeKalb var. XL82 and a hypothetical later-maturing cultivar at Katherine, Northern Territory (—) and Douglas Daly (....).*

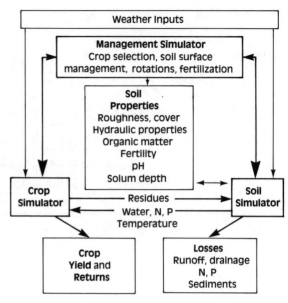

Figure 18.7 *The structure of a model for agricultural operations research.*

This approach can be used to examine strategies for timing the planting of crops and for assessing differences among soils due to variation in water storage characteristics. However, such models and their simulated outputs have limitations when grossly simplifying assumptions must be made. In Figures 18.4, 18.5 and 18.6 nitrogen was assumed to be 85 per cent adequate, other nutrients were assumed adequate, and soil-surface conditions were constant in time. Many of the important problems, risks, and opportunities in farming systems concern interactions between such major components and with weather, but happily, significant progress is being made in simulating more complex systems.

Figure 18.7 depicts the required elements of a model if it is to mimic the important phenomena of the more complex cropping system. It incorporates the weather-crop model, and it facilitates the flexible substitution of one crop for another or planned rotations in the interest of managing soil properties. The soil is simulated to provide dynamic supply of nitrogen, and in some cases, phosphorus, to crops and in turn is influenced both chemically and physically by crop residues and tillage. In addition to crop yield, soil loss is predicted, providing the means to analyse the conflicts between productivity and sustainability. In the tropical case study (Figure 18.3) such a model is beginning to provide the means to evaluate the efficacy of legume leys and grain legumes to reduce fertilizer costs, of pasture and crop residues to ameliorate soil surface problems, and eventually, of no-till to control soil erosion. Final evaluation of alternatives needs to include both costs

and benefits within the context of a farm economy illustrated in Figure 18.2. This will require economic analysis of the range of alternatives that are simulated. In addition to 'which is best?', the question, 'is the best good enough to be the basis of an industry?' needs to be addressed. To be useful in the future, the final product should provide a convenient means of repeating these questions for changed economic conditions, when new technologies become available, for a probable climate change scenario, etc.

Appraisal of FSR/operations research as a methodology

As an example of FSR, the tropical dryland farming example is inadequate and abnormal in several ways. In this case, farmers have played only a minor role. The main aim was not so much to help the few farmers that were there, but more to provide a basis for judging whether they should be there. The basis of such a judgement must be largely economic, yet there was no early exploratory economic analysis. It is possible that prior consideration of such things as infrastructure needs and most probable market and price scenarios might have suggested that a search for improved technology was unwarranted.

However, there is no doubt that the FSR framework, relaxed as it is in Figure 18.1 to deal with agricultural systems at a range of scales, is appropriate to decision making in industrially developed countries in the new era of user pays/demand-driven research. Identification of needs involves producers/industry spokesmen, agricultural economists, and agricultural scientists. The potential value of a more rigorous use of the framework would seem to depend largely on how effective the operations research component can be (Figure 18.1, iii). In contrast to many developing country situations, resources are generally available for the development of simulation models that provide surrogate system performance information for economic analysis of a wide range of possible alternatives.

Operations research provides several benefits. Even without models and computers, analysis of problems in a systems framework reveals a hierarchy of dependencies that aid research planning (Figure 18.3). The construction or calibration of a model makes efficient use of quite disparate results of past research and experience (Figure 18.1, iv). The comparison of alternative actual or hypothetical alternative practices, materials, and climatic or economic circumstances provides guidance to where the biggest payoff for further component research might be (Figure 18.1, ii, v, vi). It makes efficient use of whatever climate and soil resource information is available in extrapolating existing location-specific knowledge and reduces the amount of expensive field testing (Figure 18.1, iv). Finally, system performance in response to alternative management decision rules can be compared and optimum strategies identified; these form the basis for decision-support aids for producers (Figure 18.1, viii).

The potential value of simulation models to provide surrogate performance data for agricultural systems was enthusiastically recognized by agricultural economists when such models were in their infancy twenty-five years ago, but the slow pace and high cost of development of useful models resulted in widespread disillusionment of these potential clients. It is now clear that the investment in time and resources is beginning to be recouped. Much has been learned about how to simulate crop yield effectively and affordably. After a period of wide divergence of approach, there is evidence of significant convergence on those strategies with clear advantage. A renewed interest by agricultural economists is evident, and

fruitful collaboration in operations research on agricultural systems is beginning.

An FSR/agricultural operations research (AOR) methodology clearly has great merit in a user-pays, demand-driven research environment, but there is an important danger that is easily overlooked. In Figure 18.1 pools of information and materials (vii) and expertise (vi) are presumed to exist. The speed and efficiency with which a given problem is solved or opportunity exploited depends very heavily on the wealth in one or both of these reserves. The maize and sorghum models in the case study were operational in two years only because of the availability of (a) a maize model (which had taken many person-years to develop) that was approximately suitable and (b) appropriate locally collected data (the product of several person-years) to test and calibrate it. The danger in this presentation of an AOR approach is that it treats the cost of the scientific resource on which it is based as external to the main R & D effort. Neither the pool of new applicable knowledge nor the expertise that can create it will automatically remain replenished to serve adequately the problem-solving activities. There is a continuing need for disciplinary research that, while less obviously relevant to performance of agricultural systems, ultimately is the key to both improved technology and improved operations-research tools and methods. The proportion of the research budget directed to this research, who pays for it, and the degree to which applications affect its priorities will remain topics for vigorous debate.

Further reading

Dent, J. B. and Blackie, M. J. (1979). *Systems Simulation in Agriculture*, Applied Science Publishers, London.

Muchow, R. C. (ed.) (1985). *Agro-research for the Semi-arid Tropics: North-west Australia.* University of Queensland Press, St Lucia. Chapters by Dillon and Virmani and McCown et al.

Remenyi, J. V. (ed.) (1985). Agricultural systems research for developing countries: proceedings of an international workshop held at Hawkesbury College, Richmond, NSW, Australia, 12–15 May 1985. ACIAR Proceedings No. 11, 189 pps. Section titled 'Approaches to farming systems research' and especially chapters by Conway; Anderson et al.; and Norman and Collinson.

Robertshaw, J. E., Mecca, S. J. and Rerick, M. N. (1978). *Problem Solving: a Systems Approach*, Petrocelli Books, New York.

CHAPTER 19

Comparisons of Agriculture in Countries with Mediterranean-type Climates

K. G. Boyce, P. G. Tow and A. Koocheki

SYNOPSIS This chapter examines the Mediterranean climatic zones of the world, describes the farming systems currently used, and considers the potential for successfully introducing a ley farming system. Attempts at introducing ley farming into countries of the West Asia–North Africa (WANA) region are discussed.

MAJOR CONCEPTS

1 The Mediterranean-type climate is primarily characterized by a concentration of rainfall in the cooler half of the year and drought in the warmer half. This is associated with particular crop and livestock species.

2 Farming systems of the Mediterranean basin have evolved over perhaps 10 millenia or more. Their primary characteristic is the major importance of wheat in higher rainfall areas and barley in lower rainfall areas. Grain legumes, fruits and vegetables also have their place.

3 Livestock is very important in the Mediterranean basin but is generally not integrated with crops.

4 Over the past three centuries, crop, livestock and weed species used in the Mediterranean basin have been transferred to other countries with a Mediterranean-type climate, thus having a major influence on their agriculture.

5 The composition of farming systems varies markedly among countries having a Mediterranean-type environment. In general, increasing industrialization, urbanization and per capita income lead to more specialization and intensification of agriculture.

6 The ley farming system of integrated livestock and cropping practised in southern Australia is a unique sub-set of the dryland agricultural sys-

tems used in countries with Mediterranean-type environments.

7 The adoption of the ley farming system in cereal areas of North Africa and West Asia is technically feasible provided legume varieties, associated rhizobia and management procedures are selected for adaptation to local conditions.

8 Major barriers to the adoption of ley farming in these regions are the lack of experience in managing annual pastures for regeneration within a system of integrated crop and livestock production, and lack of appropriate government incentives for farmers to change from existing systems.

I t is estimated that the world's arable crop land is about 1400 million hectares. Of this area, about 90 per cent receives its moisture supply as rain and/or snow. However, not all this rain-fed farming area is considered to be the dryland farming area. Dryland agriculture is practised in regions where the moisture supply is the limiting factor. Often these regions are referred to as 'semi–arid' as they occur between the 'humid' zone where lack of moisture is not a constraint and the 'arid' zone where agriculture is only possible with irrigation and pastoralism is the normal system.

The semi-arid zone can be subdivided into a number of sub-zones on the basis of climate: primarily the tropical semi-arid areas close to the equator, and the steppe regions of mid-latitudes. Within the steppe regions are the zones of Mediterranean-type climate characterized by a winter-dominant rainfall pattern, totalling in area less than 150 million hectares. There are a number of regions which share the Mediterranean semi-arid climate type. They are the Mediterranean basin, southern Australia, the south-west coast of the USA, central Chile and the Cape region of South Africa. Potentially these areas should be able to support the ley farming system as practised in southern Australia.

Characteristics and distribution of Mediterranean-type climates

The Mediterranean-type climate is primarily characterized by the concentration of rainfall in the winter half-year: November to April in the northern hemisphere, and May to October in the southern, with drought in summer. In California and Chile, in particular, winter rainfall may constitute 80 to 90 per cent of the annual precipitation but this is less common in the Mediterranean basin itself. A figure of 65 per cent of the year's precipitation in winter has been used as the boundary of the climatic type.

Annual precipitation values suitable for agriculture are generally considered to be from 250 to 500 mm. Production decreases with decreasing rainfall. Most areas show a wide fluctuation in total rainfall from year to year, in distribution over the season, in the time of onset of the 'opening rains', the duration of the rainy season and the intensity of rainfall during rainy periods. Table 19.1 illustrates rainfall variability in Jordan with a typical Mediterranean-type climate. The warm season is characterized by low precipitation leading to drought conditions where plant growth cannot be sustained without irrigation. The period of drought is the most important ecological factor.

Monthly rainfall data for a range of Mediterranean-type countries are shown in Table 19.2. Temperatures in Mediterranean climates are cool and mild in winter and hot and sunny in summer. Frosts are rare and although they may occur in almost all areas are normally not severe. One system of climatic classification sets two boundaries for temperature. Mediterranean climates must have an average temperature below 15°C in at least one month; and in relation to frost severity the number of hours per year at which the temperature at weatherscreen height falls below 0°C should not exceed 3 per cent of the total.

Table 19.1 *Growing season rainfall over eight seasons in dryland agricultural areas of Jordan.*

	Ramtha	Rabba	Madaba
1980–1	—	349	349
1981–2	184	276	332
1982–3	231	494	505
1983–4	193	209	231
1984–5	243	344	363
1985–6	152	187	250
1986–7	290	326	359
1987–8	295	434	441
Mean	227	327	354
Mean % Deviation from Mean	19	24	18

NOTE: Rainfall variability was high at all sites, although the severity of drier years and the likelihood of crop failure increased with decreasing mean rainfall.

Table 19.2 *Examples of monthly rainfall data in countries with Mediterranean-type climates.*

Place	Country	JAN	FEB	MAR	APR	MAY	JUN	JUL	AUG	SEP	OCT	NOV	DEC	Annual Total
Marrakesh	Morocco	28	29	32	31	17	7	2	3	10	21	28	33	241
Oran	Algeria	70	54	35	33	19	7	1	3	16	43	46	67	394
Sfax	Tunisia	18	18	25	21	12	5	1	5	26	38	26	15	210
Tripoli	Libya	62	38	19	14	3	1	1	1	10	32	41	65	287
Aleppo	Syria	68	54	42	36	19	3	1	1	1	19	25	72	341
Amman*	Jordan	64	67	37	15	4	–	–	–	–	5	30	48	272
Mosul	Iraq	67	63	69	51	25	1	tr	tr	tr	10	36	65	388
Shiraz	Iran	144	50	4	26	1	0	0	0	0	1	46	24	296
San Diego	California	51	55	40	20	4	1	tr	2	4	12	23	52	264

Monthly rainfall (mm) — *Northern Hemisphere* — Annual Total

Place	Country	JUL	AUG	SEP	OCT	NOV	DEC	JAN	FEB	MAR	APR	MAY	JUN	Annual Total
Merredin	Western Australia	55	43	23	19	15	12	8	13	17	22	43	50	320
Kadina	South Australia	49	46	37	33	22	18	15	20	20	36	49	51	396
Valparaiso	Chile	72	50	19	4	0	4	2	0	3	18	82	96	349

Southern Hemisphere

* Marqa airport.

Except for Australia, where the topography is moderately flat, countries with Mediterranean-type climates are characterized by rugged mountain ranges. Elevation, distance from the sea and aspect considerably modify the climate over sometimes quite short horizontal distances. Increasing elevation correlates positively and strongly with increasing precipitation. Areas exposed to the open sea have cool and nearly frost-free winters while interior regions exhibit much greater seasonal temperature fluctuation.

Sub-Mediterranean climates with low autumn and spring rainfall patterns occur east and to the interior of Mediterranean climate zones but west of summer rainfall regions. They experience colder winters and greater winter precipitation. Summer temperatures tend to be very high with high surface wind velocities. These continental-type climatic patterns are characteristic of the high plateau of Algeria, East Morocco, mid-latitude dry climates of the Near East including Iraq, Syria, most of Asiatic Turkey, the Iranian plateau and Afghanistan and the north-west of the USA.

As a result of the climatic variability and generally low rainfall, dryland agricultural areas of the Mediterranean climate zones are essentially zones of high risk and uncertainty for farming. Agricultural production is characterized by delayed (or missed) sowing time due to lateness of the autumn/winter rainfall, periods of drought during the growing season, and the unexpected incidence of climatic events such as hail or hot drying winds.

The worldwide distribution of Mediterranean climates is between latitudes 32° and 40° north and south of the equator on the west coasts of the continents. On the north side of the Mediterranean Sea they extend into higher latitudes and in Western Australia into lower latitudes. Toward the equator the climate becomes increasingly arid and toward the poles rainfall increases and is less concentrated in the winter period (Figure 19.1).

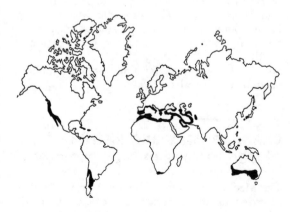

Figure 19.1 *The world distribution of countries with Mediterranean-type climate.*

Farming systems in Mediterranean-climate zones

Farming systems of the Mediterranean basin region have evolved over perhaps 10 or more millenia during the development of agriculture from the previous hunter-gatherer system of food supply. Two systems of food production developed, one based substantially on the production of grain crops and the other on the domestication of livestock. Integration of the two systems has not generally developed in many areas, except for specific purposes. Agriculture in the Mediterranean climates of the New World has developed somewhat differently to that of the Old World although the two major components are present.

Until two or three centuries ago the other four areas of Mediterranean climate were largely uninhabited. They were then colonized by Europeans, primarily from Spain, Holland and the British Isles, who brought with them plants and livestock from the homeland. A wide range of plant species of particular importance to agriculture in the Mediterranean basin was also introduced. These included cereals and tree crops such as olive, grapevines and temperate fruits and vegetables. Horses, sheep and cattle were introduced, but the goat has not become a significant livestock species in the New World despite its importance in the Mediterranean basin. Generally the endemic flora of the New World was not well adapted to the grazing habits or pressure of the introduced livestock. With time many endemic species have been replaced in the agricultural systems of these countries by the relatively more ecologically aggressive weed species of the Mediterranean basin. These species have been introduced as impurities in agricultural seed or in fodder used in transportation of livestock to the New World.

These non-crop species adapted rapidly to their new environments often radically altering the farming ecosystem during European colonization. Subsequently many of these species have become important in the current agricultural production system. This adaptive process is well illustrated by the naturalization of the annual weedy grasses of the species *Hordeum*, *Lolium* and *Bromus* and the annual forage legumes *Medicago* and *Trifolium* in Australia. In general, the same group of non-crop invaders colonized each of the New World Mediterranean-climate countries but with varying degrees of agricultural importance.

Agriculture in the Mediterranean basin

The primary feature of agricultural land use in the countries in the Mediterranean basin (Figure 19.2) is the major importance of cereal production, particularly of wheat and barley. Overall, about half of the arable land is under cereals, about 40 per cent under fallow and the remainder devoted to fruits, vegetables and other crops, particularly the pulses.

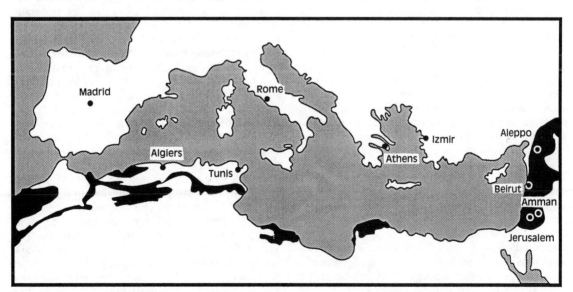

Figure 19.2 *The Mediterranean basin, including countries of the WANA region.*

There are two major types of dryland farming systems in the West Asia–North Africa (WANA) countries: the 'wheat-based' systems in wetter areas, and the 'barley-based' systems in drier areas. Although there are some variations in this broad classification due to elevation, soil depth and soil type the transition between them is considered to be the 300 mm rainfall isohyet. The generally low yields of cereals are related to low soil fertility, much of it due to soil erosion.

In the wheat-based system rainfall is generally sufficient to allow farmers to concentrate agricultural production on cropping. There is usually considerable flexibility in the types of crops grown and in the rotations used. In the wetter end of the system (>350 mm rainfall) three-course rotations of wheat, grain legume (particularly lentil and chickpea) and summer crop (particularly melons) predominate. Summer crops rely on moisture accumulated in a winter fallow. The deliberate sowing of forage legumes, particularly lathyrus and vetch, is practised in these systems in some countries and may make up 5–10 per cent of the cropping area. At the drier fringe of the 'wheat-based' area (275–350 mm) the rotation is reduced to two courses, wheat and fallow, although small areas of pulses and summer-growing melons may be sown when winter rainfall is earlier and higher than normal. Animal production in these cropping areas is relatively unimportant though some animals (mainly sheep and goats) are kept for local use (milk and meat) and for sale.

In contrast, the main goal of the barley-based farming system which predominates in the 200–300 mm rainfall areas is animal production. Reliance on animals increases with decreasing rainfall total and reliability. Animal production is an important survival mechanism in a subsistence-type agriculture where the risk of failure of crops from inadequate rainfall is high. Unlike crops, animals can be moved to sources of feed; if necessary, they can be sold to supply finance for the purchase of food for humans or feed for the most valuable animals of the flock.

The most usual cropping rotation of the barley-based system is of two courses: barley and fallow. Some summer cropping of melons is practised when soil moisture is adequate. The fallow is often not a cultivated fallow (as practised in many other dryland cereal cropping areas) but an uncultivated weedy fallow.

Animals in the barley-based zone generally graze native unimproved vegetation on non-arable land which is often communally owned, on the weedy fallows of the cropping land and the residue of crops after harvest. Supplementary feeding with grain (barley) and cut forage from irrigated areas may also be practised.

At 200–250 mm mean annual rainfall cropping ceases to be possible and agriculture is restricted to more or less nomadic pastoralism. The traditional transhumance (seasonal transfer) of grazing sheep and goats on the plains in winter and spring followed by cereal stubble and upland pastures in summer is still an important agricultural system. Transhumance, particularly long-distance transhumance, is losing its importance as social and political pressures force changes to this way of life.

The farming systems are not static but are evolving in different ways according to a variety of pressures, particularly related to human population increase, land shortage, increased mechanization, changing market forces and a variety of social and political factors. In the higher rainfall wheat-based agricultural system crop production will predominate in the future. However, there will be increasing diversification of cropping systems and rotations as technical innovations and differing economic conditions occur. In the lower rainfall wheat-based systems and the barley-based systems animal production is assuming greater importance. Barley is replacing both wheat and fallow to satisfy the increased need for animal feed production. Specifically sown fodder and forage legumes are also now under increasing use by farmers and research by scientists in an endeavour to fill the animal feed gap.

An important feature of all of the farming systems, whether wheat- or barley-based, is that cropping and livestock production are not generally integrated. Cropping land is predominantly owned and/or worked by farmers who own very few livestock and livestock owners do not usually crop the land. This has been the general pattern for centuries, if not millenia, in most of the cropping areas of the Mediterranean basin. This fact presents both a physical and social barrier to the use of forms of agriculture such as the Australian ley farming system which efficiently integrate cropping and livestock production under one management operative.

Broad area cropping and livestock production are both low productivity agricultural pursuits compared to tree crops and horticulture which are generally very labour intensive but often quite lucrative. While only consuming a small percentage of the available arable land, tree crops and vegetables contribute significantly to rural income in countries of the Mediterranean basin. The olive and grapevine are the most important of the tree crops, the fig being of little significance and the date palm confined to the arid areas of North

Africa and some West Asian countries. Olive groves are found throughout the Mediterranean basin and it is generally considered that the limits of the olive are a good indication of the range of the Mediterranean climates. The tree will grow on steep slopes and is remarkably summer-drought resistant. Grapevines show similar characteristics and are widely cultivated, particularly in the western end of the Mediterranean basin. Temperate fruits such as apples and pears are also grown as well as the subtropical species such as peach and citrus (particularly orange). A wide range of vegetables are produced, and have been an important part of Mediterranean farming since at least Roman times. Much of the area under vegetables and fruits is irrigated, in contrast to the dry-farming of cereals, vines and olives.

Agriculture in the Mediterranean basin is generally characterized by agricultural population densities significantly higher than those of countries of similar climate in other parts of the world. Farm size is small, the majority less than 5 hectares with most farmers being occupier-owner. Also, there is a tendency for farm size to decrease with time as land is traditionally divided between family members when passed from generation to generation.

Relationships between the semi-arid (agricultural) and arid (pastoral) zones in the WANA region

There has always been conflict between farmers and herders because both wish to use land favoured by reasonable rainfall. Even when separate zones for cropping and pastoralism were clearly defined on the basis of average annual rainfall, conflicts remained. One reason is that zone boundaries change annually due to rainfall variability. In some countries where marginal lands are common property, anyone can come in and grow crops. In the process, good grazing resources are destroyed and the remaining ones may be overgrazed. This problem has increased in recent times because of growing populations and an increased demand for food. The higher price of agricultural products encourages farmers to expand their cropping activities into lower rainfall areas and for herders to overgraze arid rangelands. This results in rangeland degradation and greater competition between the two groups. In many cases, farmers and nomads have had to abandon their lands due to their degradation, and move to the cities and towns.

Migratory flocks also 'invade' farm lands at certain times of the year creating some conflicts between their interests and those of the farmer. Traditional pastoral nomadism is an ecological response to variability in rainfall, temperature and forage availability. Properly practised, nomadism is in harmony with plant development and the resource is not overgrazed and destroyed. The situation is different in the case of farm lands. For instance, the grazing of crops stubble by nomadic herds often leaves the ground completely bare and prone to erosion. Fallow areas are also regarded as common ground and are heavily grazed. This applies even if the land is sown to pasture plants such as medics, which are looked on as weedy species. This is a potential cause of conflict with farmers who wish to replace fallow with pasture.

There should be co-operation between the farming and pastoral zones. This occurs to some extent in the Mediterranean-climate regions of Australia by the transfer of animals and conserved fodder between the two zones. There must also be a balance between total fodder and total livestock in the two zones, on average, if not on an annual basis.

In the developing countries discussed in this chapter there is currently an imbalance between fodder supply and livestock numbers because of the degradation of forage resources. There is also an overall deficiency in livestock products for the requirements of the individual countries. Both problems need to be overcome by increasing home-grown fodder supplies. Fortunately in WANA, there are ample naturally occurring forage species, although often in poor supply. The required increase in forage can be achieved by a combination of methods used simultaneously:

- Farmers can be encouraged to grow forage, e.g. in place of fallow, and rent it to the herders if they do not wish to use it themselves. This is already done in parts of West Asia where farmers charge for the use of cereal residues or light grazing of wheat to promote tillering and prevent lodging.
- Semi-permanent pastures should be established in marginal rainfall areas using drought tolerant native legumes and grasses (e.g. sainfoin, burnet, ryegrass).
- Rangelands should be renovated by introducing drought-tolerant shrubs such as *Atriplex*, *Kochia* and *Salsola* and by restricting grazing to allow time for regeneration of existing species.
- Some areas of irrigated pasture can be grown for special purposes, e.g. for fat lamb or milk production, and to take grazing pressure off rangelands at critical times.

Agriculture in other regions with a Mediterranean climate

There are marked variations in farming patterns in the Mediterranean basin. This is particularly true between countries of the north-west and those of the east and south. These differences may be partly ascribed to differences in climate (hotter and drier to the south and east) and to the long influence of Islamic occupation. Probably more important in the current situation is the degree of economic development in the north-west where industrialization, urbanization and incomes have progressed more, leading to more specialization and intensification in agriculture. This trend is also evident between countries of WANA where industrial development has been differentially affected by oil and mineral discoveries and the occurrence or otherwise of between-nation conflicts.

It is this degree of industrialization that accounts for many of the differences between the farming systems of the Mediterranean-climate type countries in other parts of the world although some differences in climate, geographical position and history of settlement are also evident. Thus, Chile still has around 15 per cent of its population employed in agricultural production and a low national per capita income, whilst the South African Cape region and southern Australia have lower proportions engaged in agriculture (4–5 per cent) and a high per capita income. California is one of the richest areas of the world with a very small agricultural work force (about 1 per cent). Broadly speaking, the higher the national income per capita the more intensive and specialized the farming patterns; thus the regions with Mediterranean-type climate may be ranked in an ascending scale from North Africa to West Asia, Chile, north-west Mediterranean basin, southern Australia, south-west Cape and California.

In general, cereal production is important in the four regions and all four Mediterranean-type regions outside the Mediterranean basin grow generally the same range of crops as the Mediterranean basin. However, the mix of crop and farm animals is somewhat different. In Chile, cereals occupy three-quarters of the cropping area with wheat being four-fifths of the cereal area. Some fallowing is practised and yields are generally low and variable as fertilizer input is low and little machinery is used even on larger holdings. In southern Australia and the south-west Cape, wheat and barley together occupy about one-third of the cropping area. Although yields are low and variable primarily due to rainfall deficiency, farm areas are large, highly mechanized and use little farm labour. Cereal production in California, predominantly barley-grown dryland now occupies only about one-quarter of the arable land. Cultivation is highly mechanized with minimal labour input.

Of the other forms of agriculture used in the Mediterranean basin, tree crops, grapevines and vegetables, there are significant differences in their cultivation in the New World. For instance, grapevine production for wine and dried fruit are often highly localized and specialized. Production of fruits, nuts and vegetables is considerably more important and more mechanized in California than in the other countries.

Livestock production is generally as important to the New World Mediterranean-climate countries as in the Mediterranean basin. However, the role of domestic livestock in the development of agricultural systems in the New World and their current status is different. Sheep production was the major agricultural enterprise in Australia from European settlement until the 1860s when cereal farming became increasingly possible with the passage of land tenure legislation in NSW and Victoria. Initially livestock production (principally with sheep) was completely separated from cropping and dominated the agricultural economy.

During the latter part of the 19th century pastoralism moved further out into the extensive areas of drier land in the interior of Australia as cereal production gained a foothold on the more closely settled, higher rainfall areas. Sheep remained as a relatively minor component of the cereal-based farming systems. By the 1930s fertility of cereal-growing land had declined due to soil exploitation and erosion, and annual legumes such as subterranean clover and medics had been introduced into the cereal production areas. Increasingly, the two separate agricultural systems of cropping and pastoralism became integrated and interdependent through the pasture legume. This system has been called the southern Australian ley farming system (Chapters 1, 6).

In southern Africa the history of development was different to southern Australia and has led to a different agricultural system outcome. Sheep production until 1799 was confined to a semi-nomadic pastoralism of long-legged, fat tailed animals without wool, utilized primarily for mutton for the passing sea trade and lard for tallow. Merino sheep were introduced by the British for wool production and, with the passage of land tenure and other legislation, this led to the development of large-scale livestock ranching in the Karoo region. There was no conflict with cereal production for agricultural land; hence a mixed farming system has not developed to the extent that it has in Australia.

In Chile, the tradition of southern European pastoralism has most closely been preserved. Animals, both sheep and cattle are generally raised on larger properties on the lowlands but are moved to highland pastures in the Andes and coastal ranges during summer. The majority of farmland is held in these large estates devoted to animal production. Generally no cereal cropping is undertaken but lucerne is raised for animal feed.

Animal production in California is in two forms: open range extensive pastoralism of sheep and cattle in the mountains east and west of the central valley, and intensive cattle and dairy production on irrigated fodder (principally lucerne) in the valley floor. There is no general integration of livestock and cereal production.

Transfer of ley farming technology

Potential value of ley farming in the WANA region

The potential usefulness of the southern Australian ley farming system of agriculture to farming in other regions of similar climate around the world was not fully recognized until the late 1960s. During that decade the 'green revolution' in agricultural technology transfer to developing countries had gathered momentum. The primary thrust of this programme came from the plant improvement projects for wheat and maize at CIMMYT (Centro Internacional de Mejoramiento de Maiz y Trigo) in Mexico and for rice at IRRI (International Rice Research Institute) in the Philippines. As wheat is the most important crop in countries of the Near East and North Africa much of CIMMYT's effort became centred in these countries which, coincidentally, had a Mediterranean-type climate similar to southern Australia.

It was soon recognized by CIMMYT researchers, particularly in North Africa, that the potential gains in productivity inherent in the germplasm base of the new wheats for the Mexico breeding programme may not be fully transferred, as wheat soils in the region were more or less badly degraded in structure and nutrient fertility. The status of the soils was not unlike that of Australian wheat-growing soils in the 1930s after decades of wheat-fallow rotations prior to the introduction of forage legumes to the farming system.

It was also recognized that animal production in the countries of North Africa and the Near East was as important for food production as wheat. Animal numbers were increasing rapidly with consequent risk of severe degradation of natural forage supplies and grazing land.

By the end of the decade of the 1960s it was realized by the CIMMYT organization that the general farming productivity in the countries of North Africa and the Near East might be significantly improved by the incorporation of forage legumes into the fallow phase of the cereal rotation, as had been the Australian experience. It was subsequently estimated that climatically and edaphically, some 3 million hectares of fallow land would be suited for growing annual medic in the Maghreb (Tunisia, Algeria, Morocco) and Mashreq (Jordan, Iraq, Syria) countries. Large areas in Iran could also be sown to medics. As the ecological conditions of southern Australia are similar to those of the Mediterranean basin it appeared more appropriate to use the ley farming system as a development model than other systems of farming developed in Europe and North America. It was realized that transferring the technology of the Australian ley farming system of integrated wheat and livestock production to countries of the Mediterranean basin would not be easy, since livestock production was usually separated from cropping.

Projects and demonstrations

Attempts to transfer the Australian ley farming system to the Mediterranean basin essentially began in 1971. The initial thrust came from CIMMYT, through South Australian forage scientists working with the wheat development programme in Tunisia and Algeria. A number of other, bilateral projects then followed in other WANA countries. These were: Libya (South Australia), Jordan (South Australia), Iraq (Western Australia, South Australia) and Morocco (France/Germany). FAO (the Food and Agriculture Organization of the United Nations) has also been involved in forage development projects in these countries. Since 1980 ICARDA (the International Centre for Agricultural Research in Dry Areas, based in Syria) has been active in forage research and promotion, including the use of annual forage legumes in cereal-pasture rotations.

All the CIMMYT and bilateral projects had the following features in common, based on the CIMMYT pattern and Australian experience:

- Demonstrations of the ley farming system and its management on large-scale farms set up for the purpose and/or on numbers of private farms, and using Australian cultivation and sowing machinery.
- Evaluation of Australian and local cultivars of annual legumes (predominantly medics) and associated *Rhizobium* strains.
- Research to define local limiting factors (e.g. plant nutrient deficiencies) and to adapt the introduced technology to local conditions of climate, soil and farming practices.
- Training of local staff, through project programmes and scholarships, for research and extension in the new technology.
- Economic evaluation of results.

From a technical point of view all projects were successful. They showed the feasibility of adopting a medic-based ley farming system for integrated crop and livestock production where climate and soils suited annual medics. There were several overall similarities among the projects in the results. These included:

- Medic cultivars selected in Australia were partly successful in each country. However, in each case, some locally selected strains were found to give better production. In higher altitude areas, cold tolerance was an important attribute.
- Responses were obtained to inoculation of medics with commercial strains of *Rhizobium* from Australia but better results were often obtained using locally selected rhizobia.
- Replacing fallow with a forage phase initially reduced wheat yields because of lower moisture availability to wheat, but yields eventually increased due to extra nitrogen supplied by the legume.
- The feed supplied by the forage was highly profitable in terms of livestock products.
- The use of Australian machinery for cereal and pasture establishment was successfully demonstrated.

Each project also gave particular results which added to the lessons learnt from this period of aid projects. For instance:

- The Libyan and Jordanian projects helped to persuade farmers to adopt shallow tillage methods which favour the retention of medic seed near the soil surface.
- In the Iraq-Western Australian Project it was found that the Australian agricultural machinery and bulk grain handling equipment used there could be immediately adopted into existing farming systems.
- The Iraq-South Australian Project demonstrated

superior yields from some Australian wheat varieties.
- In the Jordanian project a stocking rate trial conducted for three years compared the two species which had shown yield advantages from the beginning of work in Jordan: snail medic and common vetch. Unexpectedly, vetch gave much higher liveweight gains in Awassi lambs than snail medic. This showed the importance of assessing the value of forage species under grazing before their recommendation to farmers.

The experiences of these projects has shown that technology transfer is a step-by-step process involving adaptation and adjustment to the environment of the host country. Success in some aspect is usually accompanied by the appearance of a new limiting factor. Some of these limitations could be foreseen but others could not, or turned out to be more intractable than predicted. Thus some types of problems are overcome much more quickly than others and progress is very uneven. Along the way, new insights enable alternatives to locally unadapted technology to be devised.

The work of ICARDA

Since 1980 ICARDA has played a steadily increasing part in the research and promotion of pasture legumes for the WANA region. Originally, most of ICARDA's work was done in Syria on research centres and farms. Since 1985 the Forage, Pasture and Livestock Programme of ICARDA has strongly promoted the medic-based ley farming system. ICARDA has research bases in various parts of the region and collaborates with universities and Ministry of Agriculture research personnel. It is therefore in a position to have a very important role in solving problems associated with adapting modern technology to the needs of the region. It also collects, stores and distributes local genotypes of crop and pasture species and *Rhizobium* strains.

Because of the size of the task of improving farming over such a wide area, Aid Programmes sponsored by various developed countries will still be needed to combine with the efforts of ICARDA. Australia is the only country with first-hand experience of dryland farming systems suited to the climate and soils of the WANA region. It continues to have a role in helping to develop appropriate farming systems, although the allocation of Australian Government funds to this region remains small.

Adoption of the ley farming system

The process of technology transfer and farming system development do not end until there is adoption by farmers. Several attempts have been made by governments in the WANA region to encourage farmers to adopt the ley farming system. As early as 1975–6, Tunisia had sown 3000 ha of medic, Algeria 8000 ha and Libya 47 000 ha (where it was planned to settle 2000 farmers). In Morocco some 46 000 ha of medics were sown in the 1985–9 period under 'Operation Ley Farming', to encourage farmers to replace grazed fallow with annual legumes. Subsidized seed was provided and considerable effort spent to assist extension staff and farmers by training, field days and demonstrations.

The success rate of adoption of new farming technology is frequently low in the early years of introduction. In the case of the Moroccan 'Operation Ley Farming' a survey showed that, while initial establishment of medics was satisfactory, subsequent regeneration was often poor, perhaps due to excessive stocking rates during the first seed production period. Correct control of stocking pressures during seed production is a vital aspect of ley management.

The survey also provided some guidelines on the circumstances which may be conducive to the adoption of the ley farming system. In higher rainfall areas there is relatively little scope to introduce ley farming because of the small proportion of fallow land, competition from fodder crops and dominance of cattle in animal enterprises. There seems more scope for ley farming in the drier parts of the rain-fed areas where the predominant enterprises are cereals and sheep, and fallowing is common. Farmers most interested in medic pastures are those with larger farms (here greater than 20 ha).

The most important finding was the deficiency in farmers' understanding of the system. They considered medic pastures as a grazed fodder crop rather than an integral part of a farming system. They tried to get maximum yield by continuous heavy grazing and did not care about persistence.

Another difficulty about managing medic pastures in the WANA region is that shepherds in charge of migratory flocks regard pastures as natural grazing to which they have the right of access. Agreements would be needed between farmers and migratory shepherds to control grazing of ley pastures and to compensate farmers for any use that is made of them by outsiders.

Thus the adoption of the ley farming system is dependent not only on the availability of suitable cultivars and production technology but also on such factors as farm size, prior experience with sheep, degree of competition from alternative enterprises, adequate management training and harmony with other regional systems.

The essential goals of the ley farming system in relation to the WANA region are the replacement of fallow by a leguminous forage and the integration of crop and livestock production. For achieving this, two principles may be of use:

1 *A new system is sometimes best introduced in a step-by-step process, starting with a component that is easy to adopt and will show immediate benefits.*

A possible example of this is derived from the finding that the local Jordanian variety of common vetch has a very high feeding value. Because of this and its long history of use as a crop, it may be more successful than medic for introducing an integrated crop-livestock system to Jordanian farming. Lacking a high level of hard seededness, common vetch will not regenerate satisfactorily after a cereal crop. The less sophisticated vetch-cereal system would, however, have the advantage of simplicity of management. The traditional use of vetch as a crop may also mean that it is more likely than medic to be left ungrazed by migratory shepherds.

2 *Existing social, economic and technological structures should be used, perhaps with some modification, to introduce new technology and management.*

In one FAO project, simple labour-intensive methods have been devised for farmers to collect and thresh medic pods and thus produce their own seed. The project has also produced instructional packages, using slide photos from the region and text in Arabic, for use by extension officers. One approach to the sowing of medic seed on small areas has been to hand broadcast and have sheep trample it into the soil. The problem of lack of machinery by small farmers is also being solved by farmer co-operatives and private contractors. For this to be successful great attention has to be paid to timeliness and quality of operations. Incentives and education are required to make contract services adequate for the tasks of farming system improvement.

Another aid to introducing the ley farming system could be to use the type of village organization found in some parts of northern Syria. These have agreements whereby big blocks of crop-bearing land alternate with similar areas of fallow. Fallow could be converted to pasture and the pasture managed on a whole village basis. Fencing of larger areas is much more economical, per hectare, than fencing smaller areas. The use of a shepherd also becomes more economical as the size of their flock increases. In village enterprises, advantage can be taken of existing practices of assembling small family flocks into larger ones under the care of reliable shepherds for daily grazing. These shepherds should be trained in the objectives and requirements of ley pasture management. Because local pasture legumes have long been regarded by farmers as weeds, their image must be upgraded to the status of a valuable crop which has particular management requirements.

The wider socio-economic and political system

Those involved in attempts to improve farming systems invariably become aware of the constraints on their success imposed by social tradition, inappropriate government policies and lack of incentives to farmers and extension workers. Attempts to modify old systems or to introduce new ones will only be successful if government support is switched from the old system to the new. Examples of such support in the WANA region are the various subsidies commonly provided to farmers to encourage cereal grain production. Subsidization of barley grain for fodder imposes a very low ceiling on prices that can be asked for forage legumes or hay. The subsidy given to nitrogenous fertilizer is directly competitive with the use of legumes to improve soil nitrogen. Subsidizing locally grown wheat makes it difficult for any alternative crop to be adopted, or even for fallow areas to be used to grow pastures rather than to accumulate moisture for the next wheat crop.

Educational and research institutions also have a role in fostering systems of integrated crop and livestock production. Traditionally, animal husbandry and agronomy have been separated in different departments. Good co-operation is needed to promote a system which requires integration of the two.

Conclusions

The many investigations and projects conducted in the WANA region since 1970 have shown the relevance of the medic-based ley farming system to the edapho-climatic environments in this wide area. Common agronomic problems have emerged, relating to the need for locally adapted medic varieties and strains of *Rhizobium*. Many of these problems have been solved and those that occur can be effectively dealt with in the future by ICARDA, in collaboration with national research institutions and bilateral projects.

What has not yet been solved is how to adapt the ley farming system to a very different farming, social and economic environment from whence it developed in Australia. This includes features such as small farms, a traditional separation of crop and livestock enterprises, and a lack of understanding of pasture management. Government policies will need to change and a long period of adjustment and education will be required before an indigenous type of farming system is developed which will incorporate the benefits provided by medic-based pastures.

Very close co-ordination of research, extension, education and government policy will be required to ensure steady progress in the development and successful operation of new systems such as the ley farming system.

Further reading

Grigg, D. B. (1980). *The Agricultural Systems of the World: An Evolutionary Approach*, Cambridge University Press, Cambridge.

Nahal, I. (1981). 'The Mediterranean climate from a biological view point', in F. di Castri, D. W. Goodall and R. L. Specht (eds), *Ecosystems of the World, 2 Mediterranean-type Shrublands*, Elsevier, Amsterdam.

Osman, A. E., Ibrahim, M. H. and Jones, M. A. (eds) (1990). *The Role of Legumes in the Farming Systems of the Mediterranean Areas*, Kluwer Academic Publishers, Dordrecht.

Nordblom, T. R. (1990). 'Ley Farming in the Mediterranean Region from an economic point of view', *Proc. of a workshop on 'Introducing Ley Farming into the Mediterranean Basin'* 26–30 June 1989, Perugia, Italy.

CHAPTER 20

Improving Agriculture through Systemic Action Research

R. J. Bawden and R. G. Packham

SYNOPSIS This chapter criticizes and extends the view that dryland farming systems are real, and explores the consequences of the systems view and its extension for research methodologies. A case is made for a new 'systems science' that incorporates, rather than supplements or complements, previous scientific approaches to improving agriculture. A key feature of this new science is its focus on the learning process and the synthesis of objectivity with subjectivity.

MAJOR CONCEPTS

1 A socio-historical perspective provides insights into the dynamics of Australian agriculture.

2 Methods for analyses of agricultural systems fall into two broad categories: 'hard systems' which are computer based and have a clear, universally acceptable purpose, and 'soft' systems which are mental constructs, used to initiate debate between concerned parties on desirable and feasible change.

3 Some problems can be solved, others can only be improved.

4 The way an issue is perceived determines how it is dealt with.

5 Research cannot be conducted purely objectively but needs to be combined with subjectivity.

6 Complex, intractable problems are best researched with people, rather than for or on people. This new paradigm of research necessitates a change in research, education and extension infrastructure.

7 Sustainability is best seen as an emergent property, rather than a definable goal. It arises from the ethical resolution of conflict through this new paradigm research.

Thus far, this book has assumed that dryland farming systems are real, and can be observed, objectively measured and improved as whole entities, using the concepts of science and the tools of technology. This chapter will examine this stance and develop further dimensions to the concept of systems approaches to improving agriculture in dryland Australia. The argument is mounted that over the past 200 years there have been several major shifts in the way that agriculture has been viewed moving it from one era to another. Each of these changes in focus reflect new ways of thinking about the nature of the world and of the nature of knowledge about the world ... and as each of us holds to our own particular beliefs about these issues, changes in eras are uncomfortable times. We argue that in Australian agriculture we are in the midst of such an uncomfortable time.

Perhaps most significantly, each shift in what we might term 'an agricultural worldview' is intimately associated with questions about the nature of 'true' improvements and about the context in which such improvements are envisaged. Because these often differ from one era to another, they invariably dictate the need for new ways of doing science. New sciences and scientific methodologies emerge when the old ways of solving problems are accepted as inadequate in the face of phenomena which fail to be addressed satisfactorily by the conventional problem-solving sciences of the day.

In the face of continuing degradation in the biophysical and socio-cultural environments in the dryland zones of Australian agriculture, we posit the need for a systems science which is different from the current one with its focus on farms-as-systems. We intend to explore the idea that the new systems science is necessary because there is now an urgent need for research to be aimed at inquiry *with* people to complement that which is currently conducted *on* them, or on their behalf. We also suggest that this new research approach must deal with complexity and change, and it must combine action with inquiry—such methods that agricultural research must develop we are calling 'systemic action research'.

Socio-historic perspective

There have been four eras of Australian agriculture (Figure 20.1). Following white settlement, agriculture in Australia was developed on a traditional pattern, characterized by reliance on the agricultural practices and traditional wisdom of European farming methods. These needed adaptation to the dryland conditions of Australia, but the process was very much a synthesis of old ways with new circumstances to provide an often precarious supply of food for the colony as well as a living from the land for its farmers. The problem-solving process used to establish improvements in farming practices throughout this pioneering period was essentially that of trial and error, with experience rather than theory providing the logic for the practices.

Together with the improvement of conditions in the developing colony came a desire to improve export income through increasing the volume of production of agricultural commodities and thereby the enhancement of living standards. Trial and

error was an inefficient and uncertain way of achieving this change in goal. The need now was for concept and theory to inform the development of new practices for this emerging era of production. This context, set at the turn of the 20th century, was coincident with the emergence of sciences and technology appropriate to agriculture. The experimental research methods developed in Europe and North America allowed specific factors limiting production to be identified, and then ameliorated using technologies specifically developed for the purpose. An example is the discovery of nutrient deficiencies in crops and livestock through careful observation and experimentation, and their subsequent treatment through supplementation of fertilizers and rations respectively.

This 'production science' approach to establishing and designing improvements to agriculture can be seen as the application of the so-called 'law of the minimum'. It is characterized by rigorous methods of inquiry based on objective knowledge and particular forms of logic developed through a range of different scientific disciplines. This somewhat fragmented process of inquiry-through-problem-solving, of seeking causes for particular isolated effects, remains central to the mission of agricultural science.

Liebig's law of the minimum

Justus Liebig, a German chemist working in the early part of the 19th century, was the first to formally record the observation that the yield of crops was often limited not so much by nutrients needed in large quantites, but by elements which needed to be present in the soil only in trace amounts. His proposition that 'the growth of a plant is dependent on the amount of foodstuff which is presented to it in minimum quantity' is often referred to as Liebig's law of the minimum. Moreover, research which proceeds through the identification of simple cause/effect relationships as the basis for improvements in complex situations, is sometimes termed Liebigian. The more general term for the philosophy which underlies this approach of breaking complex issues into relatively simple ones, is reductionism.

In the period commencing just after the end of the Second World War, the application of the reductionist production sciences began to have a most significant impact on the rate of growth of the volume of agricultural products. A new challenge now emerged as the focus began to shift from volume of production to the efficiency with which it was produced. Given the importance which would be ascribed in this new era of productivity to the financial efficiency of the whole farming enterprise, the new thrust would lead to the rise of farm management and agricultural economics as sciences of fundamental importance in establishing the nature of improvements. These new social sciences enabled the uncertainties and ambiguities of human decision-making processes to be incorporated into research methods aimed at discovering the ways by which farm performance could be optimized. They also allowed the focus to be shifted away from the specific processes of production of individual parts of the farming enterprise, to embrace the performance of them aggregated together to represent the farming system as a whole: a set of relationships between animals, plants, soils and risk-taking human managers.

Given this new emphasis on the performance of the whole farming system, it is somewhat surprising that the success of the application of economic science was not duplicated in a similar application of ecological science. Indeed, it has taken the emergence of yet another era, the age of persistence or sustainability, for ecological thinking to have much of an impact. One of the reasons for the lack of progress in applying methods of ecological analysis to the issues of agriculture, was that the early agricultural ecologists generally failed to come to terms with the human dimension of farming systems. Of greater significance was the failure of ecologists to shrug off reductionism in their science and turn to ways of thinking about nature and about farming as well, in a holistic or systemic way. They still continued to place the search for relatively simple linear cause/effect relationships at the centre of their research, thus continuing with the idea of studying complexity through reducing it to smaller issues. Their logic, like those in the production sciences as well as those using socio-economic methods of analysis, continued to rely on the logic of deduction (conclusions from the general to the specific) and induction (from the specific to the general). Finally, they tended to hold as inviolate the concept of nature being organized in the form of real, self-regulating entities which they called ecosystems and which they sought to identify and describe as whole systems.

The new ecologists of this emerging era of persistence have adopted radically different positions about the nature of reality, as well as about the nature of ways of knowing about that reality. These systems scientists or social ecologists are creating a new and controversial research tradition where the emphasis is not on inquiry into systems as real *entities*, but as figments of the imagination of people which help them think about real *issues*.

Implications for research approaches to improve agriculture

The following distinctions are made about approaches to research in the three eras:
- Research in the production tradition tends to be disciplinary.
- Research in the productivity tradition tends to be multi-disciplinary or inter-disciplinary.
- Research in the persistence tradition tends to be trans-disciplinary.

All three of these research approaches are important and must be treated as complementary forms of inquiry rather than as competitive. This is not easy to achieve in practice as each research approach or paradigm is based on different beliefs about knowledge as well as how such knowledge can be legitimately gained and validated.

Production science research is oriented towards an objective investigation of natural phenomena as concrete things. The values and behaviour of people are not explicitly examined, and indeed, are assumed to be of no significance in the form of inquiry. The production approach is Liebigian (reductionistic) and is based on a belief that natural phenomena are ultimately knowable through such scientific investigation. Such a philosophical position is termed 'positivism'.

Productivity research deals with the much more abstract notion of relationships between things rather than with the things themselves. As the productivity of farming enterprises reflects to a large extent the efficiency of the human management of that system, productivity contains significant subjective dimensions. The performance of any particular farming system can never be fully predicted because of the self-determining nature of human beings who can change their minds, habits, goals, behaviours or any other aspect of being human, whenever they wish. To deal with the uncertainty of such purposefulness, those interested in researching into the productivity of farming systems have adopted a statistical perspective for their inquiry, based not on the objective probabilities of the natural sciences, but on subjective probabilities. The final issue of importance is that whilst most productivity researchers deal typically with whole farms, they inquire into the performance of the whole system by aggregating the performance of its major components as measured first in isolation from the whole system. In this sense, most farming systems research is systemic and multi-disciplinary rather than trans-disciplinary and systemic.

In a philosophical sense, we believe there can now be no one 'ultimately knowable or positivistic world'. According to this view reality can only ever be a construction in the mind of any individual viewing the world and so the philosophical tradition here is one of constructivism. Where disagreements on objectives of stakeholders were uncommon in research approaches focused on production *per se*, the differences represent a common source of dissent when the focus is shifted to productivity. Springing as much from this tension as from anything else, two interesting strategies have been developed over recent years. In the first, the so-called Rapid Rural Appraisal (RRA) approach, a team of scientists from a range of disciplines from within each of the traditions visits particular agroecological sites, particular regions or communities, or areas where co-operative inquiry is essential. Working over a period of days or even weeks, these scientists work in small sub-groups, gathering data through the use of techniques appropriate to their particular research methodology, and sharing their findings with their colleagues as well as with those living and working in the areas. The aim is to establish major problematic issues as perceived from the different viewpoints of the scientists, as well as from those people who are indigenous to the 'problem' area, as a guide to the research that could be conducted into how the prevailing situation could be improved.

Three major outcomes arise typically out of such an exercise:

1 Each participating scientist is exposed to the methods of other scientists who hail from different traditions, thus leading to a sharing of knowledge and appreciation of each other's domains and research approaches.
2 People living in rural communities and working in agriculture learn from the scientists the varying ways of science, as well as helping the scientists learn how they, as lay-people affected by the situation, perceive and construe the issues that they think are important to them.
3 The scientists and the lay-persons together identify the major problematic issues which need to be researched, and together help in the creation of the research agenda that is constructed to seek improvements.

It is important to emphasize that the participation of the lay-people in the research process from this point on is invariably minimal. The focus of the researcher remains on the identified issue of concern rather than on the way the people are learning to deal with that issue. It is a matter of research *on behalf of* the people, or even *on* the people, rather than researching *with* the people in a truly participative manner.

The second initiative is much more fundamental in that it represents a genuine attempt to create a new science which provides for a genuine synthesis of different approaches and viewpoints rather than just an amalgamation of them. Systems or systemic science is concerned with complexity. Its central thrust is based on the notion that whole entities, or systems, have properties and behaviours that are not revealed by, nor are even predictable from, the study of any of the parts of that same system. Systems are therefore said to have emergent properties and it is these properties that are the focus of research into improvements in such systems.

Emergent properties

This idea is central to systems thinking. If you think of the entity 'cow', this cannot be related to a hoof, horn or mammary gland. Similarly, the property of 'wetness' has no meaning when related to the constituent parts of water, hydrogen and oxygen. Such emergent properties enable complexity to be described in terms of a hierarchy of levels of organization in which each level is described in terms of its emergent properties rather than constituent parts — the concept of 'wholeness'.

Productivity is a measure of the performance of what can be termed 'a farming system', in that the efficiency with which the various inputs into a specific farm are transformed into outputs which leave the farm can vary enormously from farm to farm. It can also vary within the same farm over time. Such computations can be focused on different so-called currencies, such as energy or matter or money, with the flow of each of these giving a different and complementary picture of how the whole system is performing. It is possible that one particular farming system appears to be much more efficient in financial terms than another, yet less efficient than the other in the way it uses energy. This is, of course, most obvious when one compares a subsistence, pre-industrial farming system with a highly mechanized, intensive cropping system.

Table 20.1 *The comparative efficiency of two agricultural systems.*

Agricultural System	Energy Yield (*MJ/ha/yr*)	Gross Energy Productivity (*MJ/person/day*)	Surplus Energy Income (*MJ/person/day*)	Energy Ratio (*Output/input*)
Subsistence agriculture (New Guinea)	1 460	10	2.3	14.2
Industrial agriculture (south of England, 1971)	44 890	2 420	18.8	2.1

(SOURCE: Bayliss-Smith, 1982. *The Ecology of Agricultural Systems*, Cambridge University Press, Cambridge)

Under these circumstances, the notion of what constitutes an improvement, and thus how one would go about improving situations through research, is much less obvious and more complex than the simple issue of producing more product (Table 20.1). In order to deal with these multiple objective issues, a very common research strategy is the development of, and 'experimentation' on, models which attempt to simulate the performance of the system as a whole in quantitative terms for each identifiable purpose. Having built a mathematical model of the real system, a host of variables can be altered and the impact of differences on the various performances of that 'hard' system can be assessed through 'running' the model.

This approach is termed a 'hard systems approach' because it concerns itself with situations where each purpose of the system is relatively easy to identify, where the functional relationships which contribute to the performance of each purpose are quantifiable, and where it is assumed that the system actually exists in the real world. Both Farming Systems Research and Development (FSR) and Agroecosystem Analysis and Development (AA&D), are considered to be hard systems approaches although, as it transpires, neither rely on simulation modelling in the first instance. There are those that argue that FSR and AA&D are really systematic rather than systemic, in the sense that they do not really deal with either the essence of wholeness in a philosophical sense or the performance of an emergent property, in a scientific sense. Despite this argument, the support for this type of trans-disciplinary agricultural science continues to grow, particularly in the less developed parts of the world where farming systems tend to be small and complex. The approach

is also beginning to gain ground in countries like Australia and the USA as the reputation of the small cadre of scientists who are trained in its methods and contribute to its methodologies continues to grow.

For all the success of the traditional reductionist sciences in the continuing improvement of dryland farming in Australia, and for all of the emerging contribution being made by the hard systems sciences, much is still amiss. There is now a need to recognize a fourth era in Australian agriculture, an era which integrates within agriculture the objective ideas about production and productivity and profoundly subjective notions about the quality of the care and welfare of land and of those who live and work in rural communities. In this new era of sustainability, the central focus is on the persistence of relationships within farms, within farming and other rural communities, between these and their urban counterparts, and indeed between any people involved in rural Australia and their environments in all senses of that word.

In addition to the management of scarce resources and of the complexity of whole farming systems, those involved with agriculture must now learn to manage the conflicts inherent within any set of relationships which involves people interacting with each other and with issues and events in the environments around them. The degrading changes which are occurring in the quality of many of the bio-physical and socio-cultural environments in which dryland agriculture is conducted can be argued to be a function of lack of consciousness and knowledge about the relationships that people have with their environments. The situation is being exacerbated by the lack of appreciation of some people by others, be they neighbours or consumers. The focus for research and education under these circumstances is to achieve improvements in complex situations rather than solutions to readily identifiable production problems, or the optimization of universally applicable measures of performance of real systems.

In this new or soft systems approach, the models that are built are abstract sets of interrelated human activities that seem appropriate to improving the problematic situation. These models are not quantitative nor can they be 'run'. They are models for debating how change might occur in the world rather than being models of the world.

'Hard' and 'soft' systems

This distinction between hard and soft systems was made by Professor Peter Checkland at Lancaster University to differentiate the approaches which see hard systems as having clear, universally acceptable purposes. 'Hard' systems are definable and able to be manipulated in reality. Systems engineering, farming systems research and simulation modelling are examples of hard systems approaches. In contrast, 'soft' systems are mental constructs or figments of the imagination. They are used to initiate and structure debate about complex issues: Models have become means of structuring debate about desirable and feasible change in relationships within human activities. The Checkland Methodology and Systemic Action Research are examples of soft systems approaches.

The four eras of Australian agriculture are illustrated in Figure 20.1. Each new era provides

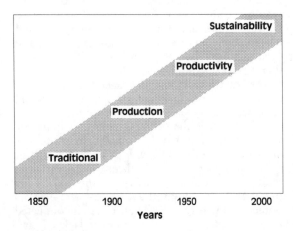

Figure 20.1 *Four eras of agriculture.* (SOURCE: Bawden, 1988)

perspectives and methods of research which complement rather than replace its predecessors, each new era dictates a fundamental shift in the way we think about the nature of nature and about the way we go about knowing it.

Unfortunately, it is extremely difficult to introduce new ways of thinking and acting no matter how inadequate the conventional approaches appear to be. In this new age of sustainability, agriculturalists must learn how to deal with both objective and subjective relationships between people, with the systems they invent, and between such systems and their environments. A new systems science for agriculture is required which focuses on such relationships. Academics of the Faculty of Agriculture and Rural Development at the University of Western Sydney, Hawkesbury, have been working on such a quest for the past decade. To understand the logic of their approach, it is important to explore some theories of learning as well as some of their philosophical underpinnings.

Experiential learning and action researching

Learning is the process by which experience (of the world) is transformed into knowledge (about the world) as a basis for adaptation (to and of the world). As a dynamic process, it involves two sets of activities aimed at finding out and at taking action. These two sets of activities are, in turn, carried out in two different contexts: in the 'concrete world of experience' as well as in the 'abstract world of the mind'. From this perspective, learning is the same as problem solving. In other words, the learner follows a systematic process in a cycle of activities (Figure 20.2). With slightly different terminology, this model could also describe the cyclical process of research. Researching is problem solving is learning!

Research into the learning process has revealed some interesting phenomena. Perhaps the most important theory that flows from this for the present purposes is that each individual develops a characteristic 'style' of learning which becomes remarkably fixed at quite a young age. Indeed, our own style becomes so fixed that we not only become aggressively resistant to change but we also do not recognize the fixity of our position. In other words, we think we are changing the way we learn when, in fact, we are learning in exactly the same way as before. This condition has been called

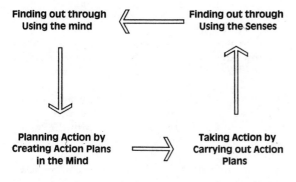

Figure 20.2 *A systematic process of learning activities.*

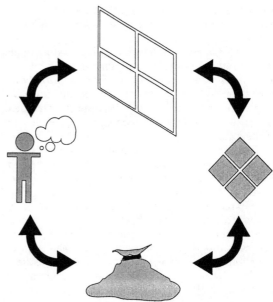

Figure 20.3 *A stylized learning cycle.*

'mindlessness' and it represents a serious impediment to fundamental reform in agricultural research, or anything else.

This matter of style and bias in the way we go about our learning is central to the whole argument being mounted here. From the model of Figure 20.2 it is obvious that an individual can develop a preference for one of the activities in the cycle over all others. Other people have preferences for different stages in the cycle and there is a myriad of possible combinations and permutations of two or more of the activities. This is a simplistic explanation for the fact that some people are very concrete, practical sort of people, whilst others are abstract, theoreticians. They often find it very hard to get along with each other just because they approach the process of problem solving from quite different perspectives. The pragmatists will want to take action almost as soon as the problem is recognized. The reflectors will want to take a long time thinking about the nature of the problem and a range of possible solutions before committing themselves to any sort of plan for action, let alone taking the real action!

The model displayed in Figure 20.3 is a stylized expansion of that in Figure 20.2, with some important new features.

A 'window' representing the particular world view of the learner is interposed between his or her 'mind', illustrated here by the person with an abstract mind, and the world which he or she experiences, illustrated by the pool of events in the concrete world. In other words, the events or issues which each of us examines in the world around us is, in many ways, predetermined by the world view we bring to bear on our senses in the first place. We only see what we want to see. Our world view, or *Weltanschauung*, is influenced by a whole host of things including our values, feelings and attitudes; our previous experiences; the knowledge we remember; even our genetic make-up. Our world view is also markedly influenced by the world views of other individuals and of the entire culture in which we live.

Nor do we just differ in learning style through preference for particular activities, or through different world views. Cognitive research reveals that we also differ in the way our brains actually seem to work and the parts of the brain that we tend to use more often than others. Popular talk of differences between left and right brain people has a serious basis in science, with the first type showing a strong preference for digital (discrete) logic and the second, for analog (continuous) logic. Partly in association with discoveries about brain function and partly drawn from observations and logical deductions, a theory of multiple intelligences has been proposed. Abilities to master language, to create and play music, and to count and do complicated mathematical computations have been regarded as all equivalent competencies of intelligence.

Finally, in our model the notion of a 'bag of tricks' representing all the practical things an individual can do has been interposed between the abstract mind and the concrete world. What we do in this world is ultimately determined by the way we see the world. Yet the way we go about seeing or perceiving is also markedly influenced by what we feel most comfortable at doing: shades of 'mindlessness' again.

A most important feature to note in this model is that rather than being a unidirectional cycle, the process is now displayed in recursive form with the direction able to be reversed at any stage. In this form, the learning cycle becomes transformed into a learning system which can be extended by bringing in other learners (Figure 20.4).

In this situation we can imagine one person trying to help another make sense out of an event in his or her world by introducing new ways of knowing, or at least by bringing different learning styles and world views to the situation. This is the essence of *action research*; the creation of a learning system with each participant relating to all others through the appreciation of their unique contributions. In its ideal form it is a truly collaborative process of joint learning where there are at least five outcomes from each researching 'project' or episode.

- Improvements occur in the original problematic situation through new insights gained through new ways of learning.
- Improvements occur in the understanding of the problematic situation that can be generalized to other similar situations.
- Improvements occur in the understanding of the processes of learning brought to bear in the 'system'.
- Improvements occur in the use of particular ways of learning.
- All of these improvements contribute to public knowledge as all of the above outcomes are presented for public critique.

In this action researching approach, the essence of systems — the systemicity — is transferred from the object being investigated to the process by which events and issues associated with the object are investigated. This is a fairly abstract concept and yet one that is vital to the understanding of the 'soft system approaches'. It is what we have called 'Systemic Action Research'.

In contemporary agriculture there is an enormous amount of controversy and conflict about a whole variety of issues associated with the future sustainability of farming in all sorts of environments, and in particular those which are subject to unpredictable climatic variations as with dryland situations. These differences of opinion have a number of different foundations, most of which are eventually grounded in differences in the way individuals go about their own learning.

It can be thus argued that conventional approaches to improving agriculture through the 'remote' generation of scientific knowledge and associated technologies, and its subsequent diffusion or transfer to the field, can only have an application limited to relatively simple problems. History has shown that it was an exceptionally successful strategy when such conditions prevailed, as in the era of production where the goal was constraint identification and removal. It can also be argued that the much less effective diffusion of knowledge about such dimensions as financial management, integrated pest management or any of the so-called expert system decision-making tools, reinforces the above submission.

Expert systems

These are computerized models that have had the knowledge of a human expert programmed into them. They function as well as the design of the computer program and the knowledge of the humans they incorporate. They do not have the capability to acquire knowledge independently and so are not truly 'artificial intelligence'. They allow the knowledge of a specialist to be available in a more general way. They are knowledge systems not learning systems.

Because people perceive the world in such different ways, and hold such different values for

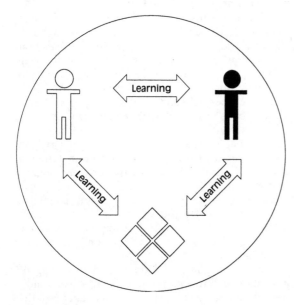

Figure 20.4 *An extended learning system.*

things and events in it, new approaches to improvements in agriculture must focus on helping people to learn to see the world in different ways; ways which are ultimately grounded in ethical considerations of what lies in the public good.

It is in this context that systemic action research, of projects which involve the collaboration of different people working together in systems of events and people, is proposed.

Implications of systemic action research

The current model of improvement in agriculture is that of research, extension and education, each separate from the other. While some universities carry out all three functions, the processes are seen to be different, and it is rare for more than two of these functions to be the responsibility of a single person or departmental group: in Australia the three functions are generally located in separate institutions, and always in separate physical facilities. Usually, large research institutes are responsible for conducting research, departments of agriculture carry out field trials and extension. Education and training is seen, depending on the level and type being discussed, as the responsibility of schools, technical colleges, government departments, universities, and agribusiness; this education and training is carried on in a mainly unco-ordinated way.

Such an approach to agricultural improvement is underpinned by a view of science as being *simplification of complexity*, with the consequent loss of focus on emergent properties; *positivism*, seeking formally laid down and definite facts; and *gradualism*, which misses out on the quantum leaps of major innovations. An example of this can be found in the debate on wool prices, which markedly affects dryland agriculture in Australia. The current approach focuses purely on the price issue—should it be price A or price B? It is believed that the resolution of this issue will lead to the correct decisions being made by farmers, including the use of relevant research findings (technical and economic) brought to farmers' attention through the current specialist-based government extension services.

A systemic action research approach would step back from the present problem and ask questions such as, why is raw wool exported? How can its value be increased, e.g. by processing greasy wool

in Australia? What was the function of the floor price — price support, equalization or what? What alternatives are available to wool production for marginal wool producers? What will be the environmental consequences of a change in the floor price flowing through to changes in wool production systems in dryland areas?

The new approach of systemic action research recognizes the strengths of the earlier paradigm but aims to overcome its weaknesses by coupling this approach to systemic methods which are able to deal with the improvement of complex agricultural issues. Examples of such complex issues that are in need of urgent attention include environmental degradation, economic distortion and community decimation.

To support this new initiative, it is not new institutes or other physical structures that are needed, but the creation of networks of collaborators engaged in the process of researching these complex issues together. It is these networks that we refer to as *action researching systems*. They include not only the people traditionally viewed as researchers, but others that are stakeholders in the issues being researched. Thus systemic action research relies on specialized knowledge, and also on the ability to help all concerned people to work in groups in a co-operative way. An example of such an approach in a dryland area is a project now being set up in NSW, in the area west of the Darling River. Problems of this region are complex and include desertification, woody shrub invasion, and soil compaction, all resulting from attempts to maintain the growth of productivity in the face of ever declining terms of trade. Further, these issues cannot be dealt with in isolation from the social issues of income security, isolation, family and community disharmony and racial tensions. The issue will be tackled initially through two subprojects: the first to carry out an agroecosystem analysis, while the second conducts a social analysis. A continued interchange will occur between the two projects to integrate relevant data and facilitate the on-going formulation and review of research strategies. This will lead to the formation of an investigation network of graziers, their families and other community representatives, together with the research staff who initiated the project, to co-investigate the implementation of strategies for improvement emerging out of the two earlier subprojects. Particular emphasis will be placed on documenting the applicability of the findings to other dryland areas, and on the improvement of the systemic action research methods used.

Further reading

Bawden, R. J. (1988). 'Problem-Based Learning — A Focus for the Quiet Revolution', *Proc. of the Society for the Provision of Education for Rural Australia*, Launceston, Tasmania.

Bawden, R. J. (1991). *Systems Agriculture: Learning to Deal with Complexity*, Kentucky University Press, Lexington.

Baylis-Smith, T. R. (1982). *The Ecology of Agricultural Systems*. Cambridge University Press, Cambridge.

Checkland, P. (1981). *Systems Thinking, Systems Practice*, John Wiley and Sons, Chichester.

Churchman, C. W. (1971). *The Design of Inquiring Systems*, Basic Books, New York.

Davidson, B. R. (1981). *European Farming in Australia*, Elsevier, New York.

Kemmis, S. and McTaggart, R. (eds) (1987). *The Action Research Reader*, Deakin University Press, Geelong, Victoria.

Reason, P. (ed.) (1988). *Human Inquiry in Action: Developments in New Paradigm Research*, Sage Publications, London.

Weil, S. W. and McGill, I. (eds.) (1989). *Making Sense of Experiential Learning: Diversity in Theory and Practice*, The Society for Research into Higher Education and Open University Press, Milton Keynes.

Wilson, J. (1988). *Changing Agriculture: An Introduction to Systems Thinking*, Kangaroo Press, Sydney.

CHAPTER 21

Future Directions — Meeting the Challenge

V. R. Squires, P. G. Tow and C. M. Boast

SYNOPSIS This chapter considers the pressures and constraints on Australian dryland farming systems now and in the future and discusses the role of the various agents for change in devising new farming systems to meet the challenges.

MAJOR CONCEPTS

1 Australian agriculture will undergo further dramatic change.

2 Modern agriculture requires unprecedented management precision.

3 Pressure is on the family farm and the trend toward fewer and bigger farms will continue.

4 The new circumstances will arise from emerging environmental issues and from global climatic changes.

5 A new approach to extension and advisory services will be needed with emphasis away from a production orientation toward a market approach for commodities and toward sustainability.

6 New farming systems will be needed. Many participants will be involved with a diversity of roles.

As indicated in Chapter 1 and elsewhere in this book, agriculture is going through a period of rapid structural change and development. New plant strains and animal breeds, new pesticides and fertilizers, more mechanized farming, new marketing arrangements and above all, vastly improved knowledge and management skills (Chapters 4 and 14) have all contributed to almost doubling overall agricultural productivity in the past thirty-five years (Figure 21.1). Perhaps the most visible impact of agricultural modernization has been a socio-economic one, demonstrated by changed rural lifestyles, notably in the decline in the number of family farms, and the progressive depopulation of agricultural areas.

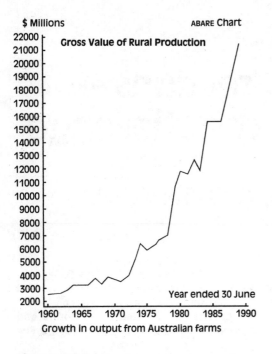

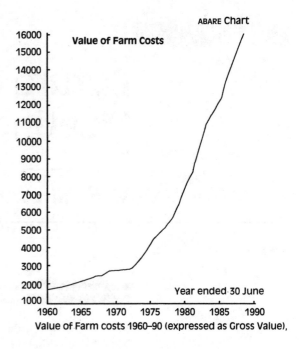

Figure 21.1 *Production has increased over the period 1960–90 but so have costs.*

Sustaining agricultural yields

As yields rise so must the effort devoted to sustaining gains. Elevating crop and pasture yields is still possible, but the easy ground has been conquered. For the foreseeable future high yields in agriculture will continue to depend on rational and effective chemical applications but the use of the integrated pest management and changes in cropping patterns can help sustain productivity gains. To a large extent, the high yields characteristic of modern agriculture depend on a steady stream of new cultivars. Commercial farmers witness a constant march of new crop varieties as the pattern in the wheat belt of NSW shows (Figure 21.2).

The dynamic nature of agroecosystems is largely responsible for this turnover among varieties. Much of Australia's wheat production depends on continuous breeding for disease resistance as new races of the pathogens evolve, but changes in climate, water or soil conditions can also force the early retirement of crop varieties.

Modern agriculture requires unprecedented management precision to avert serious shortfalls. Dramatic gains in productivity can be ephemeral unless a follow-up research programme is in place to shore up the advances. Effort to sustain produc-

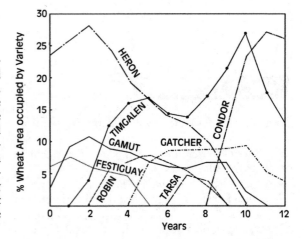

Figure 21.2 *Trajectories of selected wheat varieties in NSW 1967–79.* (SOURCE: NSW Department of Agriculture, Division of Plant Industry Annual Reports)

tivity and dampen oscillations in crop and livestock yield has been called maintenance research. As productivity rises, maintenance research is needed to uphold the gains (Chapters 6 and 17). Research on improving and sustaining yields of most major crops and pastures began about the turn of this century. In the early 1900s yields generally increased, but recently that trend has slowed. The challenge for farming systems researchers is to design new systems which can cope with changing circumstances and which also sustain the productivity gains of the past few decades.

Environmental concerns — an emerging issue

Farming has become much more mechanized and more diversified with greater regional and on-farm specialization. By increasing its reliance on mechanization and technology, developing a heavy dependence on fossil fuels, exploiting the productivity of the soil and enhancing crop yields with fertilizers and pesticides, agriculture has evolved to a state where short-term benefits are being gained at the expense of long-term costs. Such actions ignore the interdependence between agriculture and the environment. A conflict has arisen between maintaining productivity and achieving sustainability. The intensification of agricultural practices and the changing terms of trade have had an adverse effect on the environment, e.g. from pollution by pesticides and by land degradation through overcropping and overgrazing. In the cereal belt, land degradation is due to failure to apply appropriate conservation practices and to take into account processes which lead in the long term to salination, acidification etc. Agricultural inputs such as fertilizers, pesticides and feed additives have been responsible for many of the recent gains in agricultural productivity, but unfortunately a number have also had, or threaten to have, adverse side effects on the environment.

To this list of problems must be added concerns that are only now beginning to emerge. Many of the mechanisms of environmental degradation associated with agriculture have only come to be more clearly understood in the past two decades. The state of knowledge about some of the major environmental influences on agriculture is even less advanced. Indeed, many of the phenomena now considered to pose significant long-term threats to agricultural production have only recently come under intense study. Among threats which are now perceived are:

- Increasing concentrations of CO_2 which, in addition to being a factor in the 'greenhouse effect' of atmospheric heating, may also indirectly affect plant metabolism.
- Increases in ozone concentration at the earth's surface which appear to be causing declines in crop yields.
- Decreases in ozone concentration in the upper atmosphere which allow greater amounts of ultraviolet radiation to reach the earth's surface, with as yet undetermined effects on plant growth.
- Pollution of the soil by persistent harmful chemicals which has led to a decline in food quality and increased cancer risk.
- Leaching of nitrate and the washing of phosphate into surface and groundwaters.
- Global climatic changes, including prolonged periods of unseasonal weather and an apparent trend toward atmospheric warming which could have far-reaching effects on growing seasons and precipitation (Figure 21.3).

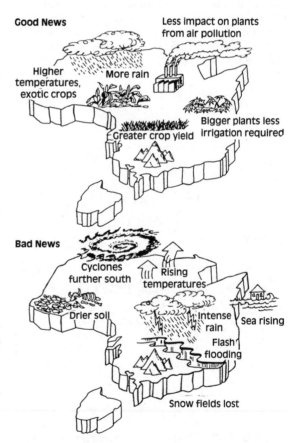

Figure 21.3 *An apparent trend towards atmospheric warming could have far-reaching effects on growing seasons and precipitation. There will be Good News and Bad News.* (SOURCE: Henderson-Sellars, A. and Blony, R., 1989. *The Greenhouse Effect: Living in a warmer Australia*, University of NSW Press, Sydney)

It has been suggested that Australia's climate will change dramatically over the next fifty years or so. It is likely to be wetter in the centre of Australia and also likely that CO_2 enrichment of the atmosphere may lead to higher yields because CO_2 enhances growth. Yield increases of up to 10–15 per cent might be expected for wheat, rice, barley and some fruits and vegetables, and increases from 0–10 per cent for corn, sorghum and sugar cane. Of course, weeds will grow better too! Climatic change may lead to sea level rises, drier soils in some areas, raised groundwater tables in others, intensified rainfall and flash flooding.

Responses to the greenhouse effect and to climatic change in general will be complex, particularly as agricultural activities in Australia range in intensity. Important factors include increased CO_2, enhanced temperatures, rainfall changes, increased evaporation, and changes in the length of the growing season. In the southern part of Australia (the traditional cereal belt) winter rainfall may be reduced (Figure 21.4).

Figure 21.4 *Projected southward movement of the northern limit of cultivation in South Australia following the onset of the climatic change brought about by 'greenhouse effect'.* (SOURCE: Greenwood, G. and Boardman R., 1989. 'Climatic change and some possible effects upon the terrestrial ecology of South Australia' in *Greenhouse '88: Planning for Climatic Change*, Department of Environment and Planning, Adelaide).

Agriculture is far more sensitive to weather and weather changes than any other human activity both in relative and absolute terms. For agricultural systems, the prospects of change will prove challenging but are unlikely to prove fatal. New developments in biotechnology and selection and breeding methods should allow relatively rapid adaptations to change in CO_2, temperature, rainfall and length of the growing season. This will depend on the success of agricultural researchers and technologists to devise new farming systems (see below).

Energy inputs to Australian agriculture

An agricultural system that relies heavily on energy-intensive, purchased inputs, is a success story in terms of traditional measurements of output and productivity. One of the greatest technological achievements of history is the efficiency with which modern societies feed their populations. Despite the impressive gains and the hopes for continued or even expanded growth, the rate of increase in food productivity has been diminishing. In Australia, almost eight times more energy in the form of fossil fuel inputs is required to process, distribute and cook the food consumed than to produce it. The major requirements for crop production are energy, labour and land. Within limits, the three inputs are interchangeable. Industrialized agriculture relies heavily on energy inputs to substitute for labour and land.

Government programmes that provide incentives for high-input farming were devised in a cheap energy era and remain largely intact. The food processing and distribution system has evolved to complement the current production system and to meet the needs of masses of people in metropolitan areas. For example, the premium put on fruits and vegetables that are cosmetically appealing to consumers makes it difficult to produce and market profitably without chemicals. On the other hand, consumer attitude surveys have revealed that pesticide residues on agricultural commodities are judged to be a serious hazard to health and higher prices will be paid for 'safe' food.

Over recent decades the trends have been to even greater fossil fuel subsidy:
- Farms — more mechanization and chemicalization, larger farms and monoculture, cattle feedlots and battery poultry. There has been a decrease in labour with increase in energy.

• Factories — trend to more processing and the mechanization of processing, e.g. hi-fibre bread.

If it continues along its present course, industrialized agriculture will increase pressures on the environment, on energy supply and on Third World food problems as the rich countries will inevitably continue to use more than their fair share of the limited supplies of fertilizers and fuels.

Australia follows the industrialized pattern of food production and so faces similar energy supply problems to those faced by other industrialized nations. Changes to the Australian food system appear to be inevitable in the long run, particularly since petroleum is the major fuel used by agriculture and it is the one resource Australia has the least of. Knowledge of the energetics of agricultural systems helps determine what and where changes might be made, for instance, to take advantage of economic changes resulting from price fluctuations in energy, food, land, labour and water. Together, energy analyses and economic valuations can facilitate and strengthen agricultural decision-making processes.

Comparisons between the energetics of different food systems can provide helpful information when changes are thought necessary to a system or when alternative systems are being evaluated. In addition, for industrialized nations energy inputs do not end at the agricultural sector; far more energy is now used in post-farm aspects of the industrialized food system.

Despite sizeable expenditures of fossil fuel, Australian agriculture overall is a mechanism for obtaining a net harvest of solar energy for human purposes. The fuel value of the plant material harvested by machine and domesticated animals represents about 0.01 per cent of the potentially photosynthetically active radiation falling on the arable and pastoral lands. Firstly, 85 per cent of the total plant yield is from pastures which consist mostly of native species whose yield is limited primarily by shortage of water. The fuel value of primary plant material harvested by people and their animals in Australia is about 1.5 times the energy value of primary fuel burned for all purposes. The fuel value of the organic products sold at the farm gate is about 10 per cent that of the primary plant material. In order to get this food energy into an acceptable condition in the context of the Australian lifestyle, fossil fuel is burned equivalent to at least five times the metabolic energy value of food. This observation suggests that it is unlikely that it would be possible to convert the fossil fuel used in the food production system into nutritionally balanced food, either chemically or microbially, more efficiently than by our present system,

since the high energy consuming activities of processing, packaging, transporting and cooking the basic food materials would continue to be necessary.

Improvement in energy-use efficiency in the partial processes of the existing system would be a more viable approach. At least 40 per cent of the total energy consumed in getting the food to the dining table is used in machines powered by internal combustion engines. Thus, any improvement in the efficiency of such machines could have large effects on the efficiency of food production. It is clear that cooking stands out as a major source of inefficiency in the whole system. Another large item is the energy required for the distribution of food, particularly in getting it from the retail store to home in the private automobile. For each joule (J) of digestible food eaten in Australia, at least 5 J of fuel are expended in making it available: $c.0.6$ J to the farm gate, 2 J from the farm to the retail store, and $c.2.8$ J from the retail store to the dining table. Hence although farming is, on balance, an energy-generating process, the food production-processing system as a whole is far from being so. A mere 29 per cent of the total farm inputs is involved in getting it to the farm gate. In the course of getting this material to the dining table almost 50 per cent of its total energy value is lost in various ways. Thus, for each joule of prepared food on the Australian dining table over 5 J of primary fuel is needed to put it there, 11 per cent of this places it at the farm gate, 38 per cent takes it from the gate to the retail store and 51 per cent takes it from the store to the table.

The fossil energy required to place the food on the table is only about 10–15 per cent of the total consumption of primary energy in Australia, and farm inputs account for only about one-tenth of that amount. These figures could be taken to indicate that the food-producing processes in general, and in farming in particular, have little potential for energy saving. The most direct use of fuel in farming is for tillage and some scope exists there for fuel savings through better design of implements and via changes in the tillage practices (Chapter 7).

Sustainable agriculture — a dream or an economic imperative?

Pressure to modify agricultural practices for environmental reasons is a relatively recent phenomenon, and the impact is only beginning to be felt. Awareness of the environmental consequences of different agricultural practices is a necessary prerequisite to the development of more sustainable agriculture. Fundamental perceptions about agricultural activity and about the farmer—as beneficiary and protector of the natural environment, as an economic factor living in symbiotic harmony with the land in an increasingly urbanized world—are beginning to alter as the significant, long-term environmental impact of agricultural activities on the landscape, on the water, the air we breathe, the food we eat is increasingly recognized.

Land degradation is a growing problem and many have come to realize that the off-farm costs are generally greater than the on-farm costs. A key environmental problem is that farmers only have incentives to deal with on-farm costs, and could in future be forced to absorb the costs of off-farm damage from soil erosion, salinity and chemical pollution. Many, if not most, of the farms that are environmentally sensitive are marginal in an agricultural production sense. Climatic and edaphic factors may make for marginality, as in the Eyre Peninsula of South Australia where periodic drought and sandy, erosion-prone soils combine to make cropping a risky business.

There is a continuing role for farmers in maintaining agricultural landscapes amid changing circumstances and conditions. Farmers play an important role in maintaining and providing wildlife habitat, maintaining rural landscapes which are appreciated for their aesthetic, touristic and recreational value, and in many areas farmers contribute to regional infrastructure and development. Most people accept that inevitably there is some trade-off between environmental quality and otherwise essential and beneficial agricultural activity. However, questions are being raised about the logic of sustained pressure on the environment to produce food and fibre from marginal lands when prime agricultural land is being submerged beneath urban sprawl.

There are a number of opportunities for better integration of sustainable practices into agriculture. These include:

- Developing research and advisory programmes with a view to giving greater emphasis and broader consideration to sustainability objectives.
- Strengthening the provision of education and advisory services to improve the use of agricultural inputs and modify agricultural practices to minimize damage to the environment.
- Encouraging or requiring farmers to prepare management plans which indicate how they will use inputs and adopt practices so as to protect

and enhance the environment and ensure sustainability.

- Entering into management agreements and other arrangements with farmers to improve landscape amenity and nature conservation values, e.g. heritage agreements.
- Removing impediments and constraints to the adoption of more sustainable practices.
- Introducing charges on inputs such as pesticides, also, on practices such as bare fallowing to contribute to the cost of advisory, research and other activities designed to prevent and control land degradation.
- Enforce existing laws and regulations more stringently, e.g. soil conservation and land management provisions.
- Further harmonizing the standards and procedures used so that contradictions are avoided.
- Making income, capital and land taxation policies neutral to agricultural and environmental objectives.

Realization of these changes in the way the world views agriculture will require a new approach to delivery of advisory and extension services.

Advisory and extension services

A significant expansion in rural debt and declining profit margins have resulted in a move away from production-oriented extension advice. Instead counselling and farm management advisory work has been in demand. Farm management advisers combine whole-farm planning, financial budgeting, performance monitoring and evaluation of specific enterprises. Demand for farm management advisory services will continue to grow—provided the rural community has the capacity to pay for them.

There can be problems in ensuring that different advisory activities are integrated so that farmers are not receiving contradictory messages. In recent years this has led to a rationalization of extension services so that the greatest effort is targeted to priority problem areas. In some cases it has been found that by targeting extension and advisory resources towards areas where, for example, land degradation is greatest, extension becomes a very effective method of controlling it. Targeting also tends to increase the skills and effectiveness of the advisory officers involved.

Farmers are starting to develop their own conservation schemes (Chapters 5 and 11). Government staff are increasingly seen as specialist collaborators rather than authoritarian 'imposers' of land protection schemes. Their role is becoming more

Component Researched

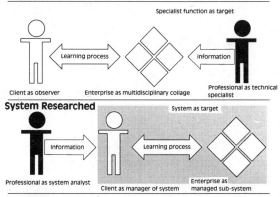

System Researched

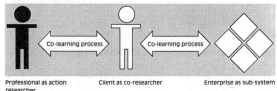

Researching System

Figure 21.5 *The roles of extension workers, researchers and the client (principally the farmer) can be viewed in different ways. The recognition that the client is a co-researcher is growing.*

and more one of motivating individuals and communities and, once interest is aroused, to act as specialist advisers.

There are three basic types of professionals required for the future: the scientist/technologist who develops innovative technologies that encourage greater effectiveness within systems; the systems analyst who is able to examine and design measures to improve the efficiency of whole systems; and a new type of professional who is called an 'action researcher' (Chapter 20). Action researchers are able to help in the creation of participative learning systems which accept the ethical responsibility for farming practices which are at once productive, scientifically sound, ecologically enduring and socially desirable and feasible. There has been a history in Australia of the research, extension and advisory services to follow a pattern such as that outlined in Figure 21.5. These three scenarios can co-exist. In the more traditional approach the client (usually the farming community) was a mere observer and the technical specialist was seen as the person who could provide an answer. As understanding of the complexity of farming systems grew the client entered the equation as a manager of the system. In some situations the benefits of co-learning were realized early and

the pattern was more like that depicted in the bottom of the diagram.

Maintenance of economic viability

Small changes in agricultural policy can have a big impact on the viability of agriculture in a region. The whole fabric of rural society, its *raison d'être*, is the farming base. Townships service the rural population and provide schools, health facilities, local cultural and religious support. Change in the rural society, such as amalgamation of farms or change of farming system, can have a negative impact. These factors must be considered when designing new farming systems.

The maintenance of economic efficiency (and viability) requires farmers to adapt their agricultural production systems so that they remain efficient. The data illustrated in Figure 21.6 reveal the structural adjustments in the Australian rural sector which have occurred in response to the continued pressure of the declining terms of trade (Chapter 4).

In essence, productivity growth and viability of farming has been achieved so far through:
- The substitution of capital for labour.
- Increases in enterprise scale, or intensity.
- Marked increases in crop production.
- The development of complex infrastructures for support, e.g. marketing boards, government-funded research and extension services.
- Development of new land areas through clearing of native vegetation, and draining of swamps.

Australian agriculture has been notoriously unstable with regard to prices received, output levels and incomes. This instability has worsened over the past few decades because of the volatility of world trade and our inability to influence world prices. Fluctuations are cyclic but the long-term trend is downwards. This downward trend in rural commodity prices probably began before the First World War. The long-term trend toward lower agricultural commodity prices is a world-wide trend and is called, by some commentators, the 'farm problem' since it reflects a fall in farm incomes relative to those enjoyed by other sectors of the economy.

The long-run world-wide farm problem is an inevitable consequence of 'economic growth', where economic growth is defined as a general rise in living standards accompanying rapid technological progress. In essence, economic growth, coupled with a slow adjustment of resource allocation, results in a surplus of farmers (especially of small farmers). Such changes obviously affect farm

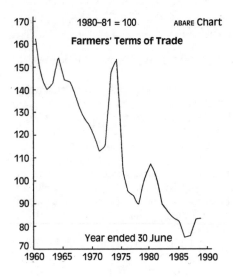

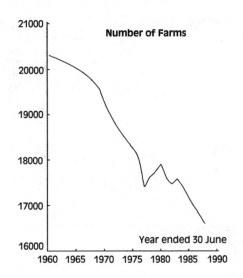

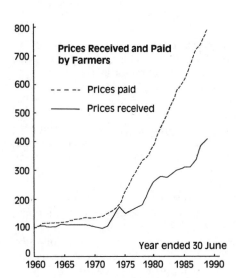

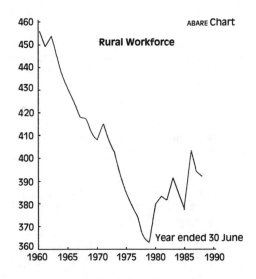

Figure 21.6 *Structural adjustments in Australian agriculture 1960–90 in response to a) changing terms of trade and b) prices received and paid. Principal changes are c) a reduction in the number of farms and d) a decline in the rural workforce.* (SOURCE: Australian Bureau of Agriculture and Resource Economics, AGPS, 1990)

incomes. The technological advances which are the mainspring of economic growth result in a rapidly increasing agricultural productivity (Figure 21.6).

Many of the new techniques call for larger farms and less labour. However, the demand for food is price-inelastic and income-inelastic, i.e. it does not respond very much to changes in price and income. As growth occurs and people's incomes rise, they spend a smaller proportion of

their incomes on food and a larger proportion on the products of secondary and tertiary industry. Thus, the rapidly increasing supplies of agricultural commodities encounter a demand which is increasing only slowly. Large reductions in price might be expected if surplus stocks are to be cleared. This will inevitably lead to the demise of many farm enterprises as farmers and their resources (capital, labour and management skills) move out of agriculture. An alternative is to remain in agriculture but suffer considerable financial disadvantage. In some countries, e.g. France, farmers receive large subsidies.

The fact that farming is not just a business but a way of life complicates the adjustment process. Resource allocation and rural adjustment has not taken place at the same rate as technological change. The proportion of the workforce required in farming tends to fall as living standards rise. The pattern of demand is the key factor. The longer term solution is to encourage resources in agriculture (labour, management and capital) to transfer to other industries where rewards are higher. The net effect of this transfer would be that average income and value of resources in agriculture will rise. Eventually a parity will be reached.

Why, then, have successful Australian family farms been able to survive? Mainly because farmers have a remarkable capacity to restrict family expenditure and reap benefits during boom times. This is largely because they have followed an ongoing strategy in investment and asset growth. Over the years, investment in machinery, land, livestock and new technology has allowed farmers to achieve productivity gains and an increase in overall farm equity and thus consolidate their financial positions. This consolidation process has allowed producers to enhance their production base and improve equity levels.

During the 1980s the key reason for farm business failure was lack of productivity due to low output, poor prices and rapidly escalating costs. Other factors included a decline in on-farm investment, land degradation and high levels of indebtedness. These factors will continue to affect the family farm in the 1990s and will lead to the inevitable consequence — a decline in farm profits and an inability to service debts. During the past decade, successful farmers have enhanced their productive base. This is an important ingredient for success and such actions will continue to be important into the 1990s and beyond. As productivity and profits grow, cash will be available to meet debt servicing, family needs, and investment in sustaining farming practices and technologies. In addition, the agricultural industries will need to adopt a marketing approach to business and shift away from the traditional production orientation to business management. This change in emphasis will require producers to meet a specific demand from consumers. In other words, if customers want assured quality and continuity of supply, e.g. wheat grain which meets a specification for high protein, the Australian farmer must supply it to retain market share. Clearly, to achieve these new objectives will require the development of new farming systems.

Need for new systems of dryland farming

Pressure for new systems will come from two directions:
- Market pressures to produce the products which the consumers want.
- Pressures from society at large for agriculture to be environmentally responsible and sustainable.

To be sustainable farming systems should minimize the use of external inputs and maximize internal inputs, but the inescapable fact is that if mineral elements are removed in farm produce, they must be replaced. Scientists and agricultural technologists are charged with improving agriculture by designing new farming systems, selecting the most effective from an array of existing and potential alternatives, and anticipating problems that may arise as agriculture changes. Agroecosystem research can help to design locally adapted farming systems that are reasonably high in yields, low in risks, and sustainable under changing conditions. Whether there is any value in designing farming systems at a level beyond that of an individual farm is debatable. However, agriculturalists are being challenged to give deep thought to the design of more sustainable systems. The need to make more efficient use of limited land and water resources and the problems arising from excessive use of agricultural chemicals make for a daunting task. Added to this is the need to allow for climatic change.

These forces, as well as social, economic and political pressures, may increase the need for agriculturalists to design better farming systems sooner. These new systems, in association with appropriate technology, would provide farmers with a basis for more price-competitive and sustainable systems on their own farms.

In considering the design of systems, there may be two possible approaches: modify an existing system or design a new system. Farming and especially farm systems should be goal-oriented, and several general goals have been mentioned in this book. These are summarized in Table 21.1. There may also be some short-term or medium-term goals such as the need to reduce the incidence of soil-borne disease organisms. Priorities such as taking an opportunistic approach to short-term market trends or maximizing cash returns must also be taken into account. These short-term prior-

Table 21.1 *General goals of farming and farm systems.*

Goal	Definition
High productivity	Close to potential as determined by climate.
Profitability	Must provide a good living and capital for reinvestment after costs of inputs are deducted.
Flexibility	The composition of the system (eg crop-livestock composition, crop-pasture sequences) must be able to be changed without serious disruption to the system.
Stability	Buffered against effects of fluctuations in climate, biological-hazards and market prices.
Viability	Must be able to survive as a business, especially in the immediate future.
Adoptability	Adoption of the system must be within the capability of relevant farmers (in terms of requirements of physical and management skills and infrastructure).
Sustainability	Must maintain the physical, biological and economic resource bases over the long term.

ities may not change longer-term goals or the type of system selected, but they may influence the direction and rate of progress towards the adoption of the system.

Agricultural research will probably yield many new technologies for expanding food production while preserving land, water and genetic diversity. The real trick will be getting farmers to use them. The odds will be greater if the systems are perceived to be relevant at the level of the individual farm.

Farming system designs have three major parts:
• The crop and livestock species to be used as components of the farming system.
• The arrangements of the components in space and time.
• The quantity and nature of the inputs and outputs.

Who designs the new systems?

Given that we need to devise new systems, the question arises as to who will design them, or at least participate in the process. In the past, farmers, advisers, economists and scientists have been obvious participants in the process of developing new farming systems but, of course, a much wider array of participants may be involved (Table 21.2).

As discussed throughout this book, farming systems comprise key components and relationships, all of which must be present and operating satisfactorily for system goals to be achieved. A century or so ago the major features of Australian dryland farming systems were:
• The availability of virgin land.
• Good overseas markets for farm produce.
• Cheap transport for wool, meat and wheat.
• Appropriate farm machinery, such as stump-jump ploughs and harvesters.
• Inputs, such as superphosphate and improved wheat varieties.

Nowadays, opening up of new agricultural land has ceased to be an option but the list is otherwise similar, if more extensive. Both the opportunities and constraints on farmers have increased. Maintaining the productivity of the existing farmlands has become of the utmost importance. The implications of this fact of life must be faced by all who take part in the planning of new farming systems. It must become part not only of government policy but also be reflected in a land ethic adopted by society as a whole. Then all who participate in designing or developing systems will share a major goal and the farmer will have the necessary support at all levels to achieve sustainability.

This emphasis on conservation of resources for use in perpetuity still requires some adjustment in the thinking of Australians who are used to the idea of plentiful land, despite the fact that the limits of settlement were reached about 100 years ago. Although considerable progress has been made in such areas as conservation farming and integrated pest management it seems likely that further radical changes to farming systems design and operation will be necessary.

Major pressures will come from the environmental lobby, the consumer whose desire for safe food will come to predominate, the overseas customer who will demand higher quality produce, the animal welfare lobby who will be seeking more humane methods of farming and the financial market which will be seeking a better return on investments. The many participants in this process of change will include: the consumers, farmers, farm advisers and consultants, the economists, agricultural scientists, plant and animal breeders, sociologists and the agribusiness community (Table 21.2). Whatever the mix of participants it is

Table 21.2 *Agents of change in devising new farming systems.*

1 *The consumer*	Sometimes dominated by government policies and goals of multinational corporations, but often instrumental in influencing farming systems by demanding such things as low prices for essential foodstuffs and organically produced foodstuffs. Ironically enough, it was the consumer who, through seeking cosmetically perfect produce, led to the increase in pesticide use. Attitudes can be changed.	results are incorporated into simulation and other models (Chapter 18). • Management Research. This is needed to define and overcome limitations to the successful operation of systems. Management constraints imposed by soil and climate also need to be understood in designing new systems.
2 *The farmer*	The user and often the originator and/or refiner of the system. Thus should be consulted by all others who participate in the design of systems.	• Development Research. This is an area where the researches of many participants (including the farmer) overlap. It is inter- or trans-disciplinary and oriented towards the integration of the human, biological, technical,
3 *Farm advisers and consultants*	These people circulate information and ideas among clients and help to translate them into profitable and efficient practice. In systems design, their special task is to help to fit particular components into a whole farm plan, using both research and farm experience.	economic and environmental aspects of system design. It is also involved in the verification of the utility of systems. However, it will be location-specific unless accompanied by processes of formulating concepts and generalizations which can be applied in other situations (Chapter 20).
4 *Farm management economists*	They are a hybrid of accounting, agriculture and economics who draw on physical, biological and social sciences and consider farm organisation in terms of efficiency and continuing profitability. At the farming system level, they must relate the goals, opportunities and constraints of farming system design to market forces and other aspects of the economy, thus often modifying the system. At the farm level, their task is to select optimum land use strategies and management practices.	
5 *Plant, animal and soil scientists*	The scientist may work in three modes: • Process Based Research. This is needed so that, in the designing of new systems, the consequences of changing components, inputs or management can be predicted. It can be more effective if the	6 *Plant and animal breeders and geneticists* — These are separated out from other scientists because of their essential role in systems development. Viable systems can be developed only when environmentally adapted and disease/pest resistant genotypes are available (Chapters 6 and 9). 7 *Social scientists* — These have a role in understanding the human factors involved in systems design and are complementary to biological and soil scientists and economists. 8 *Agribusiness* — As providers of inputs and services (machinery, fertilizer and chemical manufacturers, seed merchants, etc.) their services often provide the means or impetus for the improvement of existing systems and they will (inevitably) be involved in the design of new systems.

clear that any new farming system will need to fulfil the following criteria:
- Technical feasibility.
- Economic viability.
- Social acceptability.
- Ecological sustainability.

The challenge is to put the farming family (the consumer of the new approach) at the centre of the proposed new system. The criteria used in developing improved strategies should reflect the needs of farming families, provided that these are compatible with the needs of society, e.g. provided there is no increase in rates of soil loss, levels of pesticide residue etc.

Strategies developed need to ensure convergence between the rather short-term private interests of farmers and those of society in the long term. Designing improved systems may involve incremental or single component changes, or incorporate packages of practices. The major advantage of packages is the complementary or synergistic effect between the various components. Disadvantages of packages are the complexities of putting them together and the likelihood of them being partly inappropriate for some farmers. Packages are effective when components are assembled from an understanding of the farming system into which they will be introduced. Farmers will be influenced by two factors. Firstly, whether they are able to adopt the strategies. This raises questions about whether the required inputs can be obtained, whether the resource levels implied are within reach of the farmer, and whether the changes will be socially acceptable to the community. Secondly, whether they are willing to adopt the strategies. Will they be better able to meet their goals by using the improved system? The criteria on which farmers decide their willingness to adopt are often obscure. The challenge for the future is to identify a focus to give best leverage to the improvement of the system as a whole.

Further reading

Campbell, K. O. (1980). *Australian Agriculture: reconciling change and tradition*, Longman Cheshire, Melbourne.

Crosson, P. R. and Rosenberg, N. J. (1989). 'Strategies for agriculture', *Scientific American*, September 1989.

Plucknett, D. L. and Smith, N. J. H. (1986). 'Sustaining agricultural yields', *Bioscience* 36, (1), 40–5.

Reeve, I. J., Lees, J. W. and Hammond, K. (1990). *Meeting the Challenge — a future perspective on agricultural education and agricultural educators*, Rural Unit, University of New England, Armidale, NSW.

Reganold, J. P., Papendick, R. I. and Parr, J. F. (1990). 'Sustainable agriculture', *Scientific American*, June 1990.

Remenyi, J. V. (1985). 'Agricultural systems research for developing countries', international workshop held at Hawkesbury Agricultural College, Richmond, NSW, 12–15 May 1985, *ACIAR Proc.* no. 11, 189pp.

Squires, V. R. (1990). 'Major issues in land management in Australia with particular reference to the role of educators', *Proc. of the Ecological Society of Australia*, 16, 419–25.

V

Case Studies

Dryland Farming on the Darling Downs in Queensland

J. Gaffney

This case study outlines the farming system presently used by brothers-in-law Wayne Newton and Glenn Pumpa on their farm on the Darling Downs and analyses factors affecting crop performance.

Farm location and environment

The farm is near Dalby, about 170 km west of Brisbane, where the annual rainfall is around 650 mm per year, most of it occurring as storms in summer. The summer days are hot and dry while the winter ones are mild and sunny. Late frosts at the end of winter and hail in summer are a fact of life. Variability in the climate around this average seasonal picture is very high.

The farm comprises two blocks, 'Greenvale' and 'Weroona'. 'Greenvale' is 260 ha of gently sloping, black, treeless plain cultivation and 80 ha of cleared and cultivated poplar box country. 'Weroona' is a very flat brigalow scrub soil block with 360 ha of cultivation. All the soils are deep cracking clays. Relative to the black soils the box soils are of lower productivity and are difficult to manage. The brigalow scrub soils have a very fine friable top soil that make them a joy to farm. They rank behind the plain soils but ahead of the box soils in productivity (Table A1).

Table A1

	Black Soil	Brigalow Soil	Box Soil
wheat t/ha	3.5	3.5	2.5
sorghum t/ha	4.5	3.5	3.5

Both blocks are subject to some overland flooding and waterlogging; 'Weroona' more so because of its pronounced flatness. A feature of this farm is 50 km separating the two blocks: 'Greenvale' is 20 km east of Dalby, while 'Weroona' is 30 km south-west of the town.

Farmers' goals and present measures for their attainment

The goals are the usual ones of survival, profitability, sustainability and lifestyle. The measures for their attainment are:

- An overall asset structure of $1.4 m comprising:
  land and building   1 010 000
  machinery   320 000
  working capital   80 000
- Farming system enterprises. Summer crops — sorghum, mungbeans. Winter crops — wheat, barley, chickpeas.

The following seven-year rotation indicates the overall cropping pattern: /–wheat /–wheat /–barley / mungbean– / sorghum– / sorghum– / sorghum chickpeas– / (the / separates the years, the – signifies a six months fallow).

In switching from winter to summer and back again roughly 50 per cent of the ground is long fallowed, the other 50 per cent is double cropped.

Mungbeans and chickpeas typically fill the double crop roles (as in the above examples).

Types of machinery

Reduced tillage farming is used throughout. To do this one tractor, cultivator and airseeder are permanently assigned to each block, while a slasher, blade plough, and a very high capacity spraying unit service the whole farm. As well, there is the usual equipment to harvest, handle and store grain.

Operations and management strategies

Most crops are grown using six-monthly fallows. This leads to an overall annual cropping rate of one crop per hectare of cultivation per year. The balance of summer and winter crops grown each year is roughly 50/50. This balance can be adjusted to a 35/65 within eighteen months should market prospects warrant it. The farming system is the same for all soils. Atrazine and Roundup are used in sorghum for weed control. Because of the need for flexibility Glean is used on no more than 20 per cent of the winter cereal ground. Roundup (sometimes Tillmaster) at higher rates is used on the rest. A high fertilizing policy is used; around $65 per hectare with sorghum, wheat and barley. The crops are grown in 120 metre wide strips. This equals twelve passes per strip when planting, and six passes when spraying.

Evaluation of technical and economic performance

The evaluation of technical and economic performance is summarized in Table A2. These figures (before-tax return to the owners' labour and management, and to the total invested capital) represent the expected result in a reasonable year. Singular, extremely unfavourable events, e.g. hail, heatwaves and wet weather can occur, easily reducing yields by the equivalent of 20 per cent across board and taking $60 000 off the expected income. The order of productivity presently being achieved is indicated in Table A1.

Analysis of factors affecting crop performance

Species and varieties

In a scarce moisture environment, the crops selected and the particular varieties planted are those that will hang on and yield in a dry time. Standability is also very important in sorghum; lodged sorghum is lost sorghum.

Crop sequence

The 50/50 balance of summer and winter crops, and the average of one crop per year, because of

Table A2 *The order of annual costs and returns for the whole farm are:*

Crop	Area	Av. Yield	Price	Total	
	ha	t/ha	$/t	$	$
Wheat	200	3.40	135	91 800	
Barley	100	3.20	120	38 400	
Mungbean	50	0.80	400	16 000	
Sorghum	300	3.80	120	136 800	
Chickpea	50	1.70	280	23 800	306 800
Variable costs	700 @ $130/ha			89 180	
Administration				32 000*	
Labour				8 000	
Depreciation				48 000	179 000
Operating Profit					127 800

* General overheads (rates etc.) $15 000, property maintenance and general running-around costs $12 000, provision for replanting and top dressing with zinc and gypsum $5000.

moisture constraints, reduces risk. This allows a smaller complement of machinery to be used than if the whole farm were devoted to one season's crops. The three and a half years of summer crop followed by the same period in winter crop also obviates, to some extent, the need for expensive weed and disease control measures if a monoculture is pursued. Second and third wheat crops are especially susceptible to the root lesion nematode, so three wheat crops in succession is not a viable option. Three or four sorghum crops in a row are not a problem. Mungbeans in summer, and chickpeas in winter not only counter balance the preponderance of cereal crops in the system but are relatively profitable and convenient change-over crops.

Tillage practices

Reduced tillage has enhanced the availability of soil moisture through better topsoil moisture management. Furthermore, higher levels of nitrogen fertilizer (50–100 kg/ha/yr) and better varieties, mean that relative to a decade ago the farm now has the capacity to produce almost twice as much grain per year as it could under the traditional, conventional tillage, winter-crop dominant farming system. The main gains have been through enhanced reliability of sorghum cropping on the brigalow and box soils.

Overall evaluation of the system in terms of aims

Over time the farmers' aims have remained stable. The means of achieving them have changed from conventional tillage, low input winter cropping to a reduced tillage, high input system of summer and winter cropping. However, even with reduced tillage, dryland farming on clay soils in an area that can experience continuous wet as well as dry weather remains tricky. Continuous wet weather with slow drying clays means getting on the ground to plant is always a problem. Reduced tillage only makes matters worse. On the other hand, hail can occur throughout the summer so shifting the balance of crops in favour of summer crops does nothing to reduce this risk. Finally, the threat of irrational community attitudes toward reduced tillage chemicals is a concern.

On balance, while reduced tillage tackles the major problem facing dryland farmers on the Downs, others remain. The farmers are definitely better off. By how much they are not sure.

Sustainable Dryland Farming Systems for the South-Western Slopes of NSW

P. Cregan

This case study examines farming practices in south-western NSW, discusses the concept of production potential and considers the key limitations to achieving it.

Before a sustainable farming system can be developed and implemented, a farmer must answer the question: what level of production do I wish to sustain? If the farmer wishes to produce at a low level (say 50 per cent potential) the level of management and technology required is much less demanding than if the farmer wishes to operate close to the biological potential.

The concept of production potential also needs to be defined. Plant and animal production are most commonly limited on the south-west slopes of NSW by available water, and it is sensible to describe potential production in terms of water-limited yield. This is particularly relevant for this environment since biological yield potential is linearly related to available water and thus input:output relations favour an economic optimum that is close to the biological optimum. Furthermore, for farming to be truly sustainable at a near potential level of production, farming practices must be directed to achieving minimum critical levels of soil chemical, physical and biological characteristics as well as plant and animal management practices. Sustainability comprises both economic and ecological concepts based on optimizing yield without detriment to the environment. For it to be achieved, biological and management requirements must first be satisfied; otherwise production will decline in time and sustainability will not be achieved.

Farm location, environment and practices

The Hart Brothers' farm at Junee Reefs, 50 km north of Wagga, is typical of much of the country on the south-west slopes. The family-owned farm of c. 1000 ha has a mix of enterprises which results in most of the farm being evenly divided between grazing (a medium wool self-replacing Merino flock) and half being cropped (wheat, grain-only oats, canola and lupins). The average rainfall for the area of 550 mm has a slightly winter dominant distribution and results in an 'average season' potential wheat yield of 4–4.5 t/ha.

The red earth soils are naturally slightly acid with a surface pH_{ca} as low as 4.2. The pastures grown by the Hart Brothers, Adrian and Bernard, are exclusively based on lucerne and subterranean clover. This pasture mix, which is rotationally grassed on a four paddock system, supports a stocking rate of about twice the district average (7.5 DSE/ha) as well as providing the opportunity in late spring/early summer for haymaking.

Like many farmers in this region, Bernard and Adrian have changed their practices significantly over the past twenty years. These changes are reflected in the farm machinery used by these farmers who are striving for high production that is sustainable. Gone are disc ploughs, long fallows and repeated cultivation. Today the equipment of the successful farmer centres round an efficient boom-spray, a scarifier, wideline harrows (used for breaking down and dispersing stubble as well as for tillage) and a rigid-tine combine or an air-seeder. Increasingly farmers using direct drilling and/or stubble retention techniques are changing to combines with six rather than four rows of cultivating/sowing tines as this configuration gives better trash clearance. Most farms also have a header but often contractors are employed to harvest 'specialty' crops like canola.

Farmers' goals

The challenge for farmers is to integrate their enterprises and practices to achieve their production goal and then to be able to sustain that goal through time. Innovative farmers on the southwest slopes like the Harts have been able to develop farming systems which approach this ideal. Table B1 is a summary of the major limitations to productive farming and how local farmers have overcome these problems.

Farmer-developed sustainable farming systems

Below are details of a successful 7–9 year farm rotation developed by farmers. Other rotations used successfully within the region are shown in Table B2.

Rotation

Year 1 Pasture: Lucerne/subclover — 10 to 20 per cent greater animal production than from annual pasture alone. Rotationally grazed.

Year 2 Pasture: Nitrogen accretion under pasture 70 to 100 kg N/ha/yr. Spray grazed to reduce broadleaf weeds.

Year 3 Pasture: Winter cleaned with paraquat[R] to remove annual grasses, alternate hosts to cereal crown and root diseases. This is particularly important prior to the use of lime and the growing of wheat as the raised pH favours the development of take-all. Heavily set-stocked late spring–early summer then killed with herbicide to start crop fallow.

Year 4 Canola: Lime application to raise pH to 5.5 — sufficient to balance addition of acid during the rotation. Minimum tillage seedbed preparation to incorporate lime, kill residual lucerne plants and prepare a fine, even seedbed for the small seeded canola. Can use heavy rates of grass control herbicides if grass weeds a problem.
All crops fertilized to balance P removal from potential yield level plus an allowance of an extra 5 kg P/yr for the pasture phase.
Canola can use efficiently the high nitrogen level built up under the pasture. It provides an effective, additional break to cereal crown/root diseases and its root system can enhance soil macrostructure.

Year 5 Wheat: Fallow management starts with stubble being heavily grazed with dry sheep. Sheep supplemented with lupins in late summer to increase utilization of stubble and to avoid excessive weight loss. Crop direct drilled. A winter wheat is sown early, using a starter fertilizer to encourage better seedling growth by minimizing nitrogen deficiency. Crop may need to be top-dressed with N, depending on leaf tissue N. Broadleaf weeds can be effectively controlled if necessary.

Year 6 Lupins: Fallow management as for wheat sowing but mechanical assistance may be required if wheat crop heavy, e.g. use stubble buster, slasher, etc. Crop direct drilled. An effective disease break that can improve soil structure and has no nitrogen fertilizer requirement. Another opportunity if necessary for grass weed control with herbicides.

Year 7 Wheat or barley or linseed undersown with pasture.

The use of lucerne confers major production advantages to those farmers who can successfully establish and then manage it within their farming system. Increases in profitability of up to 20 per cent following the across-farm use of lucerne have been recorded.

Depending on the relative prices of wool, meat and grain, the ratio of pasture:crop can be varied but consideration should always be given to the sequencing of the components so that the potential complementary relationship between them can be maximized.

When evaluating a rotation it is necessary to evaluate the production from the whole rotation rather than its individual parts. Evaluation of the components in isolation will not validly apportion production because of the interaction between these components.

Table B.2 *Some rotation used by farmers on the south-west slopes capable of achieving potential yields.*

| | Year | | | | | | | |
	1	2	3	4	5	6	7	8
crop/pasture sequence	P	P	P	C	W^3	L	W	B/P
	P	P	P^1	W	L	W/P		
	P	P	C	W	L	W/P		
	P^2	P	P	P	C	W	L	W
	P	P	P	O	W	L	W	P^2
	W^4	L	W	L	W	L	W	L

P = pasture
W = wheat
B = barley
B/P = barley undersown with pasture
C = canola

NOTES:
1 Pasture winter cleaned (grass control by herbicides).
2 Pasture sown separately, not undersown.
3 Field peas are sometimes substituted for lupins.
4 The lime requirement for this system may be higher than for other rotations.
5 Barley and triticale can be substituted for wheat. Substitution usually depends on commodity prices.

Table B.1 *Management used by farmers on the south-west slopes of NSW to overcome limitations restricting the achievement of potential yield.*

Limitation	Management Practices
Available soil water	Modify tillage to: increase infiltration, reduce loss during cultivation and reduce loss from fallows. Maintain ground cover to reduce runoff. Control weeds.
High soil strength	Deep tillage and/or use of deep rooting species to establish macropores.
Soil acidity	Lime to ensure minimum soil profile pH >5.2. Use perennial pastures to reduce acidification rate.
Soil fertility	Use legume rotations and fertiliser management programmes that supply sufficient nutrient for predetermined production goal.
Plant disease	Control with rotations/herbicidal control of alternate hosts, resistant varieties, and pesticides when threshold level reached.
Soil salinity	Increased water use by higher yielding crops and pastures. Use of trees to reduce additions to the water table.
Insect pests	Use of rotations and cultural manipulations, resistant varieties and pesticide use when threshold levels reached.
Weed competition	Herbicides and crop/pasture rotation. Pasture species manipulation with grazing and/or herbicides. Tillage.
Plant species	Use of perennial pastures and trees to increase water use. Selection of more persistent varieties and species. Sow crops that can be sown earlier to increase crop water use, improve flexibility of sowing time, and possibly improve yield potential.
Stock type	Select stock with greater efficiency of pasture use and minimum care requirements.
Stock management	Grazing management strategies to increase forage utilization and minimize health problems.

Putting Farm Systems to Work in the Mallee District of Victoria

J. B. Griffiths

T his case study is about a dryland cereal-livestock farm in the Victorian Mallee. A decline in productivity and a desire to retain family ownership have forced the farmer to examine options for future actions. This study discusses his decisions, encouraged by the results of recent crop and pasture research.

Farm location and environment

The farm, which is 1500 ha, is in the Mallee Region of Western Victoria. Annual rainfall is 330 mm, of which 230 mm falls as effective rainfall between April and September. Summer rainfall is of little value except for perennial pastures or germinating summer-growing weeds. Seventy per cent of the farm comprises sandy loam soils while the remainder are light sand rises. Under sound management the sandy loams have good fertility while the rises are mainly lower in fertility.

Farmer's objectives

The overall aims of the farmer are to ensure long-term ownership of the farm for himself, his wife and one son. This involves changes to the size or structure and management of the farm and he is currently involved in the process of change.

Evaluation of the farm enterprise

Fallowing in July, in preparation for cropping the following May has been an integral part of the farming system in the Victorian Mallee since it was opened up for farming. Consequently, a three-course rotation has been operated on the case-study farm for a number of years. The rotation has been: medic-based pasture, fallow, wheat or barley.

The farm carries 600 Merino ewes which are crossed with Poll Dorset rams in November for prime lamb production. The ewes lamb in April and the lambs are sold for slaughter in July, August and September, depending on seasonal conditions. The farmer has been able to market up to 110 per cent of prime lambs into the early spring market, but the figure is often closer to 95 per cent.

Cropping and livestock production have been integrated, however, less attention has generally been given to the sheep than to the cropping programme. Livestock have utilized the medic-based pastures and cereal stubbles as a source of feed. Supplementary fodder has been provided from hay made in years of heavy pasture production and from grain held on the farm.

Significant problems are arising with the present system. These are:

- Average yields over the farm have declined from 1.8 t/ha to 1.1 t/ha.
- Wheat quality, as measured by grain protein content, has fallen consistently over time.
- Cereal diseases, particularly root diseases, are being recorded in cropping paddocks almost every year. The major diseases are Take-all (*Gauemannomyces tritici*) and Cereal Cyst Nematode (CCN) (*Heterodera avenae*).
- Pastures are largely dominated by grass species, namely Brome grass (*Bromus diandrus*) and Barley grass (*Hordeum leporinum*).

The need for change

A critical review of the farm activities has shown that, at the present rate of decline in productivity, the income from the farm will be unable to sustain a livelihood. Significant changes must be made.

There are several options open to deal with this problem. The main ones are:

1 Buy a further 300 ha, or multiples of this amount, from neighbouring farmers as land becomes available. This would require borrowing more than $500 000 with a minimum of an additional $200 000 to replace capital equipment to farm the additional land. The existing workforce available to the farm can handle the extra land and workload. However, land may not be available in the immediate future.

2 Diversify the range of crops and inputs for livestock production on the present area of land. There is a much wider range of crops available and pasture production can be improved, with more attention to management.

There are greater risks involved in growing crops with which the available labour is not familiar, but the productivity of the available land can be much improved. There will need to be a development of the skills required to manage such crops as grain legumes and oilseeds and to give full attention to managing pastures resown to improved annual medic species.

There will be no need to buy new machinery in the early stages of the development, but modifications to headers, purchase of grain handling equipment and upgrading existing spray equipment will require about $150 000.

As the farm improves through each phase of the rotation, and as the fallow is replaced by an extra pasture each year or a grain legume or oilseed crop, there will be a need to finance these operations. This may require about $80/ha for additional cropped areas and up to $50/ha over two years for pasture improvement.

Modifications of the present cereal/sheep-based system, to include grain legumes, oilseeds and improved pastures, will place demands on staff for time and effort. This time may be provided by existing staff, or substituted by the use of contractors for a range of specialist jobs such as spraying for weeds, crutching sheep, harvesting grain legumes and contract carting of grain.

3 The third significant option is that the farm could be sold because the prospect of farming in a more complex and significantly different way is not acceptable.

Both options 1 and 3 are rejected as being too expensive and unmanageable in the long term, or too drastic a change for the farm family, if the alternatives offer a viable solution. The advantages of option 2 have been widely reported in the local press and at field days attended by the farmer. He considers that there is a significant improvement to be made in some paddocks by giving greater attention to management and substituting a wider range of crops and pastures for the fallow phase. He wishes to avoid excessive use of agricultural chemicals, and any use of nematicides to control CCN. Department of Agriculture experimental results (Table C1) show that rotations which include a two-year break, free from susceptible cereal crops or grass weed hosts, have dramatic effects on yields. If the rotation includes grain legumes or medic pastures then grain protein is also improved. Experimental results show that Take-all can also be controlled very successfully by the adoption of non-host species in the rotation for one or two years. However, the farmer realizes that it is not possible to switch immediately from his traditional rotation to a three- or four-course rotation containing no fallow, without causing management difficulties at peak work times. The change in rotation needs to occur gradually, and the paddocks at greatest risk of disease should receive attention first.

Wheat yields in the paddock selected have been about 60 per cent of the district average, and the grain protein has been less than 10 per cent. The medic pasture is dominated by grasses and broad-leaf weeds and CCN has produced a SIRONEM rating of 4 (i.e. badly infested). The paddock will be slashed after harvest, grazed heavily with sheep, and dry sown in April, using a prickle chain, broadcast distributor with 8 kg of medic/ha. After medic emergence on the opening rains, red-legged earthmite will be sprayed, as required. Grazing pressure will be increased as the grasses and self-sown cereals become established but by mid-July they will be dominating the first-year pasture. Use of a grass specific herbicide at this time ensures that the CCN cysts cannot reach maturity and thus over-summer into the next crop. This treatment will also control Take-all. Grazing pressure on the medic pasture in year 1 will be adjusted to maximize seed set and survival. In year 2, grazing pressure will be maintained from the break of the season until well into the summer dry period or until there is adequate rainfall to commence land preparation for sowing. Crop establishment after pasture requires low inputs for working since there is no fallow, and relatively good soil conditions for sowing.

In another paddock a different rotation will be adopted. Galleon Barley will be sown into wheat stubble in year 1 and peas or lupins will be sown in year 2 before returning to wheat. The paddocks will thus be free of susceptible cereals for two years. The effect of such a rotation on SIRONEM ratings has been very impressive in experiments (Table C2) and the quality of grain was also much improved. Similar experiences are now being reported by farmers. On one southern Mallee farm, the SIRONEM rating was reduced from 4 to 1 after two years of medic pasture.

Improved pastures need to be efficiently utilized by livestock if overall benefits to the farm are to accrue. The Department of Agriculture had shown that under experimental conditions the performance of a ewe flock could be improved by more than 12 per cent if joining took place in February and cross-bred lambs were dropped in mid-winter. Management of sheep is much less demanding when there is an abundance of feed. Pregnant ewes or young lambs do not require heavy hand feeding to survive in this situation. Using this strategy with sale of lambs in October, the percentage of lambs sold at market can be increased to 130 per cent. Grain residues, cereal or grain legume stubbles and dry improved pastures provide an excellent over-summer fodder source for all classes of stock.

The farmer has been further encouraged to change his rotation practices by the estimates of gross margins (Table C3) for the various rotations and by the experiences of other farmers.

Without exception, those farmers controlling grasses in their pastures in year 1 are finding that there is very little grass in year 2 pasture, that they have no need to work their land in the spring of

Table C1* *Effect of rotation sequences on the wheat crop in the third year.*

Rotation Sequence		Wheat Yield (t/ha)
Year 1	Year 2	Year 3
Wheat	Wheat	0.64
Wheat	Medic pasture	2.68
Medic pasture	Medic pasture	3.81

*Annual Report, Mallee Research Station, Walpeup.

Table C2* *Effect of cropping sequence on Sironem rating and wheat yield.*

Rotation Sequence		Sironem Rating	Wheat Yield (t/ha)
Year 1	Year 2	Year 3	
Wheat	Wheat	4.0	0.76
Wheat	Fallow	2.0	1.41
Galleon[†]	Ryecorn	2.5	2.05
Galleon[††]	Peas	1.0	2.87

**pers. comm.* M. W. Ferguson, Mallee Research Station, Walpeup.
[†]A CCN-resistant barley cultivar.
[††]SIRONEM rating based on a test devised by CSIRO; see text.

Table C3 *Estimates of average gross margins for five rotations in the Mallee.*

Rotation	Average Gross Margins $/ha
Wheat, lupin, wheat, pea	167
Wheat, grass-free medic pasture	129
Wheat, barley, grass-free medic pasture, lupin	110
Wheat, medic pasture, fallow	85

year 2 in preparation for sowing in year 3, and that total cultivation for wheat production is reduced 50–80 per cent. Most farmers are cultivating prior to sowing to minimize the effects of rhizoctonia. They have found that there are almost no weeds growing following the autumn break. Furthermore, there is an additional 2–4 months of grazing available because fallowing has been discontinued.

Thus, after careful consideration of the challenges posed by the cereal diseases and new management systems, those parts of the farm at greatest risk from disease will be introduced to new rotations and a strategic management plan undertaken for overall farm improvement.

Dryland Farming in the Yorke Peninsula of South Australia

T. Dillon

This case study describes the farming business of Rod and Margie Davies on their farm 'Rolleston' on the Yorke Peninsula and analyses factors contributing to their success.

Farm location and environment

'Rolleston' is a 491 ha farm in a 380 mm annual rainfall area, 16 km south of Kadina, on northern Yorke Peninsula. The property is situated 25 km from a grain port and fertilizer works at Wallaroo. It is slightly undulating and totally cleared of the original mallee scrub. The soil types range from clay loams, mallee loams to stony mallee. All these soils are limestone based and have a pH of 8.0 to 8.7 at the surface with a lime subsoil. They suit the growing of medic but not subterranean clover.

Rod and Margie have two sons, one studying at the Roseworthy campus of the University of Adelaide, and the other at a college in Adelaide. The farm has been in the family for four generations and began from a lease of 48 ha in the 1880s.

Labour costs need to be kept low on a farm of this size. Rod manages the farm without permanent labour. He employs shearers and crutchers and has most of the grain, at harvest, carted by contract. Casual help is used occasionally for 'two-person' jobs and during Rod's absences from the farm.

Farming system enterprises

A sustainable system in this environment ideally includes 30–50 per cent pasture and 50–70 per cent crop. The soils are high in lime so structure is not a serious problem but higher soil organic matter levels will allow plants better access to soil minerals and moisture. High levels of organic matter are best maintained by legume dominant pastures. The livestock enterprise associated with these pastures also gives a more stable income and allows a better spread of labour requirements. This farm comprises livestock and cropping enterprises. Crops sown are 180 ha of wheat, 102 ha of barley and 36 ha of field peas. The remaining 170 ha is barrel medic pasture. The livestock enterprise consists of two flocks of Merino ewes. One flock

Table D1 *Mean monthly rainfall (mm) at Kadina, SA.*

Jan	Feb	Mar	Apr	May	Jun	Jul	Aug	Sep	Oct	Nov	Dec	Total
15	19	20	36	49	57	48	45	37	32	22	18	392

breeds replacement Merino ewe hoggets and wethers (sold as lambs). The other ewe flock is mated with Dorset rams for prime lamb production. Ram lambs have been sold entire for export. 40 per cent of ewe hoggets are culled and join the prime lamb flock. 60 per cent join the Merino mated flock thus ensuring a nucleus of high producing ewes. Wethers are bought in prior to harvest and used when excess stubble feed is available, then sold before the 'break' of season for extra profit.

Machinery

Machinery costs, like labour, must be monitored carefully and kept as low as practical. The machinery must be reliable. The cropping machinery is adequate and in good condition. The tractor is updated regularly as the key piece of machinery. It is used in most operations on the farm including harvesting and by updating regularly, reliability and comfort are maintained at a high level which is essential for a one-person operation.

A second-hand seed drill with improved trash clearance has recently been bought to improve the structure and moisture retaining ability of the soil by leaving more stubble residues during the cropping phase.

The farmer also owns a stonepicker, stone roller, mister and bale stacker in partnership with other farmers. This machinery is of a type that is used for a few weeks each year in off-peak periods.

Total market value of the machinery is $90 000, giving a machinery overhead of $280/cropped hectare or $120 per tonne of grain produced.

Operation and management strategies

Rod keeps a tight reign on the machinery overheads. He uses contractors or hires machinery for many tasks on the farm, e.g. contract sheep and grain carting obviate the necessity for an expensive truck—an elderly Bedford tipper is adequate to cart superphosphate, sow the crop and still carry the odd load of grain to the silo during busy times. These journeys amount to only about 4000 km per year. A baling contractor is used to roll up excess pasture growth and the hay cuts around crops. The pea crop is also harvested by contractor, allowing Rod to continue reaping with a medium-sized PTO header which is not subjected to the wear and tear of pea harvesting.

The basic rotation is one year of medic pasture followed by one or two years of cereal depending on fertility of the paddock. The second cereal crop has nitrogen fertilizer added at seeding. The peas are sown where medic seed reserves are low. These paddocks are resown or undersown with medic later. Medic pasture is the key to the cropping rotation. It provides much of the nitrogen, maintains the soil structure and provides an opportunity to control grasses. Removing these grasses has reduced cereal root disease. Broad-leaf weeds are controlled well in the two years of cereal and the grasses in the legume year.

The control of broad-leaf weeds in the two cereal crops gives low weed density in pastures, and it is seldom necessary to spray for brassica weeds in the pea crops or pasture. This, in conjunction with grass control in the pastures, gives maximum medic growth thus maximizing nitrogen fixation. Seed set is maximized and reserves carry over to regenerate after two years cropping. Rod believes the effects of increased fertility from improved pastures and further lowering of disease levels (Fusilade has only been used for three years on pastures) should improve crop yields further in the future.

Dense medic pastures and July lambing maintain high sheep numbers which allows effective utilization of crop residues in summer–autumn. In years where the 'break' occurs after the end of April all sheep are lot fed until the young pastures can sustain the sheep (4 to 6 weeks in feed lot). Feed lotting has other benefits besides the protection of young pastures. These are: time savings in feeding the sheep in small paddocks, the feed supply is usually nearby, the sheep are more easily fed and checked, and there is a more efficient use of fodder that is fed out since the sheep are not using energy in moving around a large paddock chasing feed. Mobs do not have to be moved constantly around the farm to preserve young pastures.

Rod believes that present management will maintain soil stability and fertility. With the farming system in use now, the complementary rotation will ensure the nutrients are replaced as they are used. Soil and tissue testing is being used to check levels of phosphorus, nitrogen, carbon and trace elements. Soil structure is being maintained, if not improved, by incorporating most residues of pastures and some cereals. A conscious effort is made not to burn pasture residues to preserve organic matter and not destroy medic seed. Occasionally wheat stubble is burnt after the 'season' break for control of snails and some diseases.

Sheep complement the cropping rotation by cleaning up crop residues whilst returning a respectable profit in most seasons.

Evaluation of technical and economic performance

Increased yields have been achieved over the last ten years from 50 per cent of potential (Chapter 17) to 73 per cent of potential by reduced cereal root disease and the incorporation of modern technology. This rate of increase will be almost impossible to sustain so the purchase or leasing of more land will be necessary. Figure D1 shows it can be seen that a dramatic lift in efficiency has occurred over the last decade coinciding with the introduc-

tion of Galleon barley (eel-worm control) and improved management of pasture grasses with various sprays. Average yields are wheat 2.4 t/ha, barley 2.5 t/ha and peas 1.4 t/ha. These cereal yields represent approximately 70 per cent of theoretical potential (Chapter 17) for an April–October rainfall of 300 mm.

Average wheat yields have increased by 0.53 t/ha over the last ten years (Table D2). This is an increase of 28 per cent, but actually an increase of 44 per cent in terms of potential, due to the decrease in average growing season rainfall over the period. The results of these management strategies are expressed in Table D3 where gross margins are based on 1989–90 season costs and prices and average yields.

In an average season total costs including taxation, loan repayments, rates and taxes amount

Figure D1 *Rolling five-year averages of wheat as percentage of R. F. yields.*

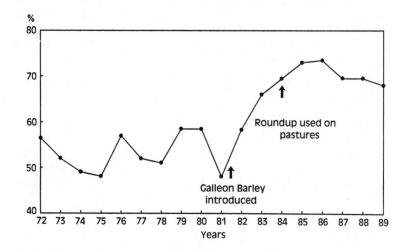

Table D2 *Change in wheat yields and percentage of yield potential between two ten year periods.*

	1970–9	1980–9	Difference −10 yrs	Result %
Average growing period rainfall	322 mm	296 mm	−26 mm	−8%
Wheat average yields	1.9 t/ha	2.4 t/ha	+.53 t/ha	+28%
% rainfall potential	45%	65%	+20%	+44%

Table D3 *Gross margins for crops and sheep, 1989–90, York Peninsula, SA.*

	Wheat	Barley	Peas	Sheep (5 DSE/pasture ha)
Gross Income/ha	450	312	308	192
Variable Costs/ha	95	90	155	70
Gross Margin $	355	222	153	122

to 60 per cent of the gross income. This increase in yield has compensated for depressed prices and rising costs during this last decade. The 'real' position has slipped and will ease further in the future without a dramatic increase in commodity prices.

The long-term viability of the family farm depends on outside influences such as grain prices, exchange rate and inflation. These will be critical factors in the 1990s. A general increase in the cash price of grain will be needed to compensate for increased costs of production and plant as well as to leave enough profit to expand and maintain a reasonable standard of living.

Philosophy for the future

The long-term aim of this family farm is to maintain it as an economic unit which will continue to be viable in the future without mining the resources. The owners realize that the farm could be cropped more frequently with cereals for more short-term profit but they have a concern that long-term overall fertility of the soil would be lowered, thus the productivity and profit would be jeopardized.

Soil structure and fertility and weed populations need to be managed to ensure long-term farming efficiency and productivity.

The owners intend to leave the farm in good heart for the next user. The farm is intended to be handed on to the next generation as a sustainable economic unit big enough to survive the troughs in the industry and to support future expansion. This will entail having the farm debt free with a modern efficient farming plant capable of handling this increased area, if the opportunity arises. They believe that off-farm investments, whether it be superannuation, separate investments or real estate, are a real part of their farming today. This is needed to fund expansion or enable a retiring member of the family to leave farming without threatening the budget with an extra home and income. Their aim is to have a mix of all three types to level out fluctuations in different sectors of the economy. They believe that they are only the caretakers of the land. It is their duty to leave the farm in a condition that is equal to, if not better than, the condition in which they received it.

Farming with a Whole Farm System Approach in the Eastern Wheatbelt of Western Australia

S. J. Trevenen and G. G. Fosbery

This study illustrates a systems approach to farming in the Merredin area of Western Australia using an average or typical farm analysed by the mathematical program MIDAS (Model of an Integrated Dryland Agricultural System).

Farm location and environment

Merredin is located approximately 250 km east of Perth and is in the heart of the Western Australian wheatbelt. The area around Merredin produces approximately 30 per cent of the State's wheat, which represents close to 10 per cent of the total Australian production. It is a mixed farming area, with approximately half the cleared area sown to crop in any year. Wheat is the major crop, followed by lupins, oats, barley, field peas and triticale. These crops have average yields of approximately 1 t/ha. Merino sheep for wool and live meat export are run on the remaining area.

Merredin has a winter-based rainfall pattern with 310 mm average annual rainfall, of which 210 mm falls during the growing season. The growing season opens in mid-May with pastures drying off at the end of September and the cereal harvest commencing in mid-November. Winter temperatures are mild but frosts are a regular occurrence from June to early August. There are irregular frosts during cereal flowering. Summer temperatures are hot and are regularly in the $30°C^+$ range. Grain moisture problems are rare at harvest.

The farm and farmer goals

The average farm at Merredin was determined to be 2300 arable hectares in size, spread over six soil classes or land management units. A description of each of these land management units and the arable area in each is shown in Table E1.

The property is owned and operated by one family unit, employing casual labour at seeding, harvest and for shearing. The plant includes a 120 kw tractor, a 7.5 m air seeder, and 5.5 m self-propelled harvester. All crop weed control spraying is carried out by the landholder. The property has normal bank borrowing limits and credit facilities from the stock firm.

The objective of the farmer is to achieve maximum sustainable profit. The MIDAS model shows how this can be done and allows the farmer to know how much profit he will forgo by pursuing other objectives.

Table E1 *Land management units on the average farm Merredin, Western Australia.*

Management units Classification	No.	Description	pH (water 0–10 cm)	Clay %	Area (ha)
Heavy	S6	Sandy clay loam	6.5+	25–35	575
Medium	S5	Sandy loam	6.0–7.0	20–25	345
Duplex	S4	Loamy sands and gravelly sands over clay	5.5–6.5	10–15	230
Gravelly sands	S3	Gravelly sands	5.5–6.0	10–15	230
Sandplain soils	S2	Deep loamy sands	5.5–6.0	10–15	460
Acid sands	S1	Loamy sands and gravelly sands — acid at depth	5.0–5.5	10–15	460
Total Arable Land					2300

Assessment of the previous, 'traditional' farming approach

During the 1970s the farms were cropped on a year-in/year-out wheat-pasture rotation with little regard to soil type. On the lighter soils, pastures were made up of annual grasses and broadleaf weed species with only a minor legume content. There were limited areas where subterranean clover dominated. On the heavy soils Cyprus barrel medic was the major pasture species. Redlegged earthmite and lucerne flea affect the early production of the pasture legume species. Crops were generally established using multiple cultivation for weed control although herbicides, particularly for annual ryegrass control, were gaining acceptance. Half the farm was in crop in any one year, almost exclusively to wheat, with some barley and oats grown for stock feed. Cereal root and leaf diseases are of minor significance in this environment. A self-replacing Merino sheep flock was run on the rest of the farm. During the late 1970s and early 1980s there was a swing into more intensive cropping rotations. This was a reaction to a run of years with low wool prices relative to wheat, coinciding with the introduction of improved herbicides and large cropping machinery. The sustainability of this approach to farming in the eastern wheatbelt was seriously in doubt as:

- Inappropriate economic analysis was being used to encourage farmers into more intensive cropping rotations. This was used out of context as the overall costs and returns of cropping and the complementarity of enterprises were not being adequately considered.
- There is a range of very different soil types on the farm, with highly variable crop and pasture production potential.
- Most soils are nitrogen responsive yet there were few legumes in the system.
- There was a reliance on multiple cultivation to establish the crop. Not only did this lead to soil structure decline and wind and water erosion but it also delayed the time of sowing of the crop. Early sowing, with good weed control, is the most important factor for a high crop yield in the eastern wheatbelt.

This led agricultural advisers and leading farmers to develop a whole farm approach to agriculture in the eastern wheatbelt. MIDAS was developed and used to analyse and promote the whole farm approach.

Whole farm analysis

Analysis by means of the MIDAS model involves dividing the farm into land management units based on soil types (of which there are several). This is important because each unit supports a different set of crop and pasture species and requires different rotations and other management practices to maximize economic profit. The analysis also uses a systems approach to the development of strategies which takes into account the positive and negative ecological interactions that occur in farming systems. For instance, values are estimated for the yield advantage gained by sowing a cereal crop after a good legume pasture and any negative effect of cropping on pasture regeneration and feed availability in the following year.

The six major land management units have very different cereal crop potentials and grain and pasture legume options. The most suitable species for

Table E2 *Most suitable species for each land management unit on the typical Merredin farm, Western Australia.*

Land management unit	Cereal	Best option Grain legume	Pasture legume
Heavy	Wheat* Barley	Field peas	Barrel/burr medic*
Medium	Wheat*	Field peas	Burr medic*
Duplex	Wheat*	Lupins	Sub. clover
Gravelly sands	Wheat*	Lupins	Sub. clover
Sandplain soils	Wheat*	Lupins	Sub. clover
Acid sands	Oats* Triticale*	Lupins	Natural

*Most suited option to that land management unit. Subterranean clover had been traditionally sown on light textured soils in the wheatbelt but will not persist in close cropping rotations.

each land management unit are shown in Table E2.

When the parameters for the typical Merredin farm were entered into the MIDAS model (see Chapter 15) it was found that profit maximization subject to biological sustainability was at 60 per cent of the cleared area in crop. However, there was only a relatively small change in net profit between 50 and 80 per cent in crop (Figure E1).

This was contrary to the gross margin analysis of individual enterprises which promoted 100 per cent cropping. These gross margins did not take into account

• The interdependencies between enterprises.
• The change in labour and plant requirements when increasing the farm cropping programme.
• The variation in yield potential between different land management units on the farm.

Evaluation of the farming system

The major finding of the farming systems analysis was the importance of farming each land management unit to its capacity. It also showed that the best rotations varied substantially between the land management units. Again, this was contrary to the traditional farming approach. From this whole farm approach, advisers developed the 'farming by soil type' concept which is being widely adopted by wheatbelt farmers in Western Australia.

The ideal rotations for each soil type to maximize farm profit subject to biological sustainability were designed for heavy, medium, duplex, gravelly sand, sand plain and acid sand.

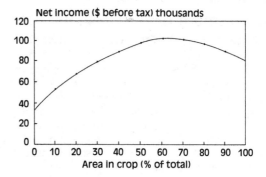

Figure E1 *Profit versus per cent of the farm in crop.*

Heavy soil type

This soil type has a high clay content and good waterholding capacity. In performs well in a wet year but very poorly in a dry season. It is subject to degradation by over-cultivation but has the potential to grow good medic pastures.

The whole farm economic analysis showed that for this soil type all options from continuous wheat through to continuous pasture were very close economically on the average farm. Farmers now tend to crop this soil type after a wet summer or early break of season, when yield prospects are likely to be above average. It is left as self-regenerating pasture in other years. The medic pasture is well adapted to this opportunity cropping. With its high hard-seed content, medic can successfully regenerate even after two or three successive crops.

Crop establishment is successfully achieved by a single shallow cultivation (2–4 cm) to stimulate a germination of weeds, particularly annual grasses, followed 7 to 14 days later by the seeding operation. Herbicides are also used in combination with the cultivation to kill weeds. Minimizing the tillage on this soil type improves soil structure, resulting in easier mechanical tillage and up to a ten-fold increase in the rate of water infiltration.

Medium soil type

This soil type has the most consistent production over a range of seasonal types. Both the burr medic legume pasture and field peas as a legume crop are well adapted to this soil type and cereals require only small additions of nitrogen fertilizer. It is also suited to a one-pass crop establishment system, with all weeds being controlled by the sensible use of herbicides in combination with the seeding cultivation. From the MIDAS analysis, this soil type is most suited to a continuous wheat rotation when wheat prices are high, relative to wool. Cereal diseases are of minor importance, but field peas can be used every three to four years to break disease cycles. With more favourable wool prices relative to wheat, a burr medic pasture-wheat rotation is most profitable.

Duplex and gravelly sands soil types

These two soil types perform similarly in the whole farm analysis. Wheat performs well on this soil type but lupins are less adapted and have a low yield potential. Subterranean clover is the most adapted pasture legume, but does not persist in intensive cropping rotations. The whole farm analysis showed that a two wheat/one lupin rotation was the most profitable option. It was replaced by a two subterranean clover/one wheat rotation only when lupin performance dropped by 30 per cent below the expected performance. Farmers are increasing the amount of lupins sown on these soil types, but avoid lupins in years when the season opens late and low yields are more likely.

Sandplain soils

This deep loamy sand soil type is ideally suited to lupin production. The lupin/wheat rotation is very profitable on this soil type and is stable over a wide variation in yield and prices. As lupins are high in protein, they are rapidly replacing oats as the major stockfeed grain retained on farms. Lupin production is therefore a major contributor to the livestock enterprise as well as having the benefits of a legume to following cereal crops. When cereal prices are high relative to wool, a two wheat/one lupin rotation is most profitable. When wheat prices fall relative to wool, a one wheat/one lupin rotation is most profitable in the whole farm analysis.

Farmers have been quick to continuously crop these sandplain soils to lupin/wheat rotations. As well as being profitable, the lupin/wheat rotation has a high water use with benefits in minimizing salinization through reduced ground water recharge.

Acid sands

These soils are naturally acid and the acidity increases with depth. The acidity (aluminum toxicity) affects the performance of most species although oats, triticale and to some extent, lupins, show some adaptability to this land management unit. Serradella is a pasture legume species which shows some potential on this soil type and research is being conducted as to its suitability.

As the acidity restricts production to relatively low levels, the whole farm analysis shows that the returns on investing inputs into this soil type are low and it should be left in natural pasture. Farmers are now tending to leave this soil type as native pasture with the occasional crop of oats.

Importance of the whole farm system approach

The eastern wheatbelt of Western Australia is considered by many to be a marginal farming area. To have a sustainable and profitable farming system farmers are changing to 'farming by soil type', use of minimum tillage and the growing of lupins, field peas and burr medics on appropriate soil types. Farmers who have adopted these programmes are reducing the major biological limitations of low nitrogen status of soils and soil structural decline. Farmers are finding they are attaining their goals through the whole farm system approach as it is synonymous with whole farm sustainability and profitability as discussed in Chapters 4 and 5.

Index